Quasi-One-Dimensional Organic Superconductors

Peking University–World Scientific Advanced Physics Series

ISSN: 2382-5960

Vol. 5 *Quasi-One-Dimensional Organic Superconductors*
by Wei Zhang and Carlos A. R. Sá de Melo

Vol. 4 *Applied Symbolic Dynamics and Chaos (Second Edition)*
by Bailin Hao and Weimou Zheng

Vol. 3 *Computer Simulations of Molecules and Condensed Matter: From Electronic Structures to Molecular Dynamics*
by Xin-Zheng Li and En-Ge Wang

Vol. 2 *Geometric Methods in Elastic Theory of Membranes in Liquid Crystal Phases (Second Edition)*
by Zhanchun Tu, Zhongcan Ou-Yang, Jixing Liu and Yuzhang Xie

Vol. 1 *Dark Energy*
by Miao Li, Xiao-Dong Li, Shuang Wang and Yi Wang

Peking University-World Scientific Advanced Physics Series

Vol 5

Quasi-One-Dimensional Organic Superconductors

Wei Zhang

Renmin University of China, China

Carlos A. R. Sá de Melo

Georgia Institute of Technology, USA

北京大學出版社
PEKING UNIVERSITY PRESS

World Scientific

Published by

World Scientific Publishing Co. Pte. Ltd.

5 Toh Tuck Link, Singapore 596224

USA office: 27 Warren Street, Suite 401-402, Hackensack, NJ 07601

UK office: 57 Shelton Street, Covent Garden, London WC2H 9HE

British Library Cataloguing-in-Publication Data
A catalogue record for this book is available from the British Library.

Quasi-One-Dimensional Organic Superconductors

The Work was originally published by Peking University Press in 2014.
This edition is published by World Scientific Publishing Company Pte Ltd by arrangement with Peking University Press, Beijing, China.

B&R Book Program

Peking University-World Scientific Advanced Physics Series — Vol. 5
QUASI-ONE-DIMENSIONAL ORGANIC SUPERCONDUCTORS

ISBN 978-981-3272-94-1

For any available supplementary material, please visit
http://www.worldscientific.com/worldscibooks/10.1142/11061#t=suppl

Desk Editor: Christopher Teo

Typeset by Stallion Press
Email: enquiries@stallionpress.com

Printed in Singapore

To Robert, Aleksandra, Jackie, and Kyle

Acknowledgements

During the writing of this book, we received constant support from Peking University Press, in particular, the excellent editorial board of this series. To name a few, we thank Xiao-Hong Chen, Xiao Liu, and Zhao-Yuan Yin, both for their invaluable help and great patience. We would also like to thank our colleagues working on quasi-1D organic superconductors who have had active discussions with us and generously shared their results over the years. Finally, we thank Wei-Qiang Yu, Min-Hui Shangguan, and Ren Zhang for reading the manuscript and giving us excellent suggestions and comments.

Beijing, November 2014

Wei Zhang
Carlos A. R. Sá de Melo

Contents

1
Overview

Just after the discovery of the first organic superconductor bistetramethyl-tetraselenafulvalene hexafluorophosphate$(TMTSF)_2PF_6$ by Jerome *et al.* in 1980 [1], it became clear that quasi-one-dimensional (quasi-1D) organic materials $(TMTSF)_2X$ and their sister compounds tetramethyltetrathiafulvalene salts $(TMTTF)_2X$ (where $X = PF_6, ClO_4 \ldots$ is an anion), known as Bechgaard salts, were very unusual materials. A couple of years later, it was found that superconductivity in these compounds was rapidly destroyed by non-magnetic defects [2, 3, 4, 5, 6]. Soon after these experiments, Abrikosov [7] pointed out that the strong suppression of the critical temperature in the presence of non-magnetic defects could be reasonably explained under the assumption of triplet pairing. This initial suggestion by Abrikosov was later reinforced by Gorkov and Jerome [8] in their analysis of upper critical field measurements by Brussetti *et al.* [9], and Ishiguro *et al.* [10]. Gorkov and Jerome suggested that the upper critical field along the easy axis $\boldsymbol{a}$ direction $H^a_{c_2}$ seemed to exceed the Pauli paramagnetic limit [11, 12] $H_P(T=0) \approx 1.84T_c(H=0)$ by a factor of 2. This opened the possibility of triplet superconductivity, because the spin-orbit coupling in these systems was too small to account for a large violation of the Clogston limit. Further experimental evidence of unusual behavior in these systems was reported in 1987 by Takigawa *et al.* [13]. They reported the absence of the Hebel–Slichter peak [14] in the proton spin-lattice relaxation rate $(1/T_1)$ data of $(TMTSF)_2ClO_4$ and a $1/T_1 \propto T^3$ temperature dependence in the region of $T_c/2 < T < T_c$. These early results were interpreted by Hasegawa and Fukuyama [15] as evidence for the existence of zeros of the superconducting order parameter on the quasi-1D Fermi surface. Hasegawa and Fukuyama also emphasized that these early experiments probed essentially the orbital part of the order parameter, and could not directly distinguish between singlet and triplet pairing states.

An early suggestion by Lebed [16], Burlachkov, Gorkov and Lebed (BGL) [17], and later by Dupuis, Montambaux, and Sá de Melo (DMS) [18], indicated that singlet and triplet pairing in quasi-1D superconductors could be tested in a measurement of upper critical fields. These authors found that the upper critical field for a triplet superconductor would exceed substantially the Pauli paramagnetic limit and produce a reentrant phase at high fields. This reentrant phase has increasing critical temperature with magnetic field, and is paramagnetic instead of diamagnetic [19, 20]. Lee *et al.* [21] started a systematic experimental search for triplet superconductivity in $(TMTSF)_2ClO_4$ via resistive measurements of $H_{c_2}(T)$ for fields parallel to the $\boldsymbol{b}'$ direction. Their results suggested (i) the survival of superconductivity beyond the Pauli paramagnetic limit, and (ii) possible reentrant superconductivity at high magnetic fields without a magnetic superconductor [22]. Intrigued by these exciting possibilities, Lee *et al.* [23] decided to explore the phase diagram of another quasi-1D superconductor, the sister compound $(TMTSF)_2PF_6$.

Unlike $(TMTSF)_2ClO_4$, $(TMTSF)_2PF_6$ is not superconducting at ambient pressure, instead it undergoes a metal–insulator transition at 12 K. The insulating phase can be suppressed through the application of hydrostatic pressure above 5 kbar. It is important to note that different groups around the world use distinct pressure calibrations, Lee *et al.* [23] use the pressure reading from the pressure cell at low temperatures, while some other groups [1] use the pressure reading of the pressure cell at room temperature and before cooling. Since the pressure medium solidifies below a critical temperature, the pressure reading is smaller at low temperatures than that at room temperature. The system then becomes superconducting with a maximum T_c around 1.2 K. Keeping an optimized pressure of 6 kbar in order to maximize the superconducting critical temperature, Lee *et al.* [23] established that the upper critical fields along the $\boldsymbol{a}$ and $\boldsymbol{b}'$ directions (i) present a strong upward curvature, (ii) well exceed the Pauli paramagnetic limit, and (iii) have an unusual anisotropy inversion at a magnetic field around 1.6 T(tesla).

After these experiments further theoretical studies were necessary to explain the experimental data. From the observation of a strong upward curvature in $H_{c_2}(T)$ for $(TMTSF)_2ClO_4$ and $(TMTSF)_2PF_6$, when the magnetic field was applied along the $\boldsymbol{b}'$ direction, it seemed natural to make a connection to the theory of Klemm, Luther, and Beasley (KLB) [24]. In this theory, the strong upward curvature of H_{c_2} when a magnetic field is aligned parallel to the layers in a layered superconductor is explained by a temperature-induced dimensional crossover (TIDC). The upward curvature is most pronounced when spin-orbit scattering is severe ($\gamma_{so} = \tau_{so}T_c \ll 1$), and the interlayer coupling is weak, $r \geqslant 1$, with $r = 2c/\pi\xi_\perp(0)$. Here, c is the interlayer lattice spacing, and $\xi_\perp$ is the coherence length along this direction. The theory of KLB was applied successfully to layered superconductors [25] such as TaS_2, $TaSe_2$, and $NbSe_2$. However, this theory may not be applicable to the $(TMTSF)_2X$ system, where the layer spacing $c \approx 13.5$ Å is smaller than the coherence length $\xi_\perp(0) \approx 20$ Å. In addition, spin-orbit effects are very small in $(TMTSF)_2X$ ($X = ClO_4, PF_6$) leading to $\gamma_{so} \approx 100$ (see Ref. [26, 27, 10]), a value at least three orders of magnitude too large [28]. Furthermore, the KLB theory does not include the effects of the quantum ("Landau") level structure in the electronic motion, which are thought to be very important for the existence of superconductivity in $(TMTSF)_2X$ [16, 17, 18, 20] at high magnetic fields and low temperatures.

Although the $H_{c_2}(T)$ measurements in $(TMTSF)_2ClO_4$ [21] and $(TMTSF)_2PF_6$ [23] suggest a triplet pairing state in quasi-1D Bechgaard salts, the singlet possibility cannot be completely ruled out, since no reentrant phase has been observed for these materials. The absence of the theoretically predicted reentrant phase in high magnetic fields could also suggest a spatially inhomogeneous s-wave singlet state discussed by DMS [18], Dupuis and Montambaux (DM) [29] and Dupuis [19], like the one discussed by Larkin and Ovchinnikov [30], and Fulde and Ferrell [31] (LOFF, or equivalently FFLO). However, Lebed [32] showed that the LOFF state in quasi-1D superconductors is paramagnetically limited for magnetic fields applied along the $\boldsymbol{b}'$ axis when a more realistic dispersion relation was used. Thus it seems that a triplet state should be favored at least at intermediate and high magnetic fields when considered in competition to singlet (s-wave) spatially inhomogeneous states like the LOFF state.

In addition to these upper critical field measurements, the triplet state is also favored by Knight shift measurements in $(TMTSF)_2PF_6$ for a magnetic field applied along the $\boldsymbol{b}'$ axis [33]. These experiments implied that the spin susceptibility is essentially the same as the normal state value. Following these experimental results, Lebed, Machida and Ozaki (LMO) [34] implicitly assumed strong spin-orbit coupling and proposed that the triplet order parameter $\boldsymbol{d}(\boldsymbol{k})$ at zero magnetic field is frozen into the crystalline lattice in such a way to produce, at zero temperature, $\chi_{b'}(0) = \chi_{\mathrm{n}}$, and $\chi_a(0) \ll \chi_{\mathrm{n}}$, where χ_{n} is the normal state uniform susceptibility. However, NMR experiments by Lee *et al.* [35] showed that there is no Knight shift change when an external magnetic field is applied along the $\boldsymbol{a}$ direction, thus indicating that $\chi_a \approx \chi_{\mathrm{n}}$. A different proposal assuming weak spin-orbit coupling was suggested by Duncan, Vaccarella and Sá de Melo (DVS) [36], and Duncan, Cherng and Sá de Melo (DCS) [37], where the spin susceptibility tensor and other thermodynamic quantities were calculated for various order parameter symmetries within the orthorhombic group [although $(\mathrm{TMTSF})_2\mathrm{PF}_6$ is really a triclinic crystal]. The support for a weak spin-orbit coupling theory in these systems is rooted in the fact that the heaviest element in the material is selenium. Based on this analysis, DVS and DCS concluded that the fully gapped triplet state ${}^3\mathrm{B}_{3\mathrm{u}}(a)$ ("p_x-wave") would produce an experimentally measured spin susceptibility always close to χ_{n}, which implies that the diagonal components of the uniform spin susceptibility tensor are $\chi_a = \chi_{b'} = \chi_{c^*} \approx \chi_{\mathrm{n}}$, for magnetic fields strong enough to overcome the pinning effect.

At a later stage, the Knight shift measurement has also been performed for $(\mathrm{TMTSF})_2\mathrm{ClO}_4$, where results seem to be more compatible with a singlet pairing scenario, at least in the low-magnetic-field regime [38]. In their experiment, Shinagawa *et al.* measured the $^{77}\mathrm{Se}$ NMR spectra of the superconducting phase of $(\mathrm{TMTSF})_2\mathrm{ClO}_4$ single crystals and observed no evident change of the Knight shift from the normal state for a magnetic field $H = 40$ kOe. Besides, thanks for the fact that $(\mathrm{TMTSF})_2\mathrm{ClO}_4$ can become superconducting at ambient pressure, thermodynamic quantities were also investigated in such material. In 1997, Belin and Behnia [39] measured the thermal conductivity at zero and low magnetic fields. The thermal conductivity seemed to be exponentially activated at low temperatures indicating the existence of a full gap in the quasi-particle excitation spectrum. However, this statement becomes debatable after 15 years due to the appearance of another experiment on heat capacity [40], where a superconducting state with lines of nodes is more favorable. That far, we believe it is fair to say that the controversy regarding the superconducting nodal structure in $(\mathrm{TMTSF})_2\mathrm{ClO}_4$ is not settled yet.

One important unsolved puzzle associated with the superconducting states of Bechgaard salts is the pairing mechanism. Considering the quasi-one-dimensionality of the underlying crystalline structure, one approach developed by early works is trying to accommodate the "g-ology" picture of a one-dimensional conductor to Bechgaard salts [41]. By treating the system as an array of weakly coupled one-dimensional conductors formed by chains of flat molecules, the electronic properties at low energy are described as a Luttinger liquid characterized by dimensionless parameters. In such a theory, at least one of the four susceptibilities, spin density wave, charge density wave, singlet superconductivity and triplet superconductivity, diverges at low temperatures, depending on the Luttinger parameters $\mathcal{K}_{c,s}$. Thus, in a three-dimensional crystal a phase with a broken symmetry corresponding to the most divergent susceptibility will be formed. However, due to the presence of a Coulomb repulsion, it has soon be proven that the region of divergent superconducting susceptibility is difficult to achieve for realistic materials [42].

Another possibility to explain the origin of superconductivity is to construct a Bardeen-Cooper-Schrieffer-type (BCS) theory, where the inter-electron attraction is mediated by phonons and/or magnons. In a recent effort, Kuroki, Arita and Aoki (KAA) [43] examined the influence of charge and anisotropic spin fluctuation exchanges in the pairing stability for $(TMTSF)_2PF_6$, and proposed an f-wave triplet state with $\boldsymbol{d}$-vector perpendicular to the quantization axis. In contrast, Shimahara [44] proposed a nodeless d-wave state for $(TMTSF)_2ClO_4$ based in part on the thermal conductivity measurements of Belin and Behnia [39]. In recent renormalization group calculations [45, 46] the ground state phase diagram of nearly one-dimensional superconductors was investigated as a function of hopping parameters and interactions, and revealed the existence of d-wave and f-wave superconductivities as well as spin and charge density orders. Other pairing mechanisms including Ising anisotropy [47] and ring-exchange [48] have also been proposed. All these proposals, combined with the suggestions of triplet superconductivity of Lebed [16], BGL [17], and DMS [18], add to the debate about the symmetry of order parameter and the mechanism of superconductivity in quasi-1D Bechgaard salts.

The issue of the angular dependence of the upper critical field is another important question. LMO [34] proposed that the anisotropy inversion in the upper critical fields seen in $(TMTSF)_2PF_6$ was due to a $\boldsymbol{d}$-vector flop transition when an external magnetic field was applied along the $\boldsymbol{a}$ direction. However, a theoretical estimate [37] gives a value of the flopping field to be about 0.22 T, nearly one order of magnitude smaller than the anisotropy inversion field $H^* \sim$ 1.6 T observed experimentally [23]. Furthermore, NMR experiment [35] also suggested that the flop transition should happen at a magnetic field $H < 1.43$ T $< H^*$. Another possibility is that the anisotropy inversion in $(TMTSF)_2PF_6$ was due to the disappearance of closed orbits present only for fields along the $\boldsymbol{a}$ direction, and to a reduction in the degeneracy of states when open-orbit contributions start to dominate. In this case, the anisotropy inversion would be purely orbital in nature, instead of a spin effect as proposed by LMO [34]. However, the angular dependence of the upper critical field of Bechgaard salts in the $\boldsymbol{a}$-$\boldsymbol{b}'$ plane at high magnetic fields has not been calculated yet.

If indeed Bechgaard salts [or at least $(TMTSF)_2PF_6$] are triplet superconductors, then why the remarkable reentrant phase originally predicted by Lebed [16] has not been observed yet? An initial attempt to address this question was made by Vaccarella and Sá de Melo (VS) [49], who analyzed phase fluctuations at high magnetic fields. They concluded that phase fluctuations can melt the magnetic-field-induced Josephson vortex lattice at a lower temperature than the predicted mean field critical temperature. The same authors also calculated the angular dependence of upper critical field within the $\boldsymbol{b}'$-$\boldsymbol{c}^*$ plane and showed that a small misalignment with the $\boldsymbol{b}'$ axis could cause a very rapid drop of the critical temperature at high magnetic fields. However, they considered the two effects separately, and did not include amplitude and quantum fluctuations, so their results are qualitative at the best.

Setting other Bechgaard salts aside, we believe it is fair to say that the observation of unusual behavior in the upper critical fields [23] and the measurement of the spin susceptibility [33, 35] in $(TMTSF)_2PF_6$ strongly suggest the existence of a non-uniform triplet pairing superconducting state in a magnetic field. This state would reduce at zero field to a triplet phase similar to the A-phase [50] of superfluid ^{3}He, apart of course from the group theoretical differences concerning lattice and time reversal in a magnetic field [36, 37]. However, since quantitative comparison between theoretical and experimental results is still lacking, and the debate between singlet and triplet pairing is still not completely settled [especially for $(TMTSF)_2ClO_4$], we discuss in this book both singlet and triplet pairing states with different order parameter symmetries, instead of confine ourselves to triplet states only.

The reader will notice that most of the experimental and theoretical developments discussed in this book occurred from the early 1990s to 2012. Our intention in reviewing these recent efforts is to provide an idea of the current directions and frontiers related to superconductivity in quasi-1D organic conductors. This book is expanded from a review article published in 2007 by the same authors [51], while substantial revision and expansion have been made to include experimental and theoretical progresses appeared afterwards. This book is prepared for advanced graduate students, post-docs, and colleagues working in the field of superconductivity. A general audience whose expertise is not in this field may refer to some tutorial reviews or books (e.g., the excellent book by Ishiguro *et al.* [52]). The topics and the papers that are included in this book are organized in such a way to give a reasonably fluent presentation. We did not try to assign credits or priorities in the order of content. If any of the readers feel that their contributions were not properly acknowledged or reviewed, we would ask them to attribute the errors and omissions to our own stupidity, ignorance, laziness and haste rather than any malicious behavior.

2
Background

The discovery of the first organic superconductor bistetramethyltetraselenafulvalene hexafluorophosphate $(TMTSF)_2PF_6$ by Jerome *et al.* [1] in 1980 stimulated a dramatic search for superconductivity in other quasi-1D Bechgaard salts family. This search has lead to discoveries of other very interesting physical effects, among these phenomena it is worth highlighting the existence of metallic, charge-ordered, charge-localized, spin Peierls, spin density wave and superconducting phases. These phases can exist in the same material depending on temperature, pressure, magnetic field, and anion ordering. A schematic pressure–temperature phase diagram for $(TMTSF)_2X$ and $(TMTTF)_2X$ is shown in Fig. 2.1. In addition, the discovery of magnetic field induced spin density waves (FISDW) [53] in these systems has stimulated an incredible surge of theoretical and experimental interest in 1980s, as can be verified in the book by Saito and Kagoshima [54]. Among the recent aspects related to these phenomena it is important to single out the quantum Hall effect and the angle-dependent magneto-resistance oscillations which have received a lot of theoretical and experimental attention in the last few years [16, 26, 55, 56, 57]. More recently, there has been some experimental and theoretical efforts in trying to decipher the nature of the metallic state in a magnetic field [58, 59]. The origin of many of these features lies in the extremely anisotropic nature of the electronic structure in these materials resulting from the weakly coupled chain-like crystalline structure.

2.1 Crystalline structure and electronic dispersion

The $(TMTSF)_2X$ family is iso-structural, where all members are triclinic crystals with very similar lattice parameters. As can be seen in Fig. 2.2, nearly planar TMTSF molecules zigzag forming chains along the so-called $\boldsymbol{a}(x)$ axis, and form sheets in the $\boldsymbol{a}$-$\boldsymbol{b}'$ (x-y) plane [52]. These sheets are separated by anions along the $\boldsymbol{c}^*(z)$ axis. Selenium (Se) orbitals play an important role on the electronic properties of these materials [Selenium corresponds to S in the chemical formula $(\mathrm{TMTSF})_2\mathrm{X}$]. Although Se–Se spacings along the $\boldsymbol{a}(x)$ axis and along the $\boldsymbol{b}'(y)$ axis are comparable, the overlap of π orbitals is strongest along the chains (x direction). As a consequence, the electronic spectrum is quasi-1D and the electron transfer energies are estimated to be of the order of 0.5, 0.05, 0.003 eV along special directions x, y and z, respectively.

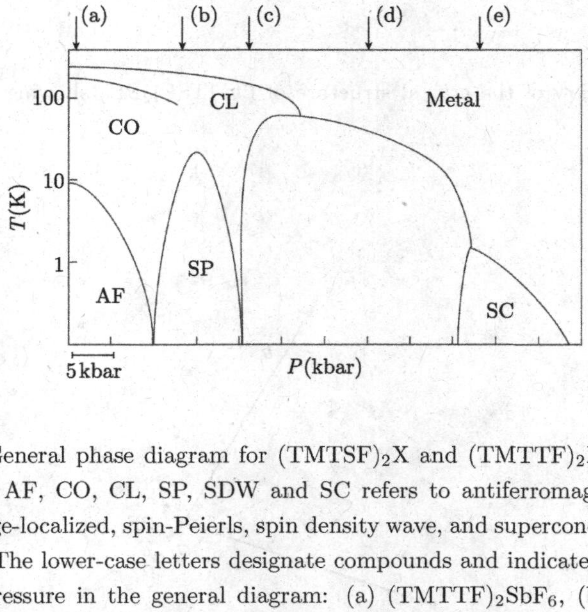

Figure 2.1 General phase diagram for $(\mathrm{TMTSF})_2\mathrm{X}$ and $(\mathrm{TMTTF})_2\mathrm{X}$ compounds. The notation AF, CO, CL, SP, SDW and SC refers to antiferromagnetic, charge-ordered, charge-localized, spin-Peierls, spin density wave, and superconducting states, respectively. The lower-case letters designate compounds and indicate their location at ambient pressure in the general diagram: (a) $(\mathrm{TMTTF})_2\mathrm{SbF}_6$, (b) $(\mathrm{TMTTF})_2$ AsF_6, (c) $(\mathrm{TMTTF})_2\mathrm{Br}$, (d) $(\mathrm{TMTSF})_2\mathrm{PF}_6$, (e) $(\mathrm{TMTSF})_2\mathrm{ClO}_4$.

In this book, we will use interchangeably $x \leftrightarrow \boldsymbol{a}$, $y \leftrightarrow \boldsymbol{b}'$ and $z \leftrightarrow \boldsymbol{c}^*$, unless specified. Furthermore, natural units are used to simplify notation where the Planck's constant $\hbar$, the Boltzman's constant k_B, and speed of light $\bar{c}$ are equal to one ($\hbar = k_\mathrm{B} = \bar{c} = 1$). We may also use the anion formulae PF_6 and ClO_4 as a shorthand notation to represent the corresponding Bechgaard salts

$(TMTSF)_2PF_6$ and $(TMTSF)_2ClO_4$, respectively.

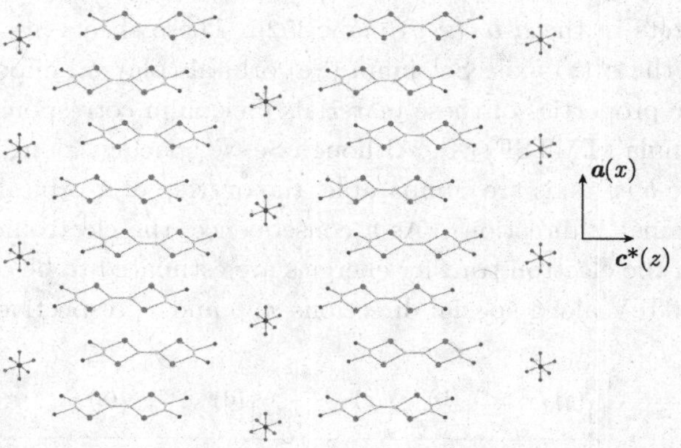

Figure 2.2 View of the crystal structure of $(TMTSF)_2PF_6$ showing the $\boldsymbol{a}$ and $\boldsymbol{c}^*$ axes.

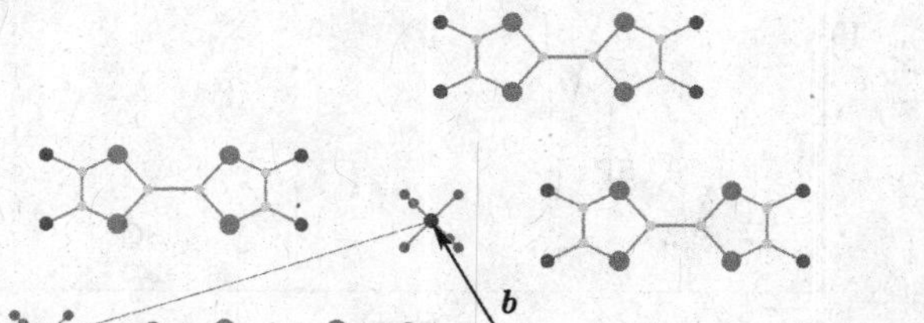

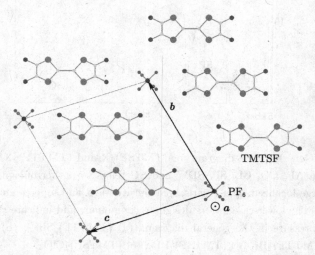

Figure 2.3 View of the crystal structure of $(TMTSF)_2PF_6$ showing the principal directions $\boldsymbol{a}$, $\boldsymbol{b}$ and $\boldsymbol{c}$. Adapted from Ref. [61].

A simple approximation for these systems is to take the dispersion relation [60]

$$\varepsilon(\boldsymbol{k}) = -t_x \cos(k_x a) - t_y \cos(k_y b) - t_z \cos(k_z c), \tag{2.1}$$

where a, b and c are the lattice constants and t_x, t_y and t_z are the transfer integrals along the x, y and z directions, respectively. The quasi-one-dimensionality is imposed by the condition $t_x \gg t_y \gg t_z > 0$. Even though this dispersion relation should be considered as a simplification, it contains some of the essential characteristics of the quasi-1D band structure near the Fermi energy. The use of this dispersion relation implicitly assumes an orthorhombic crystal structure for Bechgaard salts. However, these systems are really triclinic with typical lattice parameters at room temperature $a = 7.297\,\text{Å}$, $b = 7.711\,\text{Å}$, $c = 13.522\,\text{Å}$, with angles $\alpha = 83.39°$, $\beta = 86.27°$, and $\gamma = 71.01°$, as seen in Fig. 2.3 [61, 62] for $(\text{TMTSF})_2\text{PF}_6$. This means that all the discussions presented here in connection with triclinic Bechgaard salts should be viewed only as qualitative at best.

As a consequence of this anisotropy, the Fermi surface of quasi-1D systems is open and has two separated sheets (see Fig. 2.4), such that the electronic motion then can be classified as *right* going (with sheet index $\alpha = +$) or *left* going (with sheet index $\alpha = -$). With the limiting condition that $t_x \gg \max(t_y, t_z)$, the dispersion relation above can be linearized as

$$\varepsilon_\alpha(\boldsymbol{k}) = v_{\text{F}}(\alpha k_x - k_{\text{F}}) - t_y \cos(k_y b) - t_z \cos(k_z c), \tag{2.2}$$

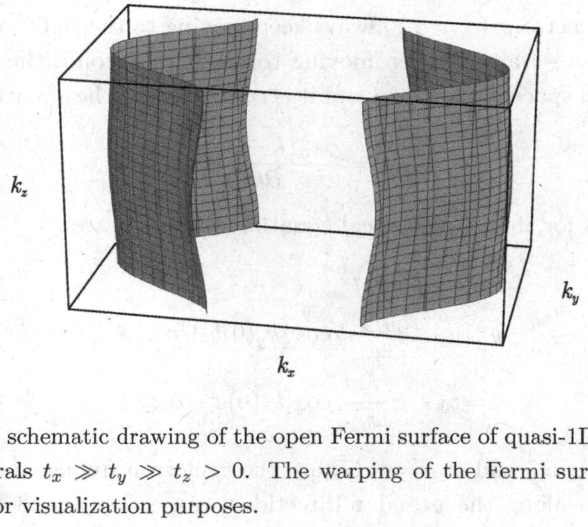

Figure 2.4 A schematic drawing of the open Fermi surface of quasi-1D systems with transfer integrals $t_x \gg t_y \gg t_z > 0$. The warping of the Fermi surface is greatly exaggerated for visualization purposes.

where the approximations $v_x \approx v_{\rm F}$ and $t_x \approx E_{\rm F}$ are used with $v_{\rm F}$ and $E_{\rm F}$ being Fermi velocity and Fermi energy, respectively. This strong anisotropy of dispersion relation leads to very interesting magnetic field effects, including the magnetic field induced dimensional crossover (FIDC).

2.2 Magnetic field induced dimensional crossover

The electronic motion in a magnetic field is substantially changed due to the strong anisotropy of quasi-1D systems. To illustrate this point we consider the effect of a magnetic field applied within the y-z plane of the material, i.e., $\boldsymbol{H} = (0, H_y, H_z)$. Using the Landau gauge $\boldsymbol{A} = (H_y z - H_z y, 0, 0)$, where α and k_x are still conserved quantities (good quantum numbers) while k_y and k_z are not, the electronic dispersion defined in Eq. (2.2) becomes $\varepsilon_\alpha(\boldsymbol{k} - e\boldsymbol{A})$, through the use of the Peierl's substitution $\boldsymbol{k} \to \boldsymbol{k} - e\boldsymbol{A}$. Thus, the semiclassical electronic motion is governed by the equation

$$\frac{d\boldsymbol{k}}{dt} = -\frac{\partial \varepsilon_\alpha(\boldsymbol{k} - e\boldsymbol{A})}{\partial \boldsymbol{r}}. \tag{2.3}$$

It is important to emphasize two major points. First, since the Fermi surface of these systems is open and composed of two separate sheets (see Fig. 2.4), the semiclassical electronic motion consists of open orbits in reciprocal space. Right going electrons ($\alpha = +$) always keep moving to the right, and left going electrons ($\alpha = -$) always keep moving to the left. Second, the semiclassical motion in real space is also open and it is controlled by the equation of motion

$$\frac{d\boldsymbol{r}}{dt} = \frac{\partial \varepsilon_\alpha(\boldsymbol{k} - e\boldsymbol{A})}{\partial \boldsymbol{k}}. \tag{2.4}$$

The solutions for this semiclassical equation of motion are

$$x = x_0 + \alpha v_{\rm F} t, \tag{2.5}$$

$$y = y_0 - \alpha \frac{t_y}{\omega_{c_y}} b \cos\left[k_y(0) b + \alpha \omega_{c_y} t\right], \tag{2.6}$$

$$z = z_0 + \alpha \frac{t_z}{\omega_{c_z}} c \cos\left[k_z(0) c - \alpha \omega_{c_z} t\right], \tag{2.7}$$

where $\omega_{c_y} = v_{\rm F} G_y$ and $\omega_{c_z} = v_{\rm F} G_z$ are the cyclotron frequencies of the electronic motion along the y and z directions, respectively, and $G_y = |e| H_z b$

and $G_z = |e|H_y c$ are the corresponding characteristic magnetic wavenumbers (inverse of the characteristic magnetic lengths). Here, $H_z = H\cos\theta$ and $H_y = H\sin\theta$, where θ is the angle between $\boldsymbol{H}$ and $\hat{\boldsymbol{z}}$.

Provided that there is no magnetic breakdown, and that $E_{\mathrm{F}} \gg \max(\omega_{\mathrm{c}_y}, \omega_{\mathrm{c}_z})$, this semiclassical analysis suggests a classification of three different regimes of confinement for the electronic motion:

(i) a one-dimensional (1D) regime, where the semiclassical amplitudes of motion are confined along both the y and z directions, i.e., $t_z/\omega_{\mathrm{c}_z} \ll 1$ and $t_y/\omega_{\mathrm{c}_y} \ll 1$;

(ii) a two-dimensional (2D) regime, where the semiclassical amplitudes of motion are confined along the z direction but not along the y direction (or *vice versa*), i.e., $t_z/\omega_{\mathrm{c}_z} \ll 1$ but $t_y/\omega_{\mathrm{c}_y} \gg 1$ (or $t_z/\omega_{\mathrm{c}_z} \gg 1$ but $t_y/\omega_{\mathrm{c}_y} \ll 1$); and

(iii) a three-dimensional (3D) regime, where the semiclassical amplitudes of motion are not confined in either y or z direction, i.e., $t_z/\omega_{\mathrm{c}_z} \gg 1$ and $t_y/\omega_{\mathrm{c}_y} \gg 1$.

Therefore, when $|t_z c/\omega_{\mathrm{c}_z}| \leqslant |t_y b/\omega_{\mathrm{c}_y}|$, or equivalently, $|\tan\theta| \geqslant t_z c/t_y b \equiv \tan\theta_{\mathrm{c}}$, the confinement (localization) first occurs along the z direction. When $|\tan\theta| \leqslant t_z c/t_y b$, the localization first occurs along the y direction. Specifically for the quasi-1D systems defined by Eq. (2.1), where $t_z/t_y \ll 1$, the localization happens first along the z direction except for a very narrow region of angles $|\tan\theta| \leqslant \tan\theta_{\mathrm{c}}$. For instance, taking $t_z/t_y \approx 0.1$, $\theta_{\mathrm{c}} \approx 5.71°$ and the angular region where localization takes place first along the z direction is very wide ($\theta_{\mathrm{c}} < \theta < 180° - \theta_{\mathrm{c}}$ or $180° + \theta_{\mathrm{c}} < \theta < 360° - \theta_{\mathrm{c}}$), except for a few degrees from the z axis. This kind of magnetic field confinement is called magnetic field induced dimensional crossover (FIDC), where depending on direction and magnitude of the field the electronic motion crosses over from 3D→2D→1D.

To illustrate this FIDC further we consider a simpler case where the magnetic field is perfectly aligned along the y direction, i.e. $\boldsymbol{H} \parallel \boldsymbol{b}'$ and $\theta = 90°$. Using the Landau gauge $\boldsymbol{A} = (Hz, 0, 0)$ where k_x, k_y and α are conserved quantities, while k_z is not, the solutions of the equation of motion in real space lead to a transverse electronic motion (motion in the x-z plane) of the form

$$x(t) = x(0) + \alpha v_x t, \tag{2.8}$$

$$z(t) = z_0 - \alpha \frac{t_z c}{\omega_c} \cos[k_z(0)c + \alpha\omega_c t], \tag{2.9}$$

where $\omega_c = v_F G$ is the associated cyclotron frequency, and $G = |e|Hc$ is the characteristic magnetic wavenumber. Although the semiclassical orbits are open both in reciprocal and real spaces, they are still periodic along the k_z (or the z) direction. Thus, ω_c can be physically interpreted either as the characteristic frequency at which electrons traverse the Brillouin zone along the k_z direction, or as the characteristic frequency of the periodic motion along the z axis. This semiclassical analysis also indicates that the amplitude of the oscillatory motion along the z direction is controlled by the dimensionless parameter $\tilde{t} = t_z/\omega_c$. When $\tilde{t} \ll 1$, i.e., $\omega_c \gg t_z$, the semiclassical amplitude $z_A = \tilde{t}c$ becomes smaller than the interplane lattice spacing c, and thus confined along the z direction. (See Fig. 2.5 for an illustration.)

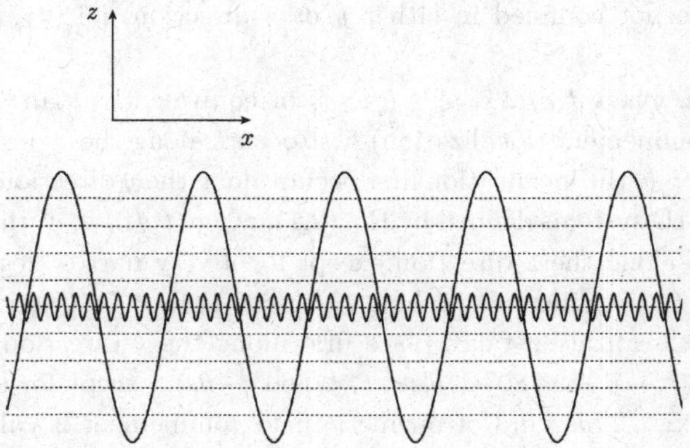

Figure 2.5 Semiclassical electronic motion in a magnetic field. Notice the magnetic field induced dimensional crossover: smaller (larger) amplitude orbits correspond to higher (smaller) magnetic fields. Doted lines indicates molecular chains along x direction. From Ref. [63].

In this particular case, the amplitude of the electronic motion z_A becomes localized in individual x-y planes as the magnetic field increases, making the electronic motion more two dimensional. This effect (FIDC) is possible mostly because of the strong anisotropy of energy scales ($t_x \gg t_y \gg t_z$) in quasi-1D materials, and thus has remarkable consequences regarding the superconducting state. Although the previous semiclassical analysis provides the correct qualitative picture, the FIDC is really a quantum phenomenon which is intimately connected to the structure of quantum levels in the presence of a magnetic field, as to be discussed next.

2.3 Quantum level structure

In order to illustrate the quantum nature of FIDC, we take the linearized dispersion of Eq. (2.2) in the simplest situation where $\boldsymbol{H} \parallel \boldsymbol{b}'$. In the gauge $\boldsymbol{A} = (Hz, 0, 0)$, the eigenfunctions of Hamiltonian

$$\mathscr{H}_0(\boldsymbol{k} - e\boldsymbol{A}) = \varepsilon_{\alpha,\sigma}(\boldsymbol{k}) + \alpha\omega_c z/c, \tag{2.10}$$

can be written as

$$\Phi_{k_x,k_y,N,\alpha,\sigma}(x, y, z = nc) = \exp\left[i(k_x x + k_y y)\right] J_{N-n}\left(\bar{\alpha} t_z/\omega_c\right) \tag{2.11}$$

with $\bar{\alpha} = -\alpha$, and the eigenvalues are

$$\varepsilon_{k_x,k_y,N,\alpha,\sigma} = \varepsilon_{\alpha,\sigma}(k_x, k_y) + \alpha N\omega_c, \tag{2.12}$$

where $J_p(u)$ is the Bessel function of integer order p and argument u and

$$\varepsilon_{\alpha,\sigma}(k_x, k_y) = v_F(\alpha k_x - k_F) - t_y \cos(k_y b) - \sigma\mu_B H \tag{2.13}$$

is a 2D dispersion. The eigenspectrum is illustrated in Fig. 2.6(a). Notice that the eigenfunctions become localized when $\omega_c \gg t_z$ in the $z = Nc$ (i.e., x-y) planes, via the argument of the Bessel function. This is in agreement with the semiclassical condition that the amplitude of motion in the z direction $z_A \ll c$. The physical interpretation is even more transparent when we analyze in addition the energetics: the energy barrier for a single electron to tunnel from plane $z = nc$ to plane $z = (n+1)c$ is ω_c, hence when the energy barrier ω_c becomes large in comparison to the hopping integral t_z ($\omega_c \gg t_z$), electrons are essentially confined to their initial planes.

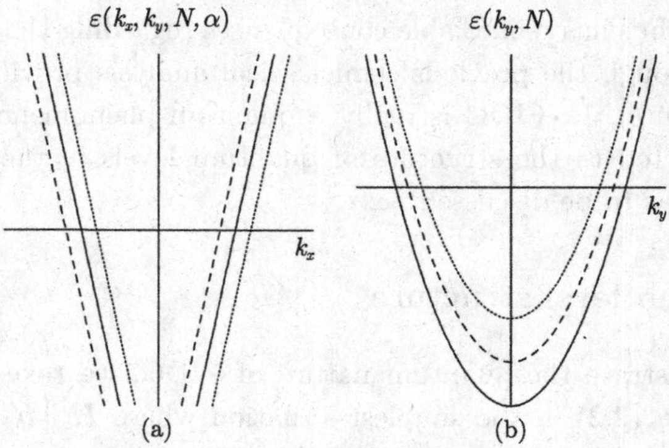

Figure 2.6 (a) Illustration of the eigenspectrum for the quasi-1D case without spin splitting. Solid lines correspond to $N = 0$, dashed lines indicate $N = +1$, and dotted lines correspond to $N = -1$; (b) Illustration of the eigenspectrum for the isotropic 3D case without spin splitting. The solid line corresponds to $N = 0$, the dashed line indicates $N = 1$, and the dotted line corresponds to $N = 2$. From Ref. [63].

At this point it is appropriate to recall the more familiar three dimensional isotropic situation with Hamiltonian

$$\mathscr{H}_0^{3\mathrm{D}}(\boldsymbol{k}) = (\boldsymbol{k} - e\boldsymbol{A})^2/2m - \sigma\mu_{\mathrm{B}}H \tag{2.14}$$

in the same gauge $\boldsymbol{A} = (Hz, 0, 0)$. The eigenfunctions of $\mathscr{H}_0^{3\mathrm{D}}$ corresponding to Landau levels are

$$\Phi^{(\mathrm{L})}_{k_x,k_y,N,\sigma}(x, y, z) = \exp\left[i(k_x x + k_y y)\right] F(z) \tag{2.15}$$

with

$$F(z) = \frac{\exp\left[-(z - z_0)^2/2a_H^2\right] \mathrm{H}_N(z - z_0/a_H)}{\pi^{1/4} a_H^{1/2} \sqrt{2^N N!}}, \tag{2.16}$$

where $a_H = \sqrt{1/(m\omega_{\mathrm{c}})}$ is the cyclotron radius, $\mathrm{H}_N(u)$ is the Hermite polynomial of order N and argument u, and N is the Landau level index. The eigenvalues are

$$\varepsilon_{k_y,N,\sigma} = (N + 1)\omega_{\mathrm{c}} + k_y^2/2m - \sigma\mu_{\mathrm{B}}H. \tag{2.17}$$

The eigenspectrum here is one-dimensional, since two degrees of freedom k_x and k_z were absorbed into the orbits characterized by the quantum number N.

Also recall that the size of electronic orbits in the perpendicular direction (x-z plane) to the applied field ($\boldsymbol{H} = H\hat{\boldsymbol{b}}$) is controlled by a_H, hence if $k_F a_H = \sqrt{2E_F/\omega_c} \ll 1$, i.e., $\omega_c/E_F \gg 1$, there is a localization effect. Since the typical size of the Landau level a_H in the x-z plane is much larger then the interparticle spacing k_F^{-1}, $\omega_c/E_F \gg 1$ is a rather difficult condition to be achieved for ordinary metals, except for gigantic magnetic fields (at the range of thousands of tesla). As a comparison, for some quasi-1D systems like the Bechgaard salts this localization effect may be achieved at moderate fields, say for $t_z \approx 5$–$10\,\text{K}$, the required fields for the onset of strong localization $\omega_c \sim t_z$ are of order of 5–10 tesla.

To emphasize further the differences and similarities between quasi-1D and 3D isotropic systems, a schematic plot of their eigenspectra is shown in Fig. 2.6 using the gauge $\boldsymbol{A} = (Hz, 0, 0)$. Notice that the quasi-1D eigenspectrum in Eq. (2.12) is a function of both k_x and k_y since only one degree of freedom (k_z) is absorbed into the spectrum due to the magnetic field dependent periodic motion along the z direction. Hence, the eigenspectrum is two-dimensional in contrast with the one-dimensional spectrum obtained in Eq. (2.17) for the 3D isotropic case. Moreover, the quasi-1D eigenspectrum is constituted of a large number of magnetic subbands labeled by the magnetic quantum number N and can be obtained from the 2D dispersion $\varepsilon_{\alpha,\sigma}(k_x, k_y)$ simply by shifting $k_x \to k_x + NG$ (see Fig. 2.6). The isotropic 3D case is also made of a large number of magnetic subbands (Landau levels) controlled by the magnetic quantum number N, but a simple shift in momentum k_y does not allow to generate the entire ladder of quantum levels since the eigenspectrum is quadratic in k_y. Despite these differences, quasi-1D systems are largely degenerate in a magnetic field and present a well defined quantum level structure like isotropic systems. As a consequence, Cooper pairing between electrons in different magnetic subbands (say N and N') is possible in analogy with Landau level pairing in the isotropic case [64, 65, 66, 67, 68]. Thus, the existence of triplet or singlet superconductivity in magnetic fields is very intimately connected to the structure of these quantum levels.

2.4 Triplet versus singlet superconductivity in a magnetic field

The key to the existence of superconductivity is that electrons can form special bound states called Cooper pairs. These Cooper pairs can propagate coherently through the system without detectable dissipation. Generally speaking (in weak spin-orbit coupling systems) there are two types of Cooper pairs, those which are in a singlet state with total spin 0, and those which are in a triplet state with total spin 1. In singlet superconductors ($S = 0$), Cooper pairing occurs with magnetic quantum numbers $m_s = 0$, corresponding to an antisymmetric linear combination of states $|\uparrow\downarrow\rangle$ and $|\downarrow\uparrow\rangle$. In triplet superconductors ($S = 1$), Cooper pairing occurs with magnetic quantum numbers $m_s = -1, 0, +1$, corresponding respectively to $|\downarrow\downarrow\rangle$ states, symmetric linear combination of states $|\uparrow\downarrow\rangle$ and $|\downarrow\uparrow\rangle$, and $|\uparrow\uparrow\rangle$ states.

Typically the application of an external magnetic field is detrimental to superconductivity due to the following reasons. First, the coupling of the magnetic field with the electron charge forces electronic orbits to bend, and as a result it becomes more difficult for two electrons to form a bound state. Therefore, the suppression of the formation of these bound states immediately undermines superconductivity. This phenomenon is called orbital frustration, and affects both singlet and triplet superconductors. Second, the coupling of the magnetic field to spins also tend to break Cooper pairs because of the Zeeman energy cost. The Zeeman splitting affects directly singlet superconductors which have magnetic quantum number $m_s = 0$, since electrons with spin up and with spin down are separated in energy. In the case of triplet superconductors, states with magnetic quantum numbers $m_s = 0$ are also directly affected by the magnetic field, thus favoring states with both spins aligned, i.e., $m_s = \pm 1$.

Therefore, in order to make the quantum level structure work in favor of superconductivity at high magnetic fields (beyond the semiclassical orbital upper critical fields and beyond the Pauli paramagnetic limit), one must envision a mechanism that eliminates or compensates for (i) the orbital frustration, and (ii) the Zeeman splitting introduced by the magnetic field. For 3D isotropic systems, such a mechanism was proposed by Tesanovic *et al.* [64] and later by Norman, MacDonald and Akera [67], building up and expanding on earlier works [69, 70]. The survival of superconductivity up to very high magnetic fields involved reaching the extreme quantum limit, where only a single Landau level is occupied. Then "Cooper" pairing can occur between electronic states in the same Landau level, thus beating the detrimental orbital frustration. In this case, superconductivity is only limited by the Pauli pair breaking effect and impurity scattering. The magnetic fields required to reach this extreme quantum limit for ordinary metals is very high, being typically in the range of thousands of tesla. The reason for such high fields is that ordinary metals are high density materials with large Fermi energies. It is very difficult to reach fields of the order of thousands of tesla with current technology, but this situation may change in the near future with the development of new facilities at the National High Magnetic Field Laboratory. With present day magnets, it is necessary to look for materials that have low density. A possible isotropic material to look for the effects suggested by Tesanovic *et al.* [64] is bismuth, which has very low density, but such isotropic metals are rare.

The advantage of quasi-1D systems is that the extreme quantum limit can be reached for magnetic fields beginning in the range from 5 to 30 tesla. This is possible for magnetic fields applied along the y axis, because of the extreme anisotropic electronic structure of these systems ($t_x \gg t_y \gg t_z$). When $\omega_c \gg t_z$ the electronic wave functions are confined along the z axis, hence they represent a highly degenerate single quantum Landau level at the Fermi energy [20, 63]. Thus pairing within the same quantum level is possible, and here it means pairing in the x-y plane. The quantum level structure has also remarkable effects upon the superconducting state of quasi-1D systems. In strong magnetic fields applied along the y axis ($\omega_c \gg t_z$) it has been shown that quasi-1D, but nevertheless 3D anisotropic superconductors, become essentially 2D anisotropic superconductors [16, 18].

On the other hand, at high magnetic fields, the confinement of the electronic motion along the z axis also tends to confine the formation and motion of Cooper pairs mostly in the x-y planes (pairing in the lowest quantum Landau level). This produces an effective coherence length $\xi_z^{\rm eff}$ which is smaller than the interplanar lattice spacing c. As a result, there exists a novel magnetic field induced dimensional crossover (FIDC) from a strongly 3D but highly anisotropic Abrikosov vortex lattice to a nearly 2D highly anisotropic Josephson vortex lattice [18]. Beyond the crossover regime, the magnetic field along the y axis can not break Cooper pairs, thus the orbital frustration disappears, and hence the Pauli paramagnetic pair breaking effect determines the final shape of the upper critical field. This FIDC mechanism then leads to the survival of superconductivity in high magnetic fields for singlet superconductors and to a reentrant behavior for triplet superconductors. For this to occur in the singlet case it is necessary to develop spatially inhomogeneous superconductivity similar to a state first proposed by Larkin and Ovchinnikov [30] and Fulde and Ferrell [31], the so-called LOFF (or equivalently, FFLO) state. In the triplet case the system would be also spatially inhomogeneous (in a magnetic field) and possibly of equal spin pairing [20, 63] similar to the A-phase [50] of liquid ^{3}He, in order to take advantage of the external field.

2.5 A minimal Hamiltonian

To study the instability of the normal state towards a superconducting state in a magnetic field, we discuss the following minimal Hamiltonian

$$\mathscr{H} = \mathscr{H}_0 + \mathscr{H}_{\text{int}}. \tag{2.18}$$

Here, the non-interaction Hamiltonian is

$$\mathscr{H}_0 = \sum_{\alpha,\sigma} \int d\boldsymbol{r} c^\dagger_{\alpha,\sigma}(\boldsymbol{r}) H_{0,\alpha,\sigma}(-i\nabla - e\boldsymbol{A}) c_{\alpha,\sigma}(\boldsymbol{r}) \tag{2.19}$$

in the presence of a magnetic field $\boldsymbol{H} = \boldsymbol{\nabla} \times \boldsymbol{A}$. The interaction Hamiltonian

$$\mathscr{H}_{\text{int}} = -\frac{\lambda_{\text{s}}}{2} \int d\boldsymbol{r} O^\dagger_{\text{s}}(\boldsymbol{r}) O_{\text{s}}(\boldsymbol{r}) - \frac{\lambda_{\text{t}}}{2} \sum_{m_s} \int d\boldsymbol{r} O^\dagger_{\text{t},m_s}(\boldsymbol{r}) O_{\text{t},m_s}(\boldsymbol{r}) \tag{2.20}$$

describes the competition between singlet(s) and triplet(t) pairing, where the coupling constants λ_{s} and λ_{t} are assumed to be independent parameters. The pair operator

$$O_{\text{s}} = \sum_\sigma \sigma c_{-,\bar{\sigma}}(\boldsymbol{r}) c_{+,\sigma}(\boldsymbol{r}) \tag{2.21}$$

corresponds to singlet pairing, and the pair operator

$$O_{\text{t},m_s} = \sum_{\sigma,\sigma'} \sigma c_{-,\sigma}(\boldsymbol{r}) \tau^{(m_s)}_{\sigma,\sigma'} c_{+,\sigma'}(\boldsymbol{r}) \tag{2.22}$$

to triplet pairing, where $\bar{\sigma}$ denotes the spin direction opposite to σ, and $m_s = +1, 0, -1$ represents the magnetic quantum number for total spin 1. The notation used above is as follows, $\boldsymbol{r} = (x, l, m)$, $\int d\boldsymbol{r} = \sum_{l,m} \int dx$, and $c_{\alpha,\sigma}(\boldsymbol{r})$ denotes the fermionic annihilation operator for particle moving on the

α sheet of the Fermi surface with spin σ. The Fermi sheet index α acts as internal quantum number that distinguishes right going electrons from left going electrons. In addition, $\tau^{(+1)} = (\sigma_x - i\sigma_y)/2$, $\tau^{(0)} = \sigma_z$, $\tau^{(-1)} = (\sigma_x + i\sigma_y)/2$, where σ_μ are the Pauli matrices with index $\mu = x, y$, or z. This representation is convenient when the direction of the magnetic field is chosen to be the quantization axis. It is also important to emphasize that when writing H_{int}, it was assumed that the instability towards singlet or triplet superconductivity are dominant with respect to charge or spin density wave instabilities. For a rather complete discussion of the competition among these instabilities (at zero magnetic field) see the review article by Solyom [71].

The order parameters for singlet (Δ_{s}) and triplet (Δ_{t,m_s}) superconductivity are respectively proportional to the average of singlet (O_{s}) and triplet (O_{t,m_s}) pair operators defined above. Thus, the singlet order parameter takes the form

$$\Delta_{\text{s}}(\boldsymbol{r}) = \frac{\lambda_{\text{s}}}{\sqrt{2}} \left[\langle c_{-,\uparrow}(\boldsymbol{r}) c_{+,\downarrow}(\boldsymbol{r}) \rangle - \langle c_{-,\downarrow}(\boldsymbol{r}) c_{+,\uparrow}(\boldsymbol{r}) \rangle \right], \tag{2.23}$$

and the triplet order parameter has three components:

$$\Delta_{\text{t},+1}(\boldsymbol{r}) = \lambda_{\text{t}} \left[\langle c_{-,\uparrow}(\boldsymbol{r}) c_{+,\uparrow}(\boldsymbol{r}) \rangle \right], \tag{2.24}$$

$$\Delta_{\text{t},0}(\boldsymbol{r}) = \frac{\lambda_{\text{t}}}{2} \left[\langle c_{-,\uparrow}(\boldsymbol{r}) c_{+,\downarrow}(\boldsymbol{r}) \rangle + \langle c_{-,\downarrow}(\boldsymbol{r}) c_{+,\uparrow}(\boldsymbol{r}) \rangle \right], \tag{2.25}$$

$$\Delta_{\text{t},-1}(\boldsymbol{r}) = \lambda_{\text{t}} \left[\langle c_{-,\downarrow}(\boldsymbol{r}) c_{+,\downarrow}(\boldsymbol{r}) \rangle \right]. \tag{2.26}$$

In what follows, the discussion will be limited to singlet pairing and to equal spin triplet pairing (ESTP). In the ESTP state, the order parameter component $\Delta_{t,0}$ with magnetic quantum number $m_s = 0$ is chosen to vanish. For those who are familiar with the A-phase of superfluid ^{3}He an immediate connection with the Anderson–Brinkman–Morel (ABM) state [50] can be made. However, it is important to emphasize that the ABM state, which is described for isotropic superconductors having p-wave symmetry, is not the ESTP state considered here due to the following reasons. First, quasi-1D systems are highly anisotropic. Second, equal spin pairing in quasi-1D systems occurs between electrons with different sets of quantum numbers, since electron pairs in the ESTP state are formed from electrons in opposite sheets of the Fermi surface, i.e., from states ($\alpha = -, \sigma$) and ($\alpha' = +, \sigma'$) in the absence of the magnetic field. In the case of zero magnetic field, the symmetry of the order parameter for a uniform singlet superconductor corresponds to positive lobes at both sheets of the Fermi surface [see Fig. 5.12(a) in Sec. 5.7]. In the case of a triplet superconductor, however, the order parameter has a positive lobe on one sheet and a negative lobe on the other [see Fig. 5.13(d) in Sec. 5.7]. Thus, the triplet state described by the minimal Hamiltonian of Eq. (2.18) corresponds to the ${}^3\mathrm{B}_{3\mathrm{u}}(a)$ weak spin-orbit coupling state ("p_x-wave") of the orthorhombic group described in Table 5.4 of Sec. 5.7. In the case of a finite magnetic field, however, the conventional Fermi spectrum is replaced by the eigenspectrum showed in Fig. 2.6(a), and pairing may occur between states $(\alpha = -, \sigma, N)$ and $(\alpha' = +, \sigma', N')$. It is this pairing of electrons in the magnetic sub-bands introduced by the magnetic field that leads to a superconducting instability.

2.6 Theoretical upper critical field along the b' direction

Since the FIDC is expected to occur first for a magnetic field applied along the y direction, we briefly discuss a calculation of the upper critical field $H_{c_2}^{b'}(T)$ for $\boldsymbol{H} \parallel \boldsymbol{b}'$. Detailed discussions including other directions are postponed until Secs. 5.3 and 5.4. Considering only singlet and equal spin triplet pairing (ESTP) superconductivity, the symbol η will be used below to indicate either the singlet case ($\eta = s, m_s = 0$) or the ESTP case ($\eta = t, m_s = \pm 1$). When the spin quantization axis is chosen to be along the direction of the applied magnetic field, the upper critical field is determined by the linearized order parameter equation

$$\Delta_\eta(x, z = mc) = \lambda_\eta \sum_N K_\eta(x, \hat{q}_x, N) \Delta_\eta\left(x, z = [m + N]c\right), \qquad (2.27)$$

where the Kernel

$$K_\eta(x, \hat{q}_x, N) = \exp\left(-2iNGx\right) F_{\eta,N}(\hat{q}_x - 2NG) \qquad (2.28)$$

is a highly nonlinear function of the operator $\hat{q}_x = -i\partial_x$ via the function

$$F_{\eta,N}(q_x) = \sum_{\alpha,\{\sigma_i\},M_1,M_2} \Lambda^{\alpha,\bar{\alpha}}_{\sigma_1,\sigma_2}(q_x + [M_1 + M_2]G) P^{\alpha,\bar{\alpha}}(M_1, M_2, N). \qquad (2.29)$$

Here, the weighting factor

$$P^{\alpha,\bar{\alpha}} = J_{M_1}(\alpha t_z/\omega_c) J_{M_2}(\bar{\alpha} t_z/\omega_c) J_{M_1 - N}(\alpha t_z/\omega_c) J_{M_2 - N}(\bar{\alpha} t_z/\omega_c) \qquad (2.30)$$

reflects the symmetry of the quantum levels in a magnetic field. This weighting factor contains the pairing structure between different magnetic sub-bands labeled by indices (M_1, M_2) and $(M_1 - N, M_2 - N)$, and behaves as a renormalization factor of the 2D generalized pair susceptibility

$$\Lambda_{\sigma_1,\sigma_2}^{\alpha,\bar{\alpha}}(Q_x) = T \sum_{i\omega_l,k_x,k_y} G_{\alpha,\sigma_1}(i\omega_l, k_x, k_y) G_{\bar{\alpha},\sigma_2}(-i\omega_l, -k_x + Q_x, -k_y), \quad (2.31)$$

where $Q_x = q_x + (M_1 + M_2)G$ and ω_l represents fermionic Matsubara frequencies. The final result for $\Lambda_{\sigma_1,\sigma_2}^{\alpha,\bar{\alpha}}(Q_x)$ is given by

$$\Lambda_{\sigma_1,\sigma_2}^{\alpha,\bar{\alpha}}(Q_x) = \frac{1}{2\pi v_{\mathrm{F}}} \left\{ \Psi\left(\frac{1}{2}\right) - \mathrm{Re}\left[\Psi\left(\frac{1}{2} + \frac{E_x}{4\pi i T}\right)\right] + \ln\left(\frac{2\omega_d \gamma_0}{\pi T}\right) \right\}, \quad (2.32)$$

where $E_x = [-v_{\mathrm{F}} Q_x + \alpha \mu_{\mathrm{B}} H(\sigma_1 - \sigma_2)]$, ω_d is the cut-off frequency, $\Psi(x)$ is the Digamma Psi function defined as the logarithmic derivative of Gamma function $\Psi(x) \equiv d \ln \Gamma(x)/dx$, and $\gamma_0 = e^{\gamma} \approx 1.7811$ with γ the Euler constant. Notice that the Cooper singularity is recovered when $E_x = 0$.

Eq. (2.27) describes a linearized Ginzburg-Landau (GL) equation, which is non-local, i.e., the gap function at $(x,\ z = mc)$ and the gap function at $(x,\ z = [m + N]c)$ are coupled. The integer N labels different x-y planes at a distance Nc away from the plane $z = mc$. This is characteristic of a Lawrence-Doniach model [72] as encountered in weakly coupled layered superconductors, where a Josephson coupling between neighboring planes exists. At low magnetic fields, where the semiclassical approximation is valid, the Josephson coupling is magnetic field *independent*. However, notice that the coefficient $K_\eta(x, \hat{q}_x, N)$ in Eq. (2.28), which controls the coupling between different x-y planes and contains information about the quantum level structure in a magnetic field, is strongly magnetic field *dependent* beyond the semiclassical regime. As a consequence, the Josephson coupling between different x-y planes has its magnitude and range controlled by the magnetic field and the quantum level structure.

The H-T phase diagram for both singlet and triplet superconductivity can be obtained by solving the order parameter equation (2.27). At non-zero, but low magnetic fields where the Zeeman splitting is not so important, the gap equation becomes

$$\Delta_\eta(x, q_z) = \left\{ \xi_x^2 \frac{\partial^2}{\partial^2 x} + \Gamma_{z,\eta} \left[1 - \cos\left(q_z c - G x\right)\right] \right\} \Delta_\eta(x, q_z) \quad (2.33)$$

after a Fourier transformation for the z coordinate. Here, the function $\Gamma_{z,\eta} = 2\left[\xi_{z,\eta}(T)/c\right]^2$ is the height of the periodic potential $\cos(q_z c - Gx)$. Notice that $\Gamma_{z,\eta}$ is independent of the magnetic field, and recall that

$$\xi_{z,\eta}(T) = \xi_{z,\eta}^{(\mathrm{GL})} \left|1 - T/T_{\mathrm{c},\eta}(0)\right|^{-1/2}, \tag{2.34}$$

where the GL prefactor is given by $\xi_{z,\eta}^{(\mathrm{GL})} = 0.1632 c \theta_\eta$. Thus, it is the parameter $\theta_\eta = t_z/T_{\mathrm{c},\eta}(0)$ that controls the strength of the coupling of the x-y planes. When $\theta_\eta \gg 1$, the GL prefactor $\xi_{z,\eta}^{(\mathrm{GL})} \gg c$, and the x-y planes are strongly coupled exhibiting typical 3D behavior. On the other hand, when $\theta_\eta \ll 1$, the GL prefactor $\xi_{z,\eta}^{(\mathrm{GL})} \ll c$, and the x-y planes are weakly coupled exhibiting a nearly 2D behavior. Notice that Eq. (2.33) is nothing but the Mathieu equation that has been previously obtained by Klemm, Luther and Beasley (KLB) [24, 73] in connection to upper critical fields for quasi-two-dimensional layered superconductors using the Lawrence–Doniach model [72]. At low magnetic fields, the maximal upper critical field $H_{\mathrm{c}_2,\eta}(T)$ always occurs when $q_z = 0$, and the suppression of the critical temperature is similar for both singlet and triplet superconductors, leading to

$$T_{\mathrm{c},\eta}(H) = T_{\mathrm{c},\eta}(0)\left[1 - \kappa \frac{t_z \omega_{\mathrm{c}}}{\{T_{\mathrm{c},\eta}(0)\}^2}\right], \tag{2.35}$$

where $\kappa = 0.0754$.

At intermediate magnetic fields the semiclassical approximation used to derive Eq. (2.35) is no longer applicable, since the magnetic field induced quantum level structure has been completely ignored. In this case, it is necessary to solve the order parameter equation (2.27) with the full Kernel, and no simplification can be made. Fortunately, for high magnetic fields ($\omega_{\mathrm{c}} \gg t_z$), the gap equation simplifies again to

$$\Delta_\eta(x, q_z) = \left\{\xi_x^2 \frac{\partial^2}{\partial^2 x} + \Gamma_{z,\eta}^{(\mathrm{QL})}\left[1 - \cos(q_z c - Gx)\right]\right\} \Delta_\eta(x, q_z) \tag{2.36}$$

provided that the condition $|\omega_{\mathrm{c}} \pm 4\eta\mu_{\mathrm{B}} H| \gg T_{\mathrm{c},\eta}(H)$ is satisfied. The eigenvalue Eq. (2.36) is also a Mathieu equation similar to Eq. (2.33) at low magnetic fields, but this mathematical similarity corresponds to a completely different physical situation. In this high field regime ($\omega_{\mathrm{c}} \gg t_z$), the single particle electronic wavefunctions become confined along the z direction, and the extreme quantum limit of very few quantum levels is reached. Thus the potential height $\Gamma^{(\mathrm{QL})}_{z,\eta}$ is now dependent on both magnetic field and temperature. If one insists in writing

$$\Gamma^{(\mathrm{QL})}_{z,\eta} = 2\left[\xi^{(\mathrm{QL})}_{z,\eta}(T,H)/c\right]^2, \tag{2.37}$$

then the magnetic field *dependent* coherence length $\xi^{(\mathrm{QL})}_{z,\eta}(T,H)$ can be defined as

$$\xi^{(\mathrm{QL})}_{z,\eta}(T,H) = \xi^{(\mathrm{QL})}_{z,\eta}(0,H)\left|1 - \frac{T}{T^{(\mathrm{2D})}_{\mathrm{c},\eta}(H)}\right|^{-1/2}, \tag{2.38}$$

where the extreme quantum limit (QL) prefactor is

$$\xi^{(\mathrm{QL})}_{z,\eta}(0,H) = \frac{t_z c}{\sqrt{2}\omega_{\mathrm{c}}} A_\eta(H), \tag{2.39}$$

and the auxiliary function $\mathrm{A}_\eta(H)$ is given by

$$A_\eta(H) = \ln\left[\frac{\gamma_0|\omega_{\mathrm{c}}|}{\pi T^{(\mathrm{2D})}_{\mathrm{c},\eta}(H)}\sqrt{\left|1-(1-\eta)\left[\frac{\omega_{\mathrm{c}}}{4\mu_{\mathrm{B}}H}\right]^2\right|}\right]. \tag{2.40}$$

The "2D" temperature for the singlet case is

$$T^{(\mathrm{2D})}_{\mathrm{c,s}}(H) = \frac{\pi T^2_{\mathrm{c,s}}(0)}{4\gamma_0\mu_B H}, \tag{2.41}$$

while for the triplet case is

$$T^{(\mathrm{2D})}_{\mathrm{c,t}}(H) = T_{\mathrm{c,t}}(0). \tag{2.42}$$

Therefore, the critical temperature is given by

$$T_{\mathrm{c},\eta}(H) = T^{(\mathrm{2D})}_{\mathrm{c},\eta}(H)\left\{1 - \left(\frac{t_z}{\omega_{\mathrm{c}}}\right)^2 A_\eta(H)\right\} \tag{2.43}$$

in the high magnetic field regime where $|\omega_{\mathrm{c}} \pm 4\eta\mu_{\mathrm{B}}H| \gg \max\{t_z, T_{\mathrm{c},\eta}(H)\}$ is satisfied.

As an example of the upper critical field analysis, we show in Fig. 2.7 numerical results of $H^{b'}_{\mathrm{c}_2}(T)$ for parameters $T_{\mathrm{c,s}}(0) = T_{\mathrm{c,t}}(0) = 1.5\,\mathrm{K}$, and $t_z = 5\,\mathrm{K}$ (i.e., $\theta_\eta = 3.33$). This choice of parameters is not too different from those describing the experimental situation encountered in $(TMTSF)_2X$. For simplicity, the triplet (λ_t) and singlet (λ_s) coupling constants are assumed to be identical in Fig. 2.7, but this doesn't need to be the case. When $\lambda_s = \lambda_t$ the critical temperatures $T_{c,s}(0)$ and $T_{c,t}(0)$ at zero magnetic field are identical. For $\theta_\eta = 3.33$ the coherence length $\xi_{z,\eta}^{(GL)} = 0.5440c$, and the x-y planes are already weakly coupled at zero magnetic field, so superconductivity is already nearly two-dimensional. In this case, the magnetic field reduces the coupling in an already weakly coupled system, making the superconductor even more 2D.

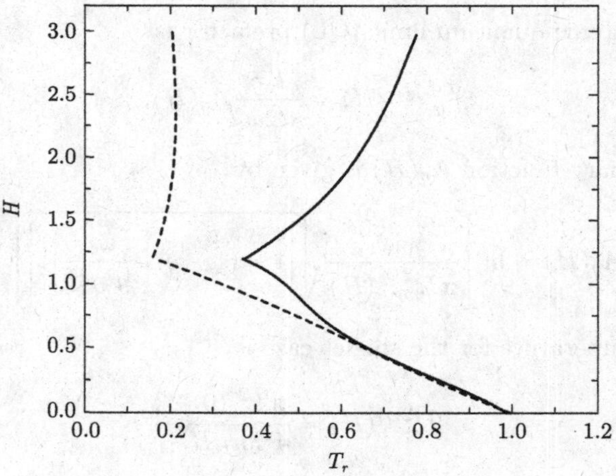

Figure 2.7 Upper critical field for parameters $T_{c,s}(0) = T_{c,t}(0) = 1.5\,K$, and $t_z = 5$ K (i.e., $\theta_\eta = 3.33$). The magnetic field H is in tesla, and $T_r = T/T_{c,\eta}(0)$. The solid line corresponds to the triplet case, and the dashed line to the singlet case. From Ref. [63].

In summary, when $t_z \gg T_{c,\eta}(0)$, i.e., $\theta_\eta \gg 1$, the GL coherence length $\xi_{z,\eta}^{(GL)} \gg c$ at low magnetic fields. Thus the x-y planes are strongly coupled exhibiting 3D anisotropic superconductivity. But at high magnetic fields, from Eq. (2.39), it follows immediately that $\xi_{z,\eta}^{(QL)}(0, H) \ll c$, when $\omega_c \gg t_z$. Thus, the increase of magnetic fields tends to make the coupling between x-y planes weaker and hence make the system be a nearly 2D anisotropic superconductor. As a result, a magnetic field induced dimensional crossover (FIDC) from

3D highly anisotropic to nearly 2D highly anisotropic superconductivity occurs. This FIDC is intrinsically a quantum effect due to the quantum level structure induced by the applied magnetic field. This effect should be contrasted with the temperature induced dimensional crossover (TIDC) proposed by KLB [24, 73] for quasi-two-dimensional superconductors in parallel magnetic fields, where the quantum effects of the magnetic field are neglected and thus play no role.

2.7 Qualitative effects on non-magnetic impurities

The effects of non-magnetic impurities is very important in order to determine the symmetry of the order parameter in Bechgaard salts, both at low and high magnetic fields. Typically non-magnetic impurities can be introduced by alloying, e.g., $(TMTSF)_2ClO_4$ or $(TMTSF)_2PF_6$ with ReO_4 or other anions [52]. Strictly at zero magnetic field, impurity scattering does not affect the critical temperature of a singlet s-wave system. This is a consequence of the Anderson's theorem which indicates that the critical temperature is independent of weak disorder for a fully gapped system with time reversal symmetry. However, the critical temperature $T_c(0)$ of $(TMTSF)_2X$ systems at zero magnetic field is sensitive to the presence of impurities [2, 3, 4, 5, 6, 74], which suggests that non-s-wave pairing may be possible [7, 8]. For instance, assuming identical critical temperatures for s-wave singlet and p-wave triplet superconductors at zero magnetic field, i.e., $T_{c,s}(0) = T_{c,t}(0)$ (as assumed in Fig. 2.7), the effect of non-magnetic impurities on the s-wave singlet state is far less dramatic than that in the case of p-wave triplet state at low magnetic fields. On the other hand, at intermediate magnetic fields the superconducting state is not formed from time-reversed pairs and it is expected to be sensitive to impurities both in the s-wave singlet and p-wave triplet cases. To leading order of $1/E_F\tau$, where τ is the elastic scattering time, impurity scattering is taken into account by including self-energy and vertex corrections into the particle-particle susceptibility K_η define in Eq. (2.28). The critical temperature in the presence of impurities $T_c^{(\text{imp})}(H)$ can be calculated using a method originally proposed to treat non-magnetic impurities in conventional superconductors [8].

In the regime of intermediate to high magnetic fields, the pair breaking parameter

$$\gamma_\eta(H) = \frac{\pi}{8\tau T_{c,\eta}(H)}(1 - U_\eta) \tag{2.44}$$

determines the critical temperature deviation $\varepsilon_\eta = |T_{c,\eta}^{(imp)}(H)/T_{c,\eta}(H) - 1|$ in the presence of impurities. The parameter U_η for the singlet case is

$$U_s = \frac{1}{2}\left(1 + \frac{\ln[2\omega_d\gamma_0/\pi T_{c,s}(0)]}{2\ln[2\omega_d\gamma_0/\pi T_{c,s}(H)]}\right)\left[1 + \left(\frac{T_{c,s}(H)}{\mu_B H}\right)^2\right], \tag{2.45}$$

while the parameter U_η for the triplet case is

$$U_t = \frac{\ln[2\omega_d\gamma_0/\pi T_{c,t}(0)]}{\ln[2\omega_d\gamma_0/\pi T_{c,t}(H)]}. \tag{2.46}$$

For $(TMTSF)_2ClO_4$ and $(TMTSF)_2PF_6$, the inverse scattering time τ^{-1} is very small ($\tau^{-1} \approx 100$ mK), indicating that these systems are very clean. Their optimal transition temperature is in the range $T_{c,\eta}(0) \approx 1.0 - 1.5$ K. If the parameter $t_z \approx 10 - 30$ K, then the pair breaking parameter γ_η may be of order ~ 1 both in the singlet and triplet cases for intermediate values of magnetic field. Thus, both singlet and triplet superconductivity may be destroyed in this field range. If the parameter $t_z \approx 5 - 10$ K, then γ_η may be of order ~ 1 only in the singlet case for intermediate values of magnetic field. Thus, only singlet superconductivity may be destroyed in this field range. Finally, if $t_z \approx 2 - 5$ K then the pair breaking parameter γ_η is always less than 0.02 for all temperatures and fields. Therefore, both singlet and triplet superconductivity at intermediate and high magnetic fields are robust with respect to weak disorder.

With this analysis about non-magnetic impurity effect, we conclude our theoretical background on singlet and triplet superconductivity of quasi-1D systems in a magnetic field. Materials covered by this introductory discussion are mostly pre-dating key experiments regarding the superconducting states, which are discussed in the next two chapters for PF_6 and ClO_4, respectively.

3

Recent Experiments on $(TMTSF)_2PF_6$

The initial suggestion by Lebed [16], Burlachkov, Gorkov and Lebed (BGL) [17], and Dupuis, Montambaux and Sá de Melo (DMS) [18] that quasi-1D superconductors like $(TMTSF)_2ClO_4$ and $(TMTSF)_2PF_6$ are good candidates for triplet superconductivity in low and high magnetic fields motivated several experimental groups to study further these materials. Some preliminary results obtained by Lee *et al.* [21, 75] on the upper critical field of $(TMTSF)_2ClO_4$ indicated very interesting behavior in these materials, in particular at high magnetic fields. After that, a wave of new experiments followed for both $(TMTSF)_2ClO_4$ and the sister compound $(TMTSF)_2PF_6$. In this chapter, we introduce some important experiments that were performed after 1997 on $(TMTSF)_2PF_6$, which are intimately related to the understanding of pairing symmetry and mechanism of this quasi-1D superconductor. In Sec. 3.1, we review measurements of the upper critical field performed by Lee *et al.* [23] and the possible connections to triplet superconductivity. Next in Sec. 3.2, we discuss the Knight shift and NMR measurements by Lee *et al.* [33, 35]. Both the upper critical fields and NMR measurements are important to clarify the pairing symmetry of the superconducting order parameter.

In addition to these experiments devoted to direct exploration of the superconducting symmetry in $(TMTSF)_2PF_6$, there are more recent experimental works concentrating on the effects of spin density wave (SDW) and normal phases, which are both adjacent to the superconducting phase. In Sec. 3.3, we analyze three transport measurements by Vuletić *et al.* [76], Kornilov *et al.* [77], and Kang *et al.* [78], which are performed in the region close to the SDW–SC phase transition line, as well as a simultaneous magneto-resistance and NMR measurements by Lee *et al.* [79]. In Sec. 3.4, we discuss a resistivity measurement in the normal state with pressure and temperature close to the superconducting transition point. The non-Fermi liquid behavior of resistivity indicates that the antiferromagnetic ordering present in proximity may have important effect on Cooper pairing.

This area of research is currently very dynamic, and it is certain that by the time we finish writing this book several new experiments would be accomplished. So we ask for an immediate apology to many of our experimental colleagues for not being able to catch up with their exciting works. For experimental works on quasi-1D superconductors prior to 1997, the reader is referred to the comprehensive reviews by Jerome and Schultz [80], Bechgaard and Jerome [81], and to the excellent books by Ishiguro, Yamaji and Saito [52] and Lebed [82].

3.1 Upper critical field

Surprisingly, prior to 1997, experimental investigations on the upper critical field and its anisotropy [3, 83, 84, 85] had not been undertaken in the temperature and field regimes where anomalous effects were theoretically predicted, i.e., $T \ll T_c$ and $(H/T_c) \gg (dH_{c_2}/dT)_{T_c}$. This regime started to be explored only after the initial experimental work of Naughton's and Chaikin's groups [23], where the upper critical field of $(TMTSF)_2PF_6$ was studied at pressure $P \approx 6.0\,\text{kbar}$. This pressure is large enough to suppress the nearby spin density wave phase and allow the existence of a metallic state at high temperatures and of a superconducting state at low temperatures [86]. (The insulator–superconductor transition pressure P_c is about 5.9 kbar in their calibrated pressure scale.) In these experiments, Lee *et al.* [23] extracted the critical temperature $T_c(H)$ as a function of magnetic field from resistance measurements for magnetic fields precisely aligned with the three principal directions $\boldsymbol{a}$, $\boldsymbol{b}'$ and $\boldsymbol{c}^*$. The phase diagram found is consistent with previous low field studies [3, 9, 83, 84, 87, 88] of $(TMTSF)_2ClO_4$, and the upper critical fields obey the relation $H^a_{c_2} > H^{b'}_{c_2} > H^{c^*}_{c_2}$ for temperatures close to $T_c(0) = 1.13\,\text{K}$. However, three new features were observed at low temperatures. First, the upper critical field displays pronounced upward curvature without saturation for $\boldsymbol{H} \parallel \boldsymbol{a}$ and $\boldsymbol{H} \parallel \boldsymbol{b}'$. Second, both $H^a_{c_2}$ and $H^{b'}_{c_2}$ exceed the Pauli paramagnetic limit [11, 12]. Third, $H^{b'}_{c_2}$ becomes larger than $H^a_{c_2}$ at low temperatures.

The temperature dependence of the resistivity in the experiment by Lee *et al.* [23] is shown in Fig. 3.1 for several values of magnetic field applied along $\boldsymbol{b}'$ axis. The normal state is metallic ($\partial\rho/\partial T > 0$) in zero applied field. However, it changes rapidly with increasing field such that $\partial\rho/\partial T < 0$ for $H > 1\,\text{T}$. This temperature dependence possibly results from a magnetic-field-induced interlayer decoupling caused by the application of an in-plane (x-y) magnetic field.

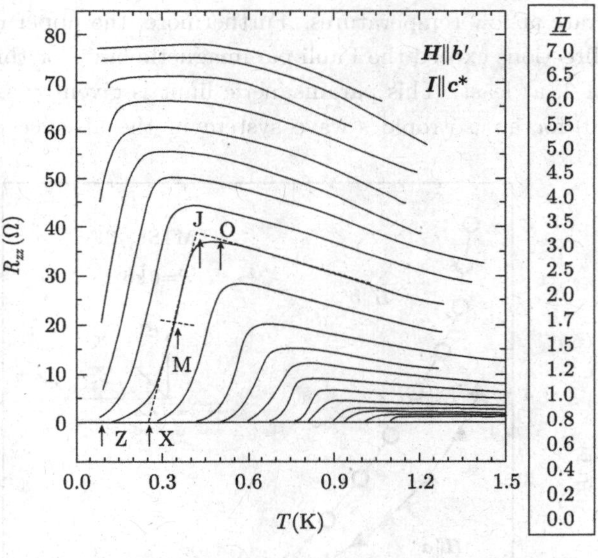

Figure 3.1 $(TMSTF)_2PF_6$ interlayer resistance versus temperature for various magnetic fields $\boldsymbol{H} \parallel \boldsymbol{b}'$ at $P = 6\,\text{kbar}$. Five different criteria for the putative critical temperature $T_c(H)$ are indicated: O (onset), J (junction), M (midpoint), X ($R \to 0$), and Z ($R = 0$). From Ref. [23].

Characterizing the superconducting transition temperature from resistivity curves in Fig. 3.1 is not a trivial task, since the transition is broad, and incomplete at high fields. Thus, Lee *et al.*[23] used five criteria to extract the critical temperature as a function of magnetic field. These criteria are defined as: (i) an onset T_O, (ii) a "junction" T_J, (iii) a midpoint T_M, (iv) a zero resistance extrapolation T_X (where the tail near $R = 0$ is ignored), and (v) a zero resistance point T_Z. Using this approach, they were able to connect the resulting curves $T_i(H)$ (for each criteria i) with the physical upper critical field $H_{c_2}(T)$. Measurements of $T_i(H)$ for $\boldsymbol{H} \parallel \boldsymbol{c}^*$ and for $\boldsymbol{H} \parallel \boldsymbol{a}$ were also performed, and the same five criteria were used to extract $H_{c_2}(T)$. Fig. 3.2 shows the H-T phase diagram using the *junction* criterion (T_J) for all three principal directions $\boldsymbol{a}$, $\boldsymbol{b}'$ and $\boldsymbol{c}^*$. Similar phase diagrams result from the application of any other criteria.

There are two unusual features embedded in the phase diagram of Fig. 3.2. The first one is that the upper critical fields for $\boldsymbol{H} \parallel \boldsymbol{a}$ and for $\boldsymbol{H} \parallel \boldsymbol{b}'$ do not seem to saturate at low temperatures. Furthermore, the upper critical fields along these directions exceed the Pauli paramagnetic limit for this compound by a factor of 2, at least. This paramagnetic limit is given by $H_P(T = 0) = 1.84T_c\,(H = 0)$ for an isotropic s-wave system in the absence of spin-orbit

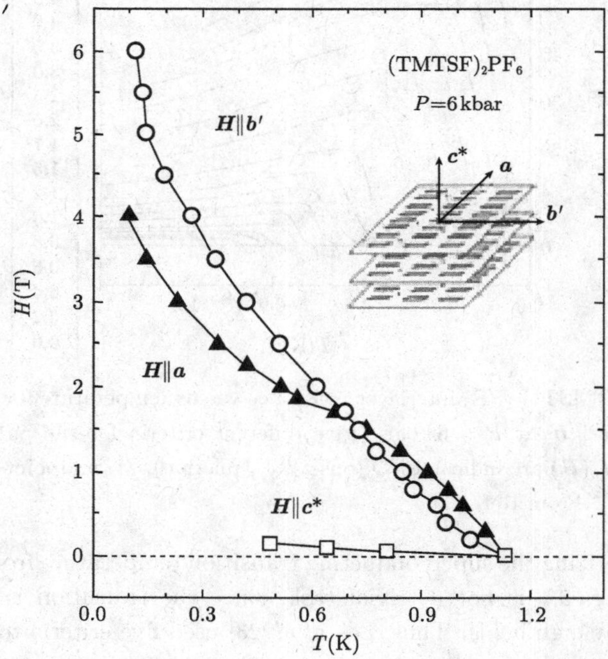

Figure 3.2 The magnetic field versus temperature (H-T) phase diagram for $(TMTSF)_2PF_6$ using the junction criterion for magnetic fields aligned along the three principal axis $\boldsymbol{a}$, $\boldsymbol{b}'$, and $\boldsymbol{c}^*$. From Ref. [23].

coupling [11, 12], or by $H_P(T=0) = 1.58T_c(H=0)$ for an anisotropic s-wave singlet pairing [15]. In the case of $(TMTSF)_2PF_6$ these estimates correspond to 2.1 T and 1.8 T, respectively. At low temperatures, $H_{c_2}^{b'} > 3H_P$ for the T_O and T_J criteria, and $H_{c_2}^{b'} > 2H_P$ for T_M and T_X.

The second unusual feature is the anisotropy inversion that occurs above the characteristic field $H^* \approx 1.6$ T, where $H_{c_2}^{b'}$ becomes larger than $H_{c_2}^{a}$. The fact that this anisotropy inversion was not seen prior to their work, is attributed to the strong sensitivity of H_{c_2} to sample alignment with respect to the external magnetic field. For instance, a tilt of 0.1° away from $\boldsymbol{b}'$ towards $\boldsymbol{c}^*$ is sufficient to bring $H_{c_2}^{b'}$ below $H_{c_2}^{a}$ at low temperatures. Another possible reason is that early upper critical field measurements in Bechgaard salts were performed on $(TMTSF)_2ClO_4$ (at $P = 0$), or on $(TMTSF)_2PF_6$ and $(TMTSF)_2AsF_6$ at higher pressures. However, the pressure used in this experiment is very close to the critical value for the suppression of the insulating SDW phase, and yields a maximized critical temperature T_c. In Fig. 3.3 the anisotropy inversion is shown in detail under four different criteria (T_O, T_J, T_M and T_X). Lee *et al.* [23] had fewer data points for the $R = 0$ (T_Z) criterion, however they found that the anisotropy inversion still occurs at $T \approx 0.35$ K.

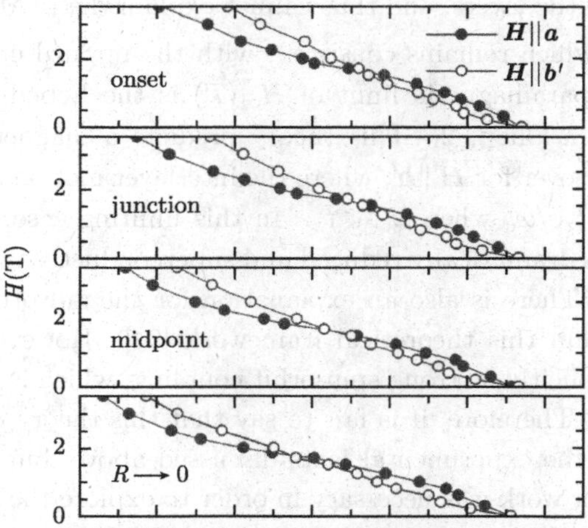

Figure 3.3 This figure shows the anisotropy inversion along $\boldsymbol{a}$ and $\boldsymbol{b}'$ for four different resistance criteria: onset, junction, midpoint and $R \to 0$. From Ref. [23].

Notice in Fig. 3.3 that the data sets look very similar. The shift of the anisotropy inversion temperature is due to the width of the superconducting transition, however the anisotropy inversion field $H^* \approx 1.6\,\mathrm{T}$ is not sensitive to the criterion used.

The unusual upward curvature and excess of Pauli limit for $H_{c_2}(T)$ can be related to interpretations from existing theories of layered superconductors. In a theory by Ovchinnikov and Kresin [89], it was shown that a upward curvature in $H_{c_2}(T)$ results at low temperatures, because the pair breaking ability of magnetic impurities is reduced as $T \to 0$. However magnetic scattering is essentially negligible in $(TMTSF)_2PF_6$ since the PF_6 ion is non-magnetic. Strong upward curvature was also discussed by Kotliar and Varma [90] in the context of cuprate superconductors for magnetic fields normal to the layers. They suggested that the proximity to a quantum critical point is responsible for the upward curvature in $H_{c_2}(T)$ at low temperatures. They found that $H_{c_2}(T) \approx 1 - t^{\alpha}$, with $t = T/T_c$ and $\alpha = 2/5$ would fit a small portion of the experimental data close to $T = 0$ for cuprate superconductors. Although the above expression agrees reasonably well with the measured $H_{c_2}^{b'}(T)$ over the entire temperature range $0.1 < t < 1$, this theory is only applicable to fields perpendicular to the layers, and thus cannot explain the $(TMTSF)_2PF_6$ data.

One theory which remains consistent with the upward curvature and the excess of Pauli paramagnetic limit of $H_{c_2}(T)$ is the Lebed–BGL–DMS scenario discussed in Chap. 2. This theory predicts a magnetic field induced dimensional crossover for $\boldsymbol{H} \parallel \boldsymbol{b}'$, where the interlayer motion along z direction is confined to $\pm t_z c/\omega_c$ when $\omega_c \ll t_z$. In this limiting case, the orbital pair breaking effect is dramatically reduced and superconductivity survives at high magnetic fields. There is also an explanation for the anisotropy inversion at $H^* \approx 1.6\,\mathrm{T}$ within this theoretical framework [34]. However, this explanation requires implicitly a strong spin-orbit coupling, which is very unlikely for Bechgaard salts. Therefore, it is fair to say that this theory can qualitatively explain some of the experimental facts discussed above, but not all of them. Moreover, further work was necessary in order to explore the symmetry of the superconducting order parameter in more detail. In the absence of phase sensitive experiments like those performed in high-T_c superconductors [91, 92, 93], Knight shift experiments are the next natural candidate to get information about the spin structure of the superconducting state.

3.2 ^{77}Se Knight shift and nuclear-magneto-resonance measurements

Knight shift is a very important method to measure electron spin susceptibility. It is defined as the shift of nuclear-magneto-resonance (NMR) resonant frequency due to interaction between nuclear and electron spins. In the presence of a magnetic field $\boldsymbol{H}$, nuclear spins in a material precess with the frequency $\omega \propto H$. At this frequency a nuclear magnetic resonance can be detected, i.e., there is a resonant absorption of energy from a radio-frequency magnetic field $\boldsymbol{H}_{\text{rf}}$ with polarization perpendicular to the constant magnetic field $\boldsymbol{H}$. Due to the finite probability of an electron to be located at the nucleus site R_n, there is an interaction between the nuclear magnetic moments and those of the conducting electrons, $H_{\text{int}} \propto \boldsymbol{m}_{\text{el}}\delta(r - R_n)$. The average value of $\boldsymbol{m}_{\text{el}}$ in this external field is given by $\langle \boldsymbol{m}_{\text{el}} \rangle \propto \chi H$, where χ is the electronic paramagnetic susceptibility. Thus, the nuclear spin energy levels due to this interaction shift the NMR frequency by $\delta\omega \equiv K_{\text{s}} \propto \chi H$, which defines the Knight shift K_{s}.

In a superconductor with singlet pairing, Cooper pairs do not contribute to the spin paramagnetic susceptibility since $m_{\text{s}} = 0$, and the entire spin magnetic moment is determined by the contribution from elementary excitations. Therefore, the electronic spin susceptibility for a singlet superconductor is

$$\chi = \frac{\partial M}{\partial H} = \chi_{\text{n}} Y(T), \tag{3.1}$$

where χ_{n} is the normal state spin susceptibility and the function $Y(T)$ is the Yosida function, which determines the fraction of normal electrons in a superconductor. In the singlet case, $Y(T)$ vanishes at $T \to 0$, which means that the Knight shift K_{s} will be strongly suppressed at low temperatures below the superconducting transition. On the contrary, in an equal spin triplet pairing (ESTP) superconductor, the Knight shift will remain unchanged upon cooling into the superconducting state, and will not vanish as $T \to 0$. Detailed discussions can be found in Leggett's review [50] for ^{3}He, and in Sec. 5.7 for a triplet superconductor [37].

There is an extensive NMR literature on both ambient pressure spin density wave [94] and pressurized metallic [95, 96] phases of $(TMTSF)_2PF_6$, prior to the experiment on ^{77}Se atoms by Lee *et al.* [33, 35]. Most of the previous works included studies of local magnetic environments of either protons in methyl groups or ^{13}C spin-labeled on various inequivalent sites. However, band structure calculations and electronic paramagnetic resonance (EPR) studies suggest that the largest spin densities associated with the conduction band are closely linked with molecular orbitals of the Se atoms [97]. Since the Knight shift is proportional to the magnitude square of the electronic wave function at the ionic nucleus, ^{77}Se is a natural choice for NMR studies. Under this consideration, Lee *et al.* [33, 35] performed NMR measurements on ^{77}Se of $(TMTSF)_2PF_6$ in the superconducting state. To ensure that the pressurized sample is superconducting while acquiring the NMR data, they conducted simultaneous transport measurements in parallel (synchronization) to the application of radio-frequency pulses. From the measured spectra, Lee *et al.* [33, 35] concluded that there is no change in the Knight shifts of ^{77}Se when $\boldsymbol{H} \parallel \boldsymbol{a}$ or $\boldsymbol{H} \parallel \boldsymbol{b}'$.

The absorption spectra is shown in Fig. 3.4 (from Ref. [33]). The lower set

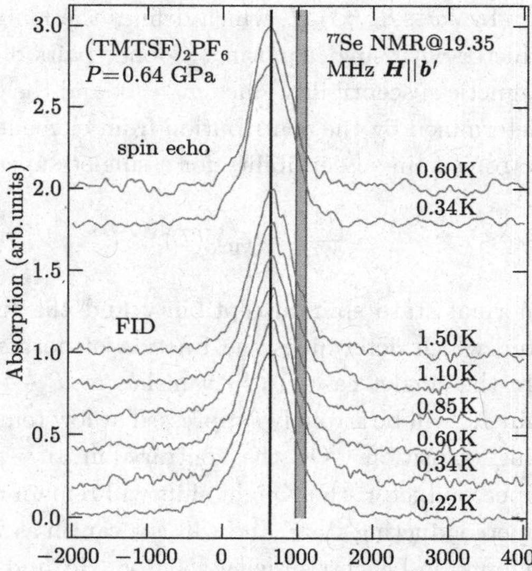

Figure 3.4 NMR ^{77}Se absorption spectra collected at temperatures below and above T_c for a magnetic field $H = 2.38\,\mathrm{T}$ applied along $\boldsymbol{b}'$ axis. The solid line indicates the measured first moment, and the shaded region indicates the expected first moment for a singlet ground state. From Ref. [33].

of spectra is collected as free induction decays (FIDs), and the upper as spin echoes. These spectra were recorded at temperatures above and below T_c, for a magnetic field aligned parallel to the layers to within 0.1° and parallel to the $\boldsymbol{b}'$ axis to within $\approx 5°$. Notice that there is no change in the absorption peak (marked by the vertical solid line in Fig. 3.4), to within the experimental error bars ($\delta K_s = \pm 20$ ppm). The lack of any observable difference between the spectra as the temperature is varied indicates that the system is not a singlet superconductor.

The vertical shaded region in the same figure corresponds to the estimated range where the center of the spectrum would have been if the spin susceptibility had vanished. This expected Knight shift window is extrapolated from the NMR frequency shift K in the normal state as a function of spin susceptibility χ_s for $\boldsymbol{H} \parallel \boldsymbol{b}'$, as shown in Fig. 3.5. The spin contribution to K (Knight shift

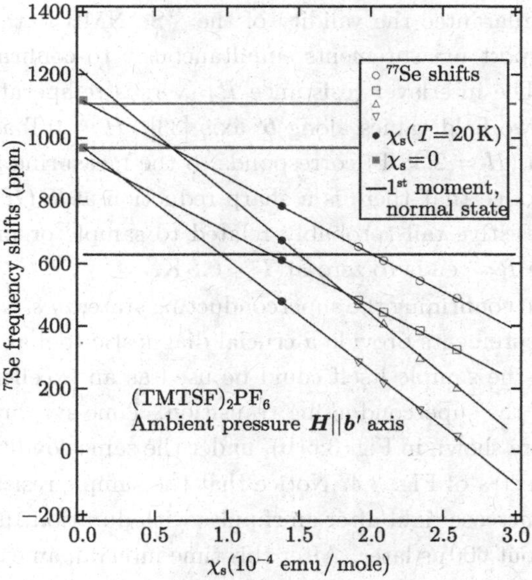

Figure 3.5 NMR shifts K vs. the spin susceptibility χ_s for magnetic fields applied along the $\boldsymbol{b}'$ axis. The bottom solid black line is the first moment in the normal state, and the top two horizontal solid blue lines indicate the window of the expected shift at $T = 0$ for a singlet superconductor, and they correspond to the shaded region in Fig. 3.4. From Ref. [33].

K_s) is proportional to χ_s, which can be extracted by comparison to the results of Miljak *et al.* [98]. The total shift K are measured relative to the NMR line (at ν_0) of ^{77}Se in $Se(CH_2)_2$, i.e., $K = (\nu - \nu_0)/\nu_0$. As shown in Fig. 3.5, the difference between the normal state shift $K(T = 20\,\mathrm{K})$ and the extrapolation values $K(\chi_s = 0)$ to $\chi_s = 0$ describes the hyperfine coupling to the electron spins, and hence corresponds to the expected Knight shift K_s for a singlet superconducting ground state.

The vertical lines that bound the shaded region in Fig. 3.4 mark the corresponding first moment at 340 – 480 ppm above the measured value. At the measuring field of $H = 2.38\,\mathrm{T}$, this corresponds to about 6 – 9kHz, with an estimated uncertainty of about 1 kHz. Lee *et al.* [35] also discovered that ^{77}Se spectroscopy with the field aligned along $\boldsymbol{a}$ axis was more sensitive. Thus, they worked at much lower fields $H = 1.43\,\mathrm{T}$ for $\boldsymbol{H} \parallel \boldsymbol{a}$ and observed similar results.

In order to guarantee the validity of the ^{77}Se NMR data, Lee *et al.* [33] performed transport measurements simultaneously to confirm the superconducting state. The interlayer resistance R_{zz} versus temperature is shown in Fig. 3.6(a) for two field values along $\boldsymbol{b}'$ axis [33]: $H = 0\,\mathrm{T}$ and $H = 2.38\,\mathrm{T}$. The second field ($H = 2.38\,\mathrm{T}$) corresponds to the measuring field in the ^{77}Se experiments. Notice that there is a sharp reduction at $T_c(H = 0) = 1.18\,\mathrm{K}$, followed by a resistive tail (probably related to sample or pressure inhomogeneities) before R_{zz} tends to zero at $T = 0.8\,\mathrm{K}$.

In addition to confirming the superconducting state, *in situ* transport along with NMR measurements provide a crucial diagnostic tool at the lowest temperatures, since the sample itself could be used as an excellent thermometer in the middle of the superconducting transition. Time-synchronous R_{zz} measurements [33] are shown in Fig. 3.6(b), under the same conditions as the data that give the spectra of Fig. 3.4. Notice that the sample resistance (temperature) begins to increase right after an rf pulse with duration 1μs, and reaches a maximum at about 600 μs later. After this time interval, an exponential decay is observed with a time constant of $\sim$ 1 ms. According to Lee *et al.* [33], the shape of the heating curve suggests that the NMR coil is the cause of heating. However, since the Knight shift measurement is completed in less than 100 μs, the sample probably does not experience any heating during this time interval.

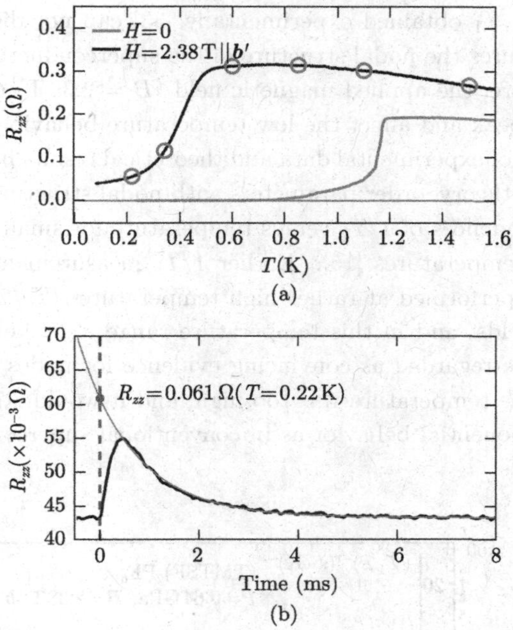

Figure 3.6 (a) Interlayer resistance R_{zz} versus temperature at zero applied magnetic field and at the ^{77}Se measuring field of $H = 2.38\,\mathrm{T}$; (b) Time-synchronous resistance measurements triggered and recorded simultaneous with the radio-frequency (rf) pulses for the NMR measurements. The base temperature is $T = 100\,\mathrm{mK}$. From Ref. [33].

To obtain an independent bulk measurement of the superconducting transition, Lee *et al.* [33] also recorded the spin-lattice relaxation rates $1/T_1$ for ^{77}Se at $H = 2.38\,\mathrm{T}$, and for the methyl group protons ^{1}H at $H = 232\,\mathrm{mT}$ and 12.8 mT. As shown in Fig. 3.7, especially in the upper inset, an enhancement of ^{77}Se spin-lattice relaxation rates $1/T_1$ is present around $T \approx 0.7\,\mathrm{K}$. However, its identification as the Hebel–Slichter (or coherence) peak needs further experimental investigation since this result needs to be contrasted with the outcome of proton $1/T_1$ measurements at $B = 232\,\mathrm{mT}$ and 12.8 mT, which are shown in the lower inset of Fig. 3.7. The enhancement of $1/T_1$ is also absent in zero field proton NMR measurements in $(\mathrm{TMTSF})_2\mathrm{ClO}_4$ [13]. Moreover, even if this enhancement of $1/T_1$ can be identified as a Hebel–Slichter peak, the temperature dependence of $1/T_1$ obtained experimentally [33] can not distinguish clearly at low temperatures the nodal structure of the superconducting order parameter. Furthermore, the applied magnetic field ($B = 2.38$ T) can suppress the Hebel–Slichter peak and affect the low temperature behavior, such that a direct comparison of experimental data and theoretical results in zero field [15] is problematic. In theory, order parameters with nodal structure should present power law dependences of $1/T_1$ versus temperature for small (zero) magnetic fields and low temperatures [15]. Earlier $1/T_1$ measurements of the proton NMR [13] were performed at rather high temperatures ($T_c/2 < T < T_c$), but low magnetic fields, and in this temperature range a T^3 behavior was seen, but could not be regarded as convincing evidence for nodes of the order parameter since the temperature was too high, and it would still be possible to have typical exponential behavior as in conventional superconductors at very low temperatures.

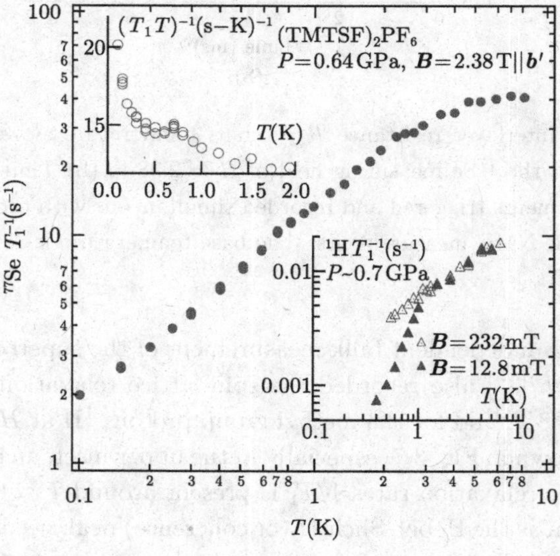

Figure 3.7 ^{77}Se spin-lattice relaxation rates $1/T_1$ versus temperature for $\boldsymbol{B} = 2.38$ T‖ $\boldsymbol{b}'$ axis. In the lower inset are shown the results of field-cycled proton $1/T_1$ measurements; at the top, the detailed temperature dependence of $(T_1T)^{-1}$ near $T \approx 0.7$ K is depicted. Adapted from Ref. [33].

In fact, Lee *et al.* performed ^{77}Se NMR measurements with a smaller magnetic field ($B = 1.43\,\mathrm{T}$) along $\boldsymbol{a}$ axis, and observed larger enhancement of $1/T_1$ below T_c. This enhancement at lower field lies between the calculated zero-field results for triplet with line nodes (where the increase of $1/T_1$ is very small below T_c), and fully gapped triplet (where the increase of $1/T_1$ is large below T_c) [15]. Thus, these $1/T_1$ measurements of ^{77}Se cannot be used directly to characterize the superconducting order parameter symmetry.

3.3 Coexistence of spin density wave and superconductivity

In the previous sections 3.1 and 3.2, we focused on experimental works attempting to determine the symmetry of superconducting order parameter in quasi-1D systems. We notice that all experiments that reveal exotic behavior of superconducting $(TMTSF)_2PF_6$ were performed in the phase diagram close to the spin density wave (SDW) state. Therefore, the effects of SDW on superconducting (SC) states within these quasi-1D systems should be addressed. In addition, the interplay between magnetic and superconducting orders by itself is a very fundamental problem in condensed matter physics. Known examples hosting neighboring magnetic and superconducting phases include high-T_c cuprates [99], heavy fermion compounds [100, 101], ferromagnetic superconductors UGe_2 [102] and $ZrZn_2$ [103], pnictides [104, 105], and Bechgaard salts [9, 80, 106].

In $(TMTSF)_2PF_6$, experiments to explore physical properties in the vicinity of the SDW-SC phase boundaries in the P-T diagram are technically difficult due to its transition critical pressure P_c. However, some earlier measurements indicated the presence of an inhomogeneous state in the vicinity of P_c [107, 108]. Specifically, Azevedo *et al.*[107] found that the quenching of the SDW state was a slow function of pressure from measurements of the Knight shift of ^{77}Se. This was interpreted as an indication of the coexistence of SDW and metallic states. Lee *et al.* [108] assumed the presence of macroscopic domains of superconducting and insulating SDW states to explain an unusual upward curvature of $H_{c_2}^{c^*}(T)$. Recently, transport measurements by Vuletić *et al.* [76], Kornilov *et al.* [77] and Kang *et al.* [78], as well as a simultaneous

NMR and electrical transport measurements by Lee *et al.* [79] in $(TMTSF)_2PF_6$ all suggest an "inhomogeneous" coexistence region of SDW and metallic (or SC) orders.

Vuletić *et al.* [76] performed resistivity measurements and suggested an inhomogeneous SDW-Metal (SDW-SC) phase due to a strong hysteretic behavior of resistivity between T_{SC} and T_{SDW} at a fixed pressure within $8.6 < P < 9.43$ kbar. This pressure range is lower than but close to the SDW–SC transition critical pressure P_c, which is 9.43 kbar in their pressure gauge. Here, T_{SC} and T_{SDW} are transition temperatures of superconducting and SDW phases, respectively.

A representative example of this hysteretic resistivity curve is shown in Fig. 3.8(a), where $P = 9.1$ kbar. In this situation, the temperature sweep starts above T_{SDW}, reverses below T_{SC} and ends above T_{SDW} again. Notice that the hysteretic behavior is strong (about 20%), and different loops appear when temperature sweeps are reversed between T_{SC} and T_{SDW}. This hysteretic behavior suggests the existence of metallic (M) domains embedded in an SDW background whose characteristics (size and shape) are strongly temperature dependent. In addition, a sharp resistance drop at $T = 1.2 \pm 0.01$ K can be observed and hence suggests the condensation of the M domains into superconducting domains. Smaller SC domains could either percolate and form large domains, or link via the Josephson effect between them with increasing pressure. Both mechanisms lead to the zero resistance state.

Vuletić *et al.* [76] also analyzed many other resistance curves and suggested a phase diagram for $(TMTSF)_2PF_6$, as shown in Fig. 3.9. Notice that there is a 0.8 kbar wide pressure region which exhibits features of coexisting SDW and M orders, such as hysteretic behavior of resistance as a function of temperature.

A similar hysteretic behavior is also observed in magneto-resistance measurements performed by Kornilov *et al.* [77] on $(TMTSF)_2PF_6$. Unlike Vuletić *et al.* [76], Kornilov *et al.* [77] used an alternative approach to cross the phase boundary by varying the magnetic field at fixed values of pressure and temperature. This technique has two advantages: (i) crossing phase boundary may be achieved almost continuously, and (ii) different phases can be determined by various magneto-resistance characteristics.

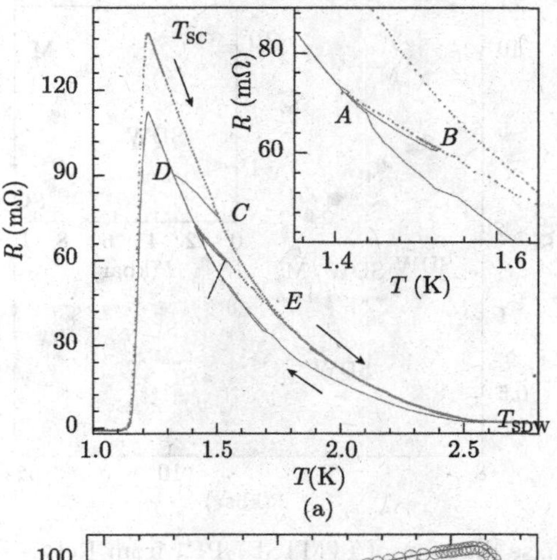

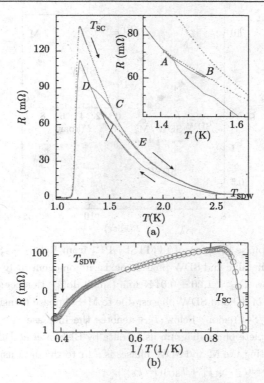

Figure 3.8 (a) Resistance versus temperature at 9.1 kbar for $(TMTSF)_2PF_6$, inside the coexistence region. Solid and dotted lines denote cooling and warming, respectively. The extremal hysteretic loop is recorded when the temperature sweep starts from above T_{SDW}, then cooling continues from points A, D, through T_{SC}, and into superconducting state. Below at least 1 K the sweep may be reversed and warming proceeds through T_{SC}, C, E and back above T_{SDW} into the SDW phase. Different hysteretic loops appear when the temperature sweep is reversed between T_{SC} and T_{SDW}. Loop 1: cooled from above T_{SDW} to A, reversed, warmed to E and further to above T_{SDW}. Loop 2: warmed from below T_{SC} to C, reversed, cooled to D and further to below T_{SC}. Loop 3: cooled from above T_{SDW} to A, reversed, warmed to B, reversed, cooled to A and further to below T_{SC}. Inset: The position of point B is shown. (b) Corresponding R versus $1/T$ plot of normalized extremal curves in cooling (solid line) and in warming (open points). From Ref. [76].

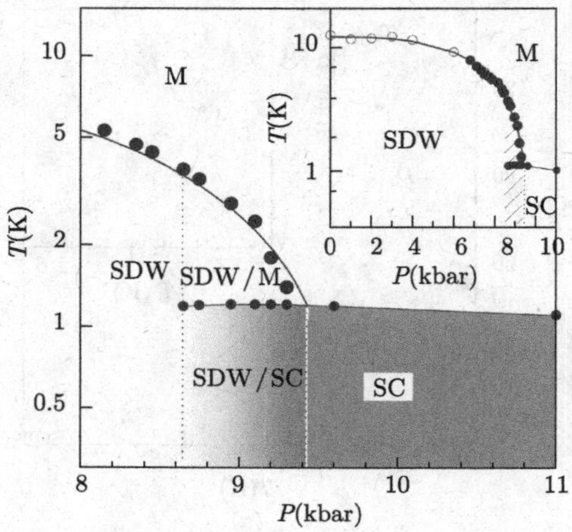

Figure 3.9 *P-T* phase diagram of $(TMTSF)_2PF_6$ from Ref. [76]. SDW/M denotes the region where metallic and SDW phases coexist inhomogeneously below T_{SDW} line (large dots). Below $T_{SC} = 1.20 \pm 0.01\,K$ line (small dots), this coexistence switches into a coexistence of SC and SDW phases, due to M–SC phase transition. A gradient in shading (SDW/SC region) below T_{SC} denotes the increase of SC volume. The inset shows a complete phase diagram using data by Biskup *et al.* [109] (circles). The solid curve separating the M and SDW phases is a fit to the data using the empirical formula $T_{SDW}(P) = T_1 - [(T_1 - T_{SC})(P/P_c)^3]$.

Three different trajectories in the *H-P-T* phase diagram are shown in Fig. 3.10. The magnetic field dependence of the resistance is qualitatively different for each of the three trajectories: trajectory 1 covers only a pure SDW phase, trajectory 2 crosses the SDW to M phase boundary and trajectory 3 covers only the M phase. In the analyzis that follows, we discuss first trajectories 1 and 3, and last the more interesting trajectory 2. For trajectory 1, resistance changes are not accompanied by hysteresis, and R increases monotonically with H in low fields. In high fields, the so-called "rapid oscillations" (RO) appear in the background of the monotonic $R(H)$ growth (figure not shown), which is typical for the SDW phase [110]. As pressure increases, but remains less than the critical $P_c(H)$, the resistance magnitude $R(H)$ decreases, but its magnetic field dependence does not change qualitatively.

When the initial (P, T) values are chosen greater than (P_c, T_c), trajectory 3 (see Fig. 3.10) lies entirely in the metallic state over the whole range of magnetic fields. As H increases the smooth growth of $R(H)$ transforms into step-like changes (figure not shown), which are related to the developing cascade of field induced spin density wave (FISDW) transitions between states with different nesting vectors [106, 111, 112, 113, 114]. In strong fields and at low temperatures, the transitions between the states with different nesting vectors are discontinuous (first order) [113, 114, 115, 116], and are accompanied by hysteresis in $R(H)$ [113, 114]. The jumps in $R(H)$ are periodic in $1/H$ [111, 112] and correspond to the phase boundaries on the known H-T phase diagram for the FISDW regime. When pressure P decreases, but remains larger than $P_c(H)$, the $R(H)$ dependence on magnetic field does not change qualitatively. Resistance jumps related to the FISDW phase transitions persist, and monotonically move to lower fields, indicating the shift of phase boundaries.

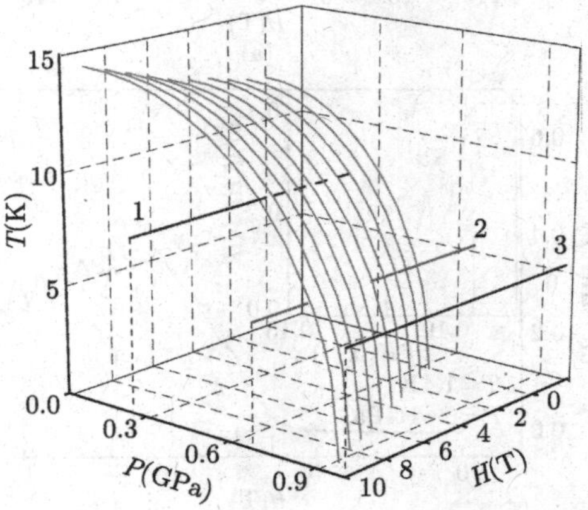

Figure 3.10 Schematic H-P-T phase diagram for $(TMTSF)_2PF_6$ without the inclusion of superconducting phase. From Ref. [77].

When the initial (P, T) values are chosen in the vicinity but slightly larger than (P_c, T_c), trajectory 2 crosses the phase boundary with increasing magnetic field. The $R(T)$ dependence measured at $H = 0$ is shown in the inset of Fig. 3.11(a), and represents the initial metallic state at $P = 0.64\,\text{GPa}$ ($P_c = 0.61\,\text{GPa}$ in their pressure gauge). When H increases, the resistance changes insignificantly up to $H \approx 7\,\text{T}$ [see Fig. 3.11(a)]. Upon further increase of H up to 16 T, the resistance sharply grows by three orders of magnitude.

This growth indicates the transition from the metallic to the SDW phase. Notice in Fig. 3.11(b) the appearance of resistance oscillations for 7 T< H <16 T, which are very uncommon features under the interpretation of a pure SDW phase.

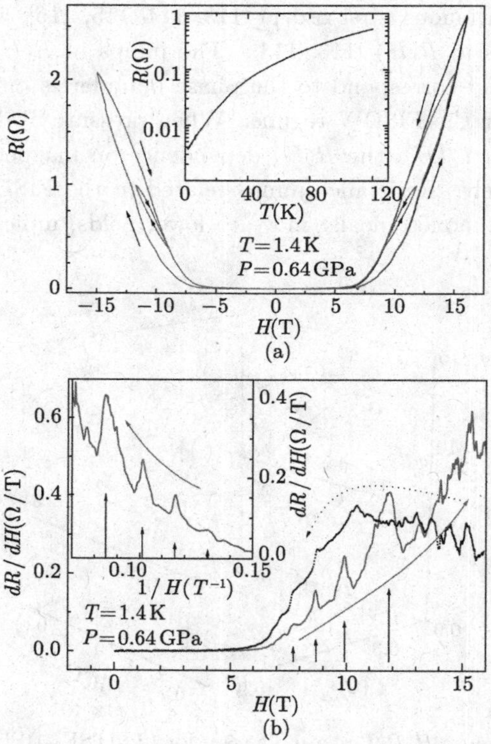

Figure 3.11 (a) Variation of the resistance with magnetic field for the case of crossing the M-SDW boundary (trajectory 2 in Fig. 3.10). The inset shows $R(T)$ changes in the initial state at $H = 0$, characteristic of the metallic phase ($dR/dT > 0$). The sharp growth of $R(H)$ at $H \approx 7$ T corresponds to the M-SDW transition. (b) Magnetic-field dependence of the derivative dR/dH, which corresponds to the $R(H)$ curves in (a) upon increasing and decreasing field. Notice that the peaks in dR/dH coincide with the boundaries between different FISDW phases. The vertical arrows show these boundaries, depicted from the experimentally determined phase diagram [113]. The inset demonstrates the periodicity of the dR/dH peaks in $1/H$. From Ref. [113].

As magnetic field is swept down (from 16 T to 7 T), a strong hysteresis ($\sim 20\%$) can be seen in the resistance, where the non-monotonic component of R practically disappears. Upon repeated magnetic field sweeps from 0 to 16T, the $R(H)$ curves described above are completely reproduced. The non-monotonic component of the resistance is more clearly seen in the derivative dR/dH [Fig. 3.11(b)]. Vertical arrows in Fig. 3.11(b) depict the locations of the FISDW phase boundaries [114]. The locations of peaks in dR/dH in Fig. 3.11(b) coincide well with the arrows (i.e., with the anticipated boundary of the FISDW phases). The inset to Fig. 3.11(b) demonstrates that these peaks are equidistant in $1/H$. For these two reasons, the observed peaks in dR/dH could be identified with crossing the boundaries between different FISDW phases. The existence of the peaks and their periodicity in $1/H$ would be quite natural for trajectory 3, but is absolutely unexpected for trajectory 2. Notice that the next peak which is expected to show up at $H \approx 14$ T is not seen. The absence of this peak at 14 T indicates that the admixture of the M phase is much reduced and, hence, the system tends to the homogeneous SDW state. However, the fully homogeneous SDW state is not achieved at this field, since the hysteresis in $R(H)$ does not disappear up to 16 T. Additional evidences supporting this conclusion are rapid magneto-resistance oscillations seen for fields $H > 14$ T, which are characteristic of the SDW phase. Therefore, at the transition from the M to the SDW phases, rather far away from the phase boundary, the magneto-resistance continues to exhibit residual signatures of the minority phase. When the field grows, the signatures of the minority phase start to disappear and are not restored when the system approaches again the same phase boundary. In order to justify the role played by H in this experiment, Kornilov *et al.* [77] also performed resistance measurements at fixed pressure and zero magnetic field. Although the resistance changes are weak and of quantitative rather than qualitative character, a hysteretic behavior is also observed ($\sim 0.5\%$) and support the interpretation of an inhomogeneous SDW/M state.

Motivated by these early results, more careful work has been done to investigate closely the underlying patterns of the possible SDW–SC coexistence region. By measuring resistivity along x, y and z directions, Kang *et al.* are able to track the evolution of the SDW-SC texture in the coexistence region [78]. The phase diagram showing this evolution is depicted in Fig. 3.12, where four different superconducting regimes can be identified as (A) 1D SC filaments along z axis, (B) anisotropic 2D SC slabs with filaments coupled along y axis via Josephson effect, (C) anisotropic 3D SC with slabs coupled along x axis, and (D) 3D homogeneous SC.

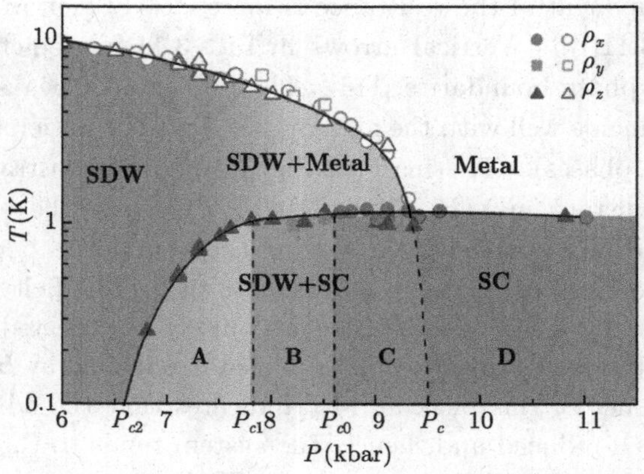

Figure 3.12 Phase diagram of $(TMTSF)_2PF_6$ as determined from resistivity measurements along x-(circles), y-(squares), and z-(triangles) axes. The SDW and SC transitions are depicted by filled and open symbols, respectively. With increasing pressure, the system undergoes an evolution from (A) 1D SC along z axis to (B) anisotropic 2D SC with Josephson coupling along y axis to (C) anisotropic 3D SC with Josephson coupling along x axis to (D) 3D homogeneous SC. (A)–(C) correspond to the coexistence region of SDW and SC. From Ref.[78].

The characteristic feature of phase A is shown in Fig. 3.13(a). With decreasing temperature, a clear SC transition can be identified when resistivity is measured along z axis, while some signature is present when measured along y axis and nothing detectable along x direction. The onset of the sudden drop of ρ_z, as indicated by T_c in Fig. 3.13(a), is confirmed to be the SC transition temperature by measuring the upper-critical field which is diverging at low temperatures [108]. Besides, the differential resistance measure along z direction gives a minimum at low voltages at low temperatures, showing a characteristic behavior of superconductivity; and exhibits a maximum at low voltages at hight temperatures, reflecting the presence of a density wave state [see inset of Fig. 3.13(a)]. This anisotropic resistivity behavior is compatible with a picture of 1D SC filaments along mainly the z direction with length approaches the sample size when $T < T_c$. As temperature further decreases, these filaments start to establish some coherence along y direction, which leads to sizable effect to ρ_y. In fact, the results shown in Fig. 3.13(a) is qualitatively similar to outcomes obtained in ultrathin superconducting nanowires [117].

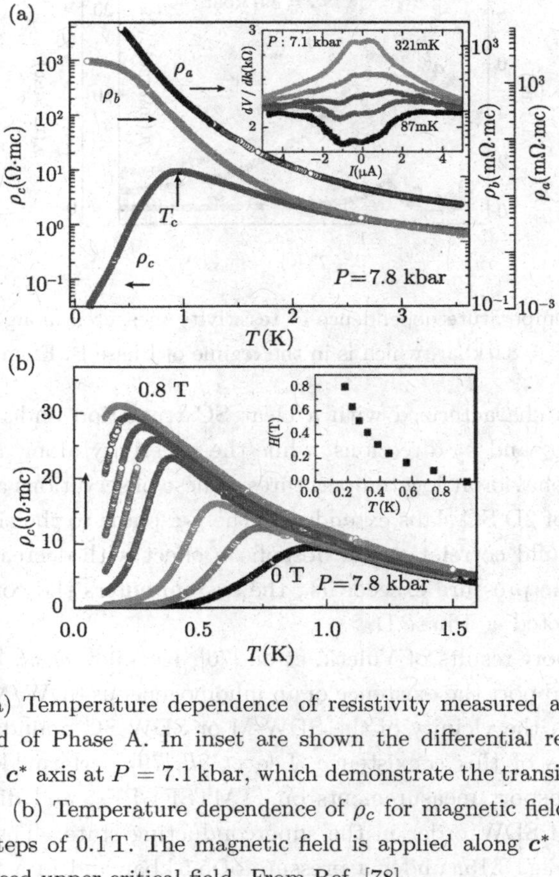

Figure 3.13 (a) Temperature dependence of resistivity measured along crystalline axes at the end of Phase A. In inset are shown the differential resistance curves measured along c^* axis at $P = 7.1$ kbar, which demonstrate the transition from SC to SDW behavior. (b) Temperature dependence of ρ_c for magnetic fields ranging from 0 to 0.8 T by steps of 0.1 T. The magnetic field is applied along c^* axis. The inset shows the deduced upper critical field. From Ref. [78].

Phase B is characterized with a clear SC transition and zero-resistivity state when measured along z direction, a clear SC transition in ρ_y, and, as in Phase A, no SC signature detectable in ρ_x. In Fig. 3.14, the resistivity along y axis also shows a double-peak behavior with decreasing temperatures, which is attributed to the evolution of Josephson coupling among the already superconducting 1D chains [78]. However, a qualitative understand of this unusual nonmonotonic behavior is still lack.

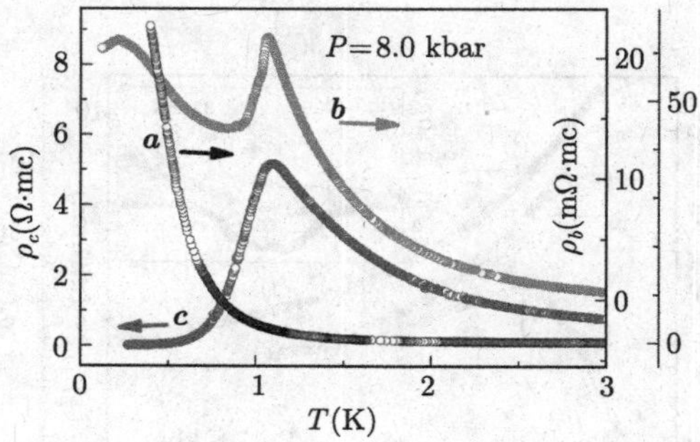

Figure 3.14 Temperature dependence of resistivity measured along the three crystalline axes at $P = 8.0\,\text{kbar}$, which is in the regime of Phase B. From Ref. [78].

Phase C is characterized with a clear SC transition and zero-resistivity state in both ρ_z and ρ_y directions, while the resistivity along x axis shows a double-peak behavior at low temperatures. These observations are compatible with a model of 2D SC slabs extended in the y-z plane to the size of samples, and start to build correlation via Josephson effect with decreasing temperature. When the pressure exceeds P_c, the system enters the conventional SC regime as depicted as Phase D.

The transport results of Vuletić *et al.* [76], Kornilov *et al.* [77] and Kang *et al.* [78] all support the existence of an inhomogeneous SDW/M or SDW/SC mixed phase in the vicinity of the SDW-M or SDW-SC boundaries. To gain further insights of this coexistence, Lee *et al.* [79] performed simultaneous NMR and transport measurements on $(TMTSF)_2PF_6$, and directly showed the presence of SDW order in the superconducting state. The main results are shown in Fig. 3.15, under a pressure of 5.5 kbar and in a magnetic field of 0.29 T. Data with triangles (circles) are obtained with field along $\boldsymbol{a}$ axis (tilted 45 degrees towards the $\boldsymbol{c}^*$ axis). The top panel shows the results from interlayer ($\boldsymbol{c}^*$ axis) electrical transport, in which the resistance is enhanced below 3 K due to the SDW transition, followed by a superconducting transition near 1 K. The middle panel of Fig. 3.15 shows the simultaneously measured proton spin-lattice relaxation time $(1/T_1)$. In the metallic state above 3 K, a single exponential curve describes fairly well the recovery of the magnetic moments. However, below 3 K, the recovery deviates significantly from an exponential behavior, as local variations of the spectral density develop. The bottom panel shows the temperature dependence of the local magnetic field at the proton site, which is essentially a measure of the full width of the NMR absorption spectra. It is important to note that the proton NMR results are dominated by the SDW signal, which is largely unaffected by the emergence of superconducting state. Lee *et al.* [79] hence argued that the SDW and superconducting regions are macroscopically separated, and each SDW and SC domain is larger than its respective correlation length. By performing an angular magneto-resistance oscillation (AMRO) study, they also concluded

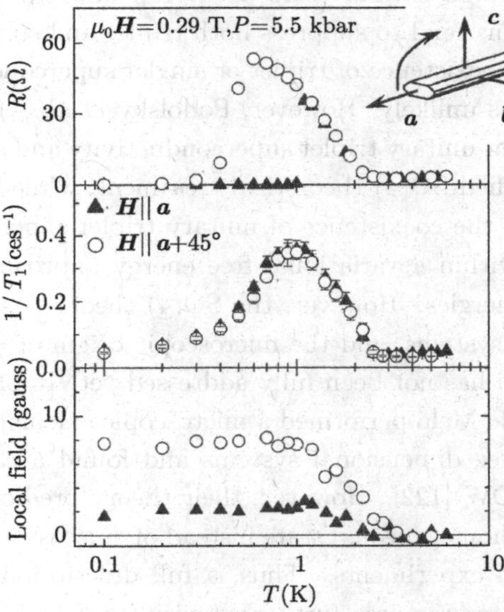

Figure 3.15 Simultaneous resistivity and proton NMR measurements. Shown here are the temperature dependence of interlayer resistance (top panel), proton spin-lattice relaxation rate (middle panel), and local field variations at the proton site (bottom panel). The data with triangles were obtained with a magnetic field aligned along ***a*** axis and circles with a 45° tilt toward the c^* axis. From Ref. [79].

that there are multiple connected SDW and SC domains, which are larger than several microns in linear dimension, and thus comparable to the mean free path of the uniform metallic phase.

Through different methods and approaches, the experiments by Vuletić *et al.* [76], Kornilov *et al.* [77], Kang *et al.* [78], and Lee *et al.* [79] suggest a similar scenario of coexisting SDW/SC and SDW/M states on the P-T diagram. The previous authors concluded that the coexistence region consists of macroscopic segregated domains of SDW and M(SC), where there may be little or no effect of any ordered phase on another due to their spatial separation. In order to explain this coexistence region, Vuletić *et al.* [76] proposed a free energy including contributions from the ordered phases and elastic energy.

The effects of nearly antiferromagnetic fluctuations in quasi-1D superconductors was investigated several years ago [118], and led to the conclusion that these fluctuations tend to suppress both triplet and singlet pairings, and that as a result the coexistence of triplet or singlet superconductivity and antiferromagnetism was unlikely. However, Podolsky *et al.* [119, 120] proposed the coexistence of non-unitary triplet superconductivity and antiferromagnetic order within a modified SO(4) theoretical treatment, while Zhang and Sá de Melo [121] proposed the coexistence of unitary triplet superconductivity and spin density wave within a variational free energy approach which includes negative interface energies. However, the SO(4) theory is strictly valid only for one-dimensional systems, and the microscopic origin of negative interface energies of Ref. [121] has not been fully addressed yet. In order to fill in this gap, Zhang and Sá de Melo performed a microscopic calculation for quasi-1D, but nevertheless three dimensional systems and found a coexistence region of triplet SC and SDW [122]. However, their theory predicts a "microscopically" mixed (but inhomogeneous) state instead of "mesoscopically" separated phases inferred from experiments. Thus, a full description of SDW/M and SDW/SC mixed states requires further experimental and theoretical investigations. Detailed discussion on these theoretical efforts to understand the SDW/SC coexistence region will be presented in Chap. 6.

Finally, it should be emphasized that all pressurized experiments done to date on quasi-1D superconductors have used pressure cells that contain liquids as the pressure medium at room temperature. However, at low temperatures it is known that the pressure cell acquires a certain amount of spatial inhomogeneity due to the solidification of the pressure medium (liquid). This pressure inhomogeneity can affect the measurements of the SC/SDW phase boundary, and consequently their putative coexistence. Therefore, further experiments should be done when more homogeneous pressure cells are produced for low temperatures.

3.4 Non-Fermi liquid behavior of the normal state

In the previous sections of this chapter, we have discussed some recent experiments on $(TMTSF)_2PF_6$, mainly focusing on the symmetry of the superconducting phase and the proximity effect of the spin density wave order. In this section, we will discuss the normal state properties at low temperatures around the critical SDW-SC transition pressure P_c. It will be seen that the normal state exhibits clear non-Fermi liquid (NFL) behavior as approaching the SDW phase. This observation suggests that the superconducting state is associated with antiferromagnetic spin fluctuations. Evidences for NFL are manifested in the ^{77}Se nuclear spin-lattice relaxation rate [123], and in a more recent resistivity measurement [124].

A understanding of the normal state of $(TMTSF)_2PF_6$, especially in the close vicinity of spin orders, not only helps tackling the mystery of the superconducting symmetry, but also shines light on a more general problem of the interplay between magnetic and superconducting orders, which is observed in a large variety of strongly correlated systems including high-T_c cuprate superconductors [126], heavy-fermion systems [127], and pnictide superconductors [104, 128]. Compared to these systems, the Bechgaard salts are considerably simpler with no complication of deformed Fermi surfaces nor complexity from underlying Kondo or Mott physics. Therefore, the NFL behavior discussed below provides us valuable information regarding the interference of antiferromagnetic spin fluctuation and pairing interaction.

In this section, we will first discuss the ^{77}Se NMR measurement on the metallic phase of $(TMTSF)_2PF_6$ performed by Wu *et al.* [123]. In the vicinity of SDW and SC, the temperature dependence of the relaxation time T_1 deviates from the canonical Korringa law of T_1T = constant, which is expected for a Fermi liquid. For instead, the relaxation time can be well described by a Curie-Weiss (CW) behavior, indicating the dominance of 2D antiferromagnetic fluctuation. Then, we will introduce results of x axis resistivity within the same parameter region by Doiron-Leyraud *et al.* [124]. A striking feature is the linear temperature dependence $\rho_x(T) = \rho_{0,x} + AT$, which is also typical NFL behavior and is compatible with 2D antiferromagnets. Based on these results, theoretical proposal has been raised to suggest a unified mechanism for both SDW and SC, where the antiferromagnetic spin fluctuation plays an essential role [46, 129]. This mechanism, however, suggesting a d-wave pairing symmetry, as will be discussed with more details in Chap. 7.

In Fig. 3.16(a), results for ^{77}Se nuclear spin-lattice relaxation rates T_1^{-1} is shown as functions of temperature for a range of pressure covering both sides of the critical pressure P_c of vanishing SDW and highest SC transition

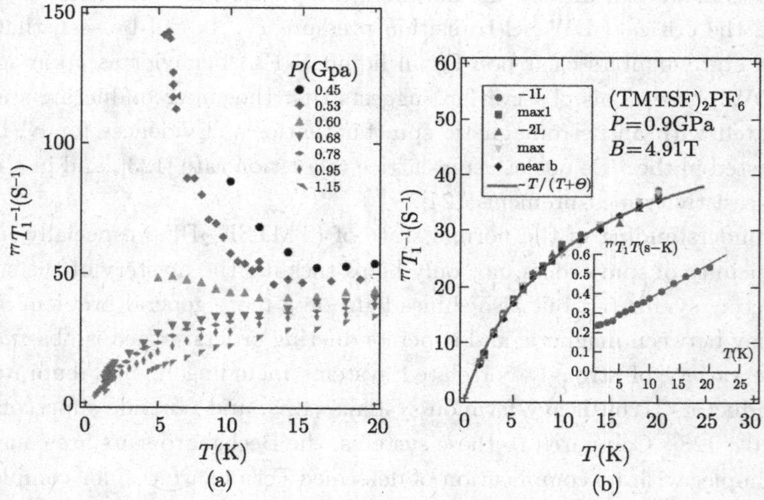

Figure 3.16 (a) Temperature dependence of spin-lattice nuclear relaxation rate T_1^{-1} at various pressures. From Ref. [125]. (b) Temperature dependence of T_1^{-1} measured with an applied magnetic field pointing along different directions within the y-z plane, including both magic angles and non-magic angles. No sizable difference can be obtained within experimental error. The inset reports the same data as T_1T v.s. T. The solid line indicates that the data are well described by a linear behavior. From Ref. [123].

temperature ($P_c \sim 0.6\,\mathrm{GPa}$ in their pressure gauge) [125]. At lower pressures, T_1^{-1} increases and tends to diverge with decreasing temperature, as one would expect for approaching the SDW ordered phase. For pressure exceeding P_c, T_1^{-1} does not show any divergence, but decreases with temperature with a pronounced downward curvature. This is in clear contrast to the linear dependence of Korringa relation, which is expected for a Fermi liquid. When temperature further decreases, the system seems to undergo a crossover such that the generic Fermi liquid behavior is regained with $(T_1T)^{-1} \to$ constant.

A closer inspection of the relaxation data for $P > P_c$ is depicted in Fig. 3.16(b), where results for $P = 0.9\,\mathrm{GPa}$ and $B = 4.91\,\mathrm{T}$ is shown, and plotted as T_1T v.s. T in the inset. The data can be well described by a Curie-Weiss (CW) relation $T_1T \propto T + \Theta$, with a pressure dependent CW constant $\Theta \approx 11\,\mathrm{K}$. The CW behavior can be explained with a spin-fluctuation model as discussed by Millis, Monien, and Pines [131], and summarized by Moriya and Ueda [132] in the context of cuprate superconductors. By assuming that the susceptibility is peaked around the antiferromagnetic wave vector, the scaling gives $T_1T \propto \xi^{-(z+2-\eta-d)}$, where $z = 2$ is expected for an overdamped response, η is the scaling factor, and d represents dimensionality. Within a mean-field theory, the correlation length for the AF order depends on temperature as $\xi \propto (T + \Theta)^{-1/2}$ and $\eta \approx 0$. Thus, the 2D fluctuating antiferromagnet would exhibit the CW relation, where Θ is linked to the amplitude of the anomaly and to the characteristic energy scale of the spin fluctuations.

The Curie-Weiss constant Θ is sensitive with changing pressure: it emerges from zero at $P \sim P_c$, increases rapidly and reaches large value at high pressure (see Fig. 3.17). The temperature dependence of Θ, especially the vanishing of this energy scale at P_c serves as a direct hint that a magnetic quantum critical point may exist below P_c, but hidden in the mesoscopic coexistence region of SDW and SC. Meanwhile, when Θ becomes large compared to temperature, the relaxation rate becomes $T_1T \propto \Theta$, indicating the recovery of the expected Korringa law as observed in experimental results with very high pressure [see, e.g., data with $P = 1.15\,\mathrm{GPa}$ in Fig. 3.16(a)].

In addition to the relaxation rate, the dominance of spin fluctuations is also suggested by a temperature dependent resistivity measurement on the normal state of $(TMTSF)_2PF_6$ by Doiron-Leyraud *et al.* [124]. By using a standard four-terminal ac technique in a nonmagnetic oil pressure cell, experimentalists manage to obtain the x axis electrical resistivity, i.e., along the chains of organic molecules, of single crystal PF_6 samples. To overcome the emergence of SC at low temperatures, a transverse magnetic field can also be applied along z direction, without bringing any complication of magnetoresistance effect. In their pressure gauge, the juncture of vanishing SDW and highest T_c SC is about 9 kbar.

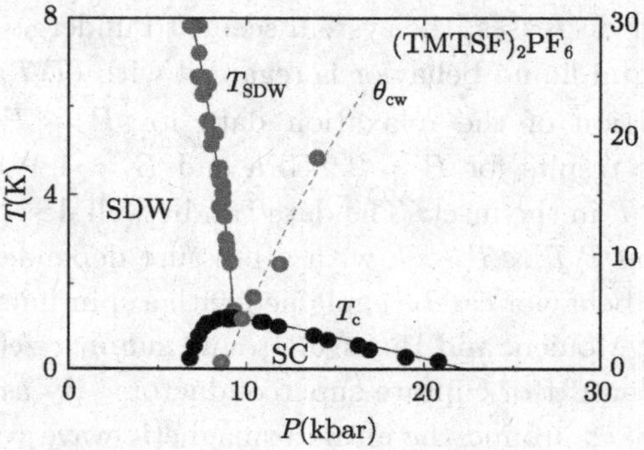

Figure 3.17 Variation of the Curie-Weiss parameter Θ as extracted from the NMR data of Ref. [125]. From Ref. [130].

The zero-field electrical resistivity data is shown in Fig. 3.18, for a range of pressure that almost cover the entire SC phase. When pressure is less than P_c (~ 9.4 kbar in their pressure gauge), the resistivity first rises when temperature is decreased below T_{SDW} and then drops sharply at the superconducting transition temperature. As $P > P_c$, the stable region of SDW is significantly suppressed, and a smooth monotonic decrease of resistivity is observed before the SC transition.

When pressure is slightly above P_c, e.g., $P = 11.8$ kbar as in Fig. 3.18, the resistivity decreases monotonically with temperature and displays a nearly strict linear temperature dependence below a certain temperature $T_0 \approx 8$ K [see, e.g., Fig. 3.19(a)]. By applying a small magnetic field of $B = 0.05$ T to overcome the emergence of SC, it is shown clearly that the non-Fermi liquid linear behavior extends to the lowest measured temperature of about 0.1 K, thus covering nearly two orders in temperature. This feature is further depicted in Fig. 3.20, where the inelastic part of the normal state resistivity is plotted on a log-log scale. The inelastic part of resistivity is defined as the difference between normal-state resistivity and the residual resistivity $\rho_0 = \rho_a(T \to 0)$,

$\Delta\rho_a = \rho_a - \rho_0$. As pressure is further increased, the low-temperature behavior of the resistivity deviates from the linear form, but acquires a upward curvature. At the highest measured pressure $P = 20.8\,\text{kbar}$ with vanishing SC transition temperature, the resistivity below 4 K shows a nearly T^2 dependence down to lowest measured temperature, reflecting a Fermi liquid nature as shown in Fig. 3.20. At intermediate pressures, a combination of linear and quadratic terms seems to best describe the low-temperature data. As a typical example, data for the case of $P = 16.3\,\text{kbar}$ is show in Fig. 3.19(b), and is well fit by a polynomial function of the form $\rho(T) = \rho_0 + AT + BT^2$. Here, a transverse magnetic field of $B = 0.03\,\text{T}$ is applied to maintain the system in the normal state.

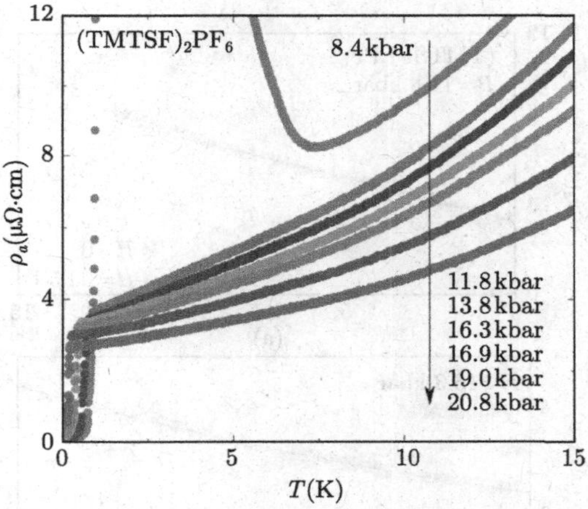

Figure 3.18 Resistivity variance $\rho_a(T)$ as a function of temperature measured at different pressures. From Ref. [124]

By fitting results with the same polynomial function, the pressure-dependent coefficients A and B can be extracted as shown in Fig. 3.21(a). The decrease of A and increase of B with pressure can be interpreted as a Fermi liquid component starts to set in from P_c, and eventually becomes dominant to resistivity at very high pressures. One interesting feature is that the coefficient A seems to acquire a linear dependence on SC transition temperature, and vanishes at the point where superconductivity disappears [see Fig. 3.21(b)]. Dorion-Leyraud *et al.* stressed that this correlation between resistivity and T_c is not fit dependent. In fact, the evolution from a pure linear to a nearly T^2 tendency can be clearly resolved from resistivity curves on a log-log plot (Fig. 3.20). This evolution can be described by the polynomial function as discussed above $\rho(T) = \rho_0 + AT + BT^2$ with pressure-dependent coefficients A and B. But it can also be tracked with a power law of the form $\rho(T) = \rho_0 + CT^\alpha$, giving an exponent α that grows from 1 to 2 with pressure. In this case, the linear resistivity behavior can be extracted by taking the inelastic part of the resistivity $\Delta\rho_a = \rho_a - \rho_0$ at 1 K, where the T^2 term is negligible. In Fig. 3.21(c), the inelastic part at 1 K is shown as a function of SC transition temperature, and acquires the same linear correlation as the coefficient A. Specifically, the extrapolation of $\Delta\rho_a$ is also close to the origin, showing that it is not the result of a particular fitting scheme.

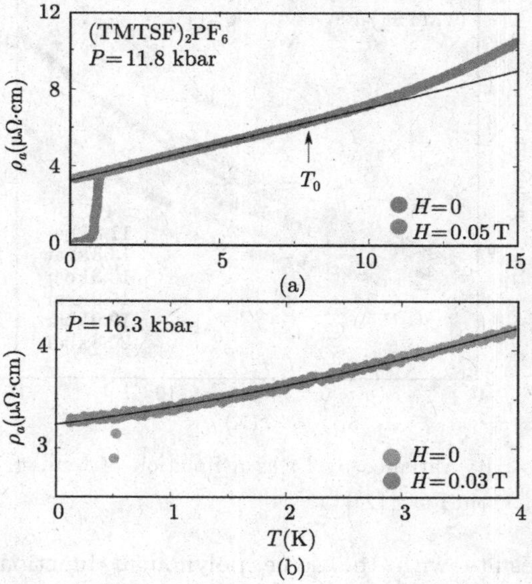

Figure 3.19 (a) Temperature dependence of resistivity $\rho_a(T)$ at $P = 11.8$ kbar, with a magnetic field of $H = 0$ (red) and 0.05 T (gray). The solid line is a linear fit to the data up to $T_0 = 8$ K. (b) Resistivity measurement $\rho_a(T)$ carried at $P = 16.3$ kbar, with a magnetic field of $H = 0$ (orange) and 0.03 T (gray). The solid line is a polynomial fit of the form $\rho_a(T) = \rho_0 + AT + BT^2$ from 0.1 K up to 4.0 K. From Ref. [124].

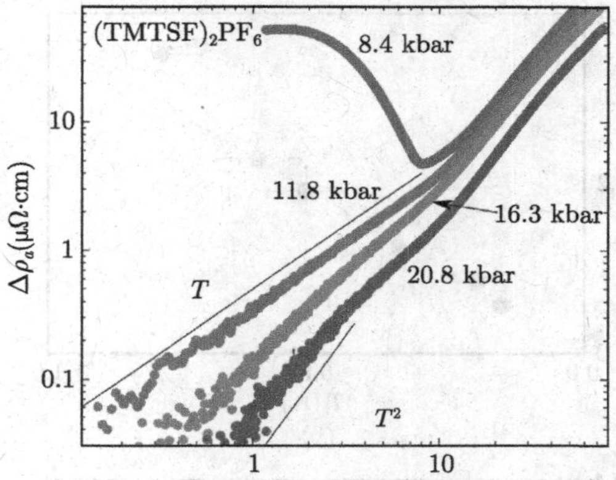

Figure 3.20 Inelastic part $\Delta\rho_a(T) = \rho_a(T) - \rho_0$ of the normal-state $\boldsymbol{a}$ axis electrical resistivity of $(TMTSF)_2PF_6$ at various pressures, where a small transverse magnetic field of typically 0.05 T is applied. The lines represent $\Delta\rho_a(T) \propto T$ and $\Delta\rho_a(T) \propto T^2$. From Ref. [124].

The observation of a strict linear resistivity as $T \to 0$ in $(TMTSF)_2PF_6$ on the merge of SDW order is another strong indication that the antiferromagnetic spin fluctuations provide major contribution to the non-Fermi liquid behavior. The linear correlation between linear resistivity and superconducting T_c discussed above suggests that the scattering and paring could share a common origin, implying that the antiferromagnetic spin fluctuations and superconductivity are intimately connected. Theoretical works exist to model

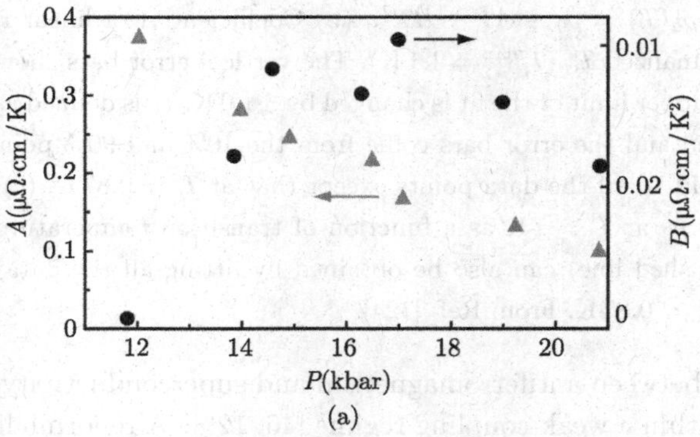

(a)

Figure 3.21 *Continued.*

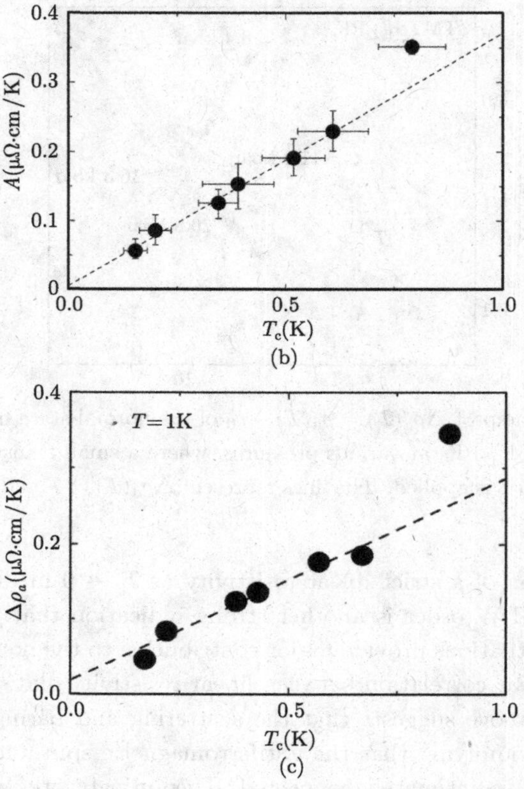

Figure 3.21 (a) Pressure dependence of the resistivity coefficients A and B of the polynomial fit $\rho_a(T) = \rho_0 + AT + BT^2$. (b) Coefficient A of linear resistivity as a function of normalized T_c ($T_c^{max} = 1.1\,K$). The vertical error bars show the variation of A when the upper limit of the fit is changed by $\pm 1.0\,K$. T_c is defined as the midpoint of the transition and the error bars come from the 10% and 90% points. The eased line is a linear fit to all the data points except that at $T_c = 0.87\,K$. (c) Inelastic part of $\Delta\rho_a = \rho_a - \rho_0$ at $T = 1\,K$ as a function of transition temperature T_c. A linear dependence (dashed line) can also be obtained by fitting all the data points except the one with $T_c = 0.87\,K$. From Ref. [124].

the interplay between antiferromagnetism and superconductivity in the Bechgaard salts within a weak-coupling regime [46, 129]. A renormlalization group approach has been adopted to explain the phase diagram of (TMTSF)$_2$PF$_6$, where a superconducting state would emerge via spin fluctuations and has d-wave symmetry [46, 129]. As discussed above, the antiferromagnetic correlation length in a 2D system acquires temperature dependence as $\xi(T) \propto (T+\Theta)^{-1/2}$, where Θ is the energy scale for spin fluctuations. Through Umklapp scattering, antiferromagnetic spin fluctuations will also provide an anomalous temperature dependence in the quasiparticle scattering rate τ^{-1}, and as a consequence modify its conventional Fermi liquid behavior $\propto T^2$. Evaluation of the imaginary part of the one-particle self-energy gives $\tau^{-1} = a\xi T + bT^2$, with a and b constants. Thus, it would be natural to expect a linear temperature-dependent component AT in the resistivity as temperature $T \ll \Theta$. If the pairing mechanism of superconductivity is indeed rooted from the same antiferromagnetic spin fluctuations, the coefficient A would be inevitably correlated with T_c. Detailed discussion regarding these theoretical efforts can be found in Chap. 7.

The linear relation between the coefficient A and SC transition temperature T_c seems to be a feature shared with a variety of materials showing SC order in proximity with SDW phase. In particular, Dorion-Leyraud *et al.* [133] observe the same correlation in the sister compound $(TMTSF)_2ClO_4$. Since the phase diagram of $(TMTSF)_2ClO_4$ is basically a high-pressure version of that of $(TMTSF)_2PF_6$, it is experimentally possible to exceed the critical pressure at which SC ceases to exist and to confirm $A = 0$ when $T_c = 0$. A detailed discussion of this experiment will be found in Chap. 4. Besides, the resistivity data of pnictide superconductor $Ba(Fe_{1-x}Co_x)_2As_2$ [104, 128] and cuprate superconductors $Tl_2Ba_2CuO_{6+\delta}$ (Tl-2201) [134] and $La_{2x}Sr_xCuO_4$ (LSCO) [135] also reveal a NFL-FL crossover behavior with $\rho = \rho_0 + AT + BT^2$, and an approximately linear dependence of coefficient A on T_c. However, whether the qualitatively similar behaviors in Bechgaard salts, cuprate and pnictide superconductors indeed share the same physical origin is still unclear.

4
Recent Experiments on $(TMTSF)_2ClO_4$

In the previous chapter, we discussed some recent experimental works on $(TMTSF)_2PF_6$. The rich phase diagram of PF_6 makes the system an interesting platform to study many quantum phenomena including possible triplet superconductivity and the interplay or coexistence of magnetic and superconducting orders. However, $(TMTSF)_2PF_6$ only becomes superconducting at a fairly high pressure $P \sim 6-9$ kbar depending on different pressure gauge, and this SC phase extends to even higher pressure of ~ 20 kbar. The requirement of high pressure brings a considerable amount of difficulties in experimental investigations on this rather fragile organic materials at cycrogenic conditions. Compared to inorganic materials, Bechgaard salt single crystals are very likely to generate cracks under pressure or thermal cycling, hence become useless for absolute measurement. This leaves no choice for experimentalists but to test a number of samples and hopefully can get at least one specimen which shows no sign of cracks. It certainly consumes much effort and time, and regrettably, many single crystal samples. Another difficulty hampers the progress of experimental investigation is to gauge pressure at low temperatures. In sample cells, the pressure transmitting medium (e.g., Daphne oil 7373) will inevitably solidify below a certain temperature, which is still above the region of interest. Thus, the pressure reading is smaller at low temperatures than that at room temperature. And to change pressure, one has to increase the temperature till the medium melts, push the piston a bit tighter or looser, and rely on experience to obtain the desired pressure after cooling the system back to low temperatures.

The complication of high pressure required for $(TMTSF)_2PF_6$ makes the sister compound $(TMTSF)_2ClO_4$ attractive to experimentalists. As discussed in Chap. 2, $(TMTSF)_2ClO_4$ is superconducting at ambient pressure. The superconducting state is also unconventional in nature. Specifically, studies on impurity effect discussed in Sec. 4.1 show that the transition temperature is drastically suppressed by the presence of non-magnetic disorders [136, 137]. This observation implies a change in sign of the superconducting gap function over the Fermi surface. Besides, the measurements on upper critical fields summarized in Sec. 4.2 confirm that the upper critical field H_{c2} can exceed the Pauli paramagnetic limit by at least a factor of two [138, 139], which gives a strong indication that $(TMTSF)_2ClO_4$ is an exotic superconductor.

Thanks to the presence of superconductivity at ambient pressure, $(TMTSF)_2ClO_4$ can be studied with some experimental probes which are impractical inside a pressure cell. We discuss in Sec. 4.3 two thermodynamic experiments on ClO_4, which draw somewhat contradicting conclusions. By extracting electronic part from measured thermal conductivity, Belin and Behnia [39] suggested that the superconducting phase of $(TMTSF)_2ClO_4$ is fully gapped with a nodeless quasiparticle excitation spectrum. On the other hand, a more recent field-angle-resolved calorimetry measurement by Yonezawa *et al.* [40] favors a superconducting state with gap nodes. Besides, we study in Sec. 4.4 some recent spin resonance experiments. In particular, a zero-field muon spin resonance measurement by Luke *et al.* found no evidence for spontaneous magnetic fields in superconducting $(TMTSF)_2ClO_4$ within experimental resolution, implying that the time reversal symmetry is likely not broken [140]. Meanwhile, an NMR and relaxation rate analysis by Shinagawa *et al.* reports a decrease of spin susceptibility upon entering the superconducting phase, hence favors a singlet pairing state, at least at low magnetic field regime [38].

In addition to the invaluable works on the superconducting phase of $(TMTSF)_2ClO_4$, experimentalists also devote great effort to gain a better understanding on the normal state, especially in the close vicinity of SDW and SC orders. Along this way, the same measurements of nuclear relaxation rate [38] and electric resistivity [133] have been done on the normal state of ClO_4 as in PF_6, and give similar non-Fermi liquid behavior as discussed in Sec. 4.5. These observations suggest that the superconducting phase in ClO_4 may also be rooted from the antiferromagnetic spin fluctuations, as will be explored in detail in Chap. 7.

4.1 Non-magnetic impurity effect

The study of non-magnetic impurity effect plays a crucial role in the understanding of the superconducting phase of Bechgaard salts. Indeed, it is the observation of the quick disappearance of superconductivity upon non-magnetic alloying that originates the early proposal of triplet pairing by Abrikosov[1] [7]. Since the two partners forming a Cooper pair of an s-wave superconductor are related to each other by time reversal symmetry, no pair breaking effect should be expected from the scattering of electrons against spinless impurities [141]. However, the condition for time reversal symmetry is not met for the case of triplet pairing, such that the transition temperature for these superconductors would be strongly suppressed by even a small amount of such non-magnetic impurities. Thus, the crucial dependence of T_c on non-magnetic disorders in Sr_2RuO_4 provides a strong support on its p-wave pairing nature.

In the context of Bechgaard salts, early investigation on non-magnetic impurity effect shows remarkable sensitivity of the transition temperature upon exposed to radiation [2, 4]. Although irradiation is recognized to be an excellent method to introduce defects in a controlled manner [142], it can generate both non-magnetic and magnetic disorders [143]. This brings some ambiguity to the explanation of T_c suppression as local moments can also act as a strong pair-breaking mechanism on conventional s-wave superconductors. Another route towards the creation of local non-magnetic defects is by chemical substitution of TMTSF for TMTTF molecules in PF_6 [144] and ClO_4 [144, 145] compounds. However, in both situations, the cationic alloying causes dramatic modification of the normal state electronic properties, hence may not be the best way to study the survival of superconductivity against non-magnetic scatterings.

Keeping the cation stack uniform, scattering centers can also be created on the anion stacks (i.e., the stacks of PF_6 and ClO_4). The role of anion stack on the ground state is enhanced as the anion, which is located at an inversion center of the lattice structure, does not possess a central symmetry [146]. In the particular case of $(TMTSF)_2ClO_4$, as temperature is decreased slowly enough across the anion-ordering temperature T_{AO}, the system enters a so-called relaxed state (or R-state for short), where the orientation of the tetrahedral

[1] This is even highlighted by the title of Abrikosov's seminal paper *Superconductivity in a Quasi-one-dimensional metal with impurities.*

anion ClO_4 becomes uniform along the stacking axis, but alternate along the y direction. This R-state then acquires a superlattice of two sites per cell, and in turn a double-sheet Fermi surface at $\pm k_F$. As the anion reorientation requires a tunneling process between two degenerate states separated by a large potential barrier, the dynamics of anion-ordering is a slow process at low temperatures. Thus, if the temperature drops fast enough, most anions are frozen at their high-temperature directions, leading to a so-called quenched state (or Q-state for short) with a large amount of orientation disorders. In this Q-state, the single-sheet Fermi surface is preserved at low temperatures, leading to a quick suppression of superconductivity [147, 148] and a $2k_F$ SDW instability of the metallic phase at $T_{SDW} = 5\,K$ [148]. Besides, since the cooling rate only modifies the ordering along the anion stacks, it has been shown that neither the Pauli susceptibility [148] nor the density of states [147] of the normal phase are affected as long as a superconducting ground state can be eventually obtained.

In addition to the controlling of cooling rate, an alternative approach for the creation of anion disorders is to synthesize an anionic solid solution involving anions of similar symmetry. For centro-symmetric anions, a suppression of the SDW state has been observed in $(TMTSF)_2(AsF_6)_{1-x}(SbF_6)_x$ [149], while effect on superconductivity has not been studied in detail, mainly due to the experimental difficulty of high pressures. As far as non-centro-symmetric anions are concerned, the early investigations by Tomić *et al.* [5] in $(TMTSF)_2(ClO_4)_{1-x}(ReO_4)_x$ show that both the low temperature conductivity and superconducting T_c are strongly affected by alloying. In this section, we discuss some extension of the analysis on $(TMTSF)_2(ClO_4)_{1-x}(ReO_4)_x$ [136, 137], where the superconducting phase diagram is systematically mapped out, and a significant suppression of superconductivity is observed by introducing non-magnetic disorders. This observation suggests that the pairing state in $(TMTSF)_2ClO_4$ is unconventional in nature.

In the experiment by Joo *et al.*, single crystal samples of solid solution $(TMTSF)_2(ClO_4)_{1-x}(ReO_4)_x$ with $x < 17\%$ are prepared with the usual electrocrystallization technique [137]. Since the actual concentration of ReO_4 can be as large as 30% smaller than the value of nominal concentration x with a large error bar [136], the nominal concentration should not be used as an indicator on the level of disorder. For instead, Joo *et al.* choose to use the residual term of the normal-state resistivity as an evaluation of the electron elastic scattering rate.

Specifically, Joo *et al.* conduct resistivity measurement along the z axis via a low frequency lock-in detection. This direction is chosen to reduce the chance of creating cracks during cooling, and to take advantage of the well-defined equipotential surface for resistance measurements provided by the x-y plane with high transport anisotropy. To obtain the relaxed (R-) state of $(TMTSF)_2ClO_4$, samples are first slowly cooled down to 4.2 K (with a rate $\leqslant$ 0.2 K/min), then warmed up to 30 K and subsequently cooled across the anion ordering temperature T_{AO} with cooling rate 0.05 K/min (i.e., 3 K/h). This rate is much lower than the rate used in pristine $(TMTSF)_2ClO_4$, for the R-state in highly defective samples requires longer annealing time close to T_{AO} [148]. By varying the cooling rate, not-fully-relaxed or even quenched samples are also obtained, and compared to R-state samples as another source of impurities.

A typical example of the temperature dependent resistivity measurement is shown in Fig. 4.1(a), where resistivity is plotted as a function of T below 40 K. Four temperature regimes can be identified from this result, including (i) a quadratic behavior regime for temperature above the anion ordering temperature $T_{AO} \sim 24\,\mathrm{K}$, (ii) another quadratic portion in an intermediate regime $T_0 \lesssim T \lesssim T_{AO}$ with $T_0 \sim 10\,\mathrm{K}$, (iii) a linear behavior regime for $T_c < T \lesssim T_0$, and (iv) a sharp drop to zero resistivity below the superconducting transition temperature T_c. In the high temperature regime (i), the quadratic temperature dependence of resistivity can be well understood within a Fermi-liquid theory, where $\rho(T) = \rho_0^+ + BT^2$ is the signature of electron-electron elastic scattering with the prefactor B determined by the density of states at the Fermi level $N(E_F)$. The onset of anion ordering leads to a folding of the Brillouin zone along the y direction, which should not affect $N(E_F)$ nor the elastic-elastic scattering amplitude. Thus, a similar quadratic behavior $\rho(T) = \rho_0^- + BT^2$ with the same prefactor B but a different residual ρ_0^- should be expected below T_{AO}, as observed in regime (ii). When temperature further approaches superconducting state, a non-Fermi liquid behavior is observed in regime (iii), with a clear linear dependence as will be discussed in detail in Sec. 4.3. Similar features have been observed for $(TMTSF)_2PF_6$, as can be found in Chap. 3.

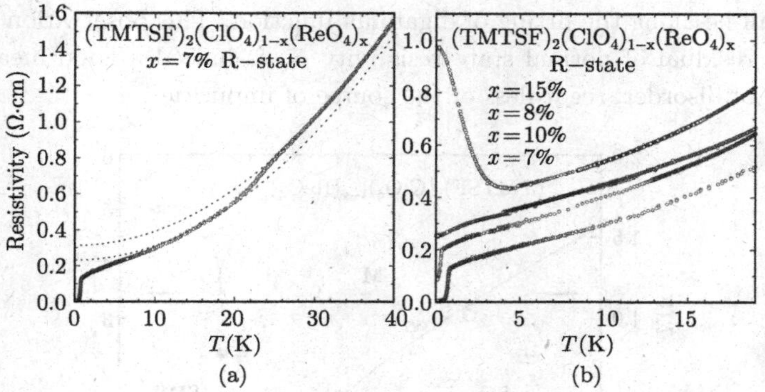

Figure 4.1 (a) Temperature dependence of the resistivity in $(TMTSF)_2(ClO_4)_{1-x}(ReO_4)_x$. Notice that on both sides of the anion ordering transition ($T_{AO} \sim 24$ K), the resistivity fits in a polynomial behavior $\rho(T) = \rho_0 + BT^2$ with the same value of prefactor B. (b) Temperature dependence of the resistivity in $(TMTSF)_2(ClO_4)_{1-x}(ReO_4)_x$ samples with different dopant concentrations. From Ref. [137].

By fitting the temperature dependence with quadratic laws in regimes (i) and (ii), Joo *et al.* [137] determine the residual resistance ρ_0^- below T_{AO} as an indicator for the elastic scattering rate, and hence the level of impurities. In Fig. 4.2, the superconducting transition temperature T_c has been plotted against ρ_0^- for different solid solutions. The transition temperature T_c is obtained either from the onset of superconductivity in the resistivity data, or equivalently, from the extrapolation of temperature where H_{c_2} reaches zero. Notice that all T_c data can be fit fairly well by the following expression [150, 151]

$$\ln\left(\frac{T_c(0)}{T_c(\rho_0^-)}\right) = \psi\left(\frac{1}{2} + \frac{\alpha T_c(0)}{2\pi T_c(\rho_0^-)}\right) - \phi\left(\frac{1}{2}\right), \tag{4.1}$$

where $\phi(x)$ is the digamma function, $\alpha = \hbar/(2\tau k_B T_c(0))$ with $\tau \propto 1/\rho_0^-$ the elastic scattering time, and $T_c(0)$ is the virtual critical temperature in the absence of any scattering. According to the data, $T_c(0)$ is fit to be ~ 1.57 K, which is still about 20% larger than the highest T_c ever observed in pristine $(TMTSF)_2ClO_4$ between 1.20 K and 1.30 K. A remarkable feature of Fig. 4.2 is that by increasing the cooling rate to some extent (say, still below 3 K/min to avoid creating inhomogeneous domains), a same sample can give different

T_c, but all lie along the fitting of digamma function. This observation implies that the residual of normal state resistivity ρ_0^- is indeed a good measure of the level of disorder, regardless of the source of impurities.

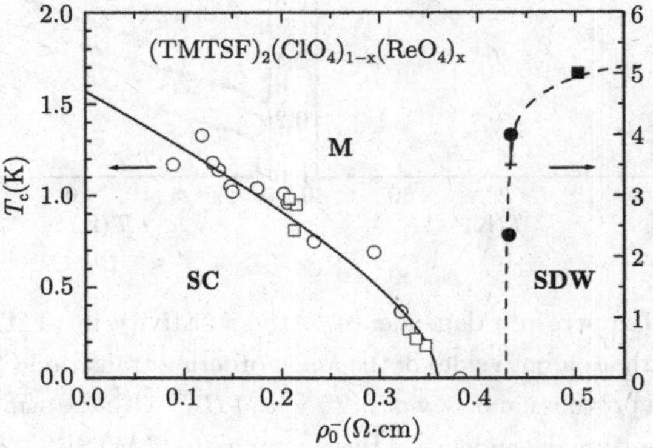

Figure 4.2 Phase diagram of $(TMTSF)_2(ClO_4)_{1-x}(ReO_4)_x$ governed by non-magnetic disorder. All open circles refer to the very slowly cooled samples in the R-state with different ReO_4^- contents. Open squares are data for the same samples corresponding to slightly faster cooling rates, while a metallic behavior above superconducting T_c is still maintained. Notice that all data can be fit fairly well by a digamma function (solid line) with $T_c(0) = 1.57\,K$ (see text). This observation indicates that the residual resistivity is a better characterization for the disorder than the nominal ReO_4^- concentration. Full dots are relaxed samples exhibiting a SDW ground state. The dashed line at the M-SDW transition is a guide for the eye. From Ref. [137].

Solid solutions created with higher nominal concentration ($x = 15\%$) show a distinct behavior on resistivity. The anions still form ordered state below $T_{AO} \sim 24\,K$ as indicated by a clear drop of residual resistivity (data not shown). But with further decreasing temperature, an upturn of the resistivity is observed in the R-state and can be attributed to the onset of an insulating ground state at 2.35 K, as shown in Fig. 4.1(b). This insulating state may be related to the occurrence of SDW [148, 152], for the high dosage of disorder breaks the anion ordering along the y axis, hence forms a single-fold Fermi surface which acquires nesting effect at $2k_F$. On the other hand, the SDW temperature of the $x = 15\%$ sample is about half the value of T_{SDW} obtained in Q-state samples. This observation implies that the partially anions ordering in the underlying system is detrimental to the SDW stability. Joo *et al.* [137] also identify a small range of impurity concentrations where neither superconductivity nor SDW phases can be stabilized at temperature as low as 100 mK. This result hence suggests a different phase diagram as compared to the *P*-*T* phase diagram of $(TMTSF)_2PF_6$, where superconductivity and SDW are neighboring or even coexisting. However, since this no-long-range-order state is obtained in only one sample (with $x = 8\%$), this possibility still needs further investigation and confirmation.

In summary, we believe it is ready to conclude that the experimental results discussed above on solid solutions $(TMTSF)_2(ClO_4)_{1-x}(ReO_4)_x$ provide a solid proof for a drastic suppression of superconductivity upon adding nonmagnetic impurities. The inclusion of ReO_4 up to a nominal concentration of $x \sim 17\%$ has little effect on the original normal-state electronic properties. This behavior cannot be reconciled with a model of conventional superconductors. For instead, the gap must change sign upon moving along the Fermi surface. However, as the impurities are local scattering sources, the scattering processes involve multi-momentum components, in particular the momentum of order $2k_F$ which can mix the + and − sheets of the Fermi surface. Thus, the impurity effect by its own cannot be used to distinguish *p* or *d* or even higher partial wave orbital symmetries for the superconducting wave function. Further information is required from analysis on the spin part of the wave function.

4.2 Upper critical field

The experimental investigations on the upper critical field of Bechgaard salts mainly focus on one central question: *Whether and how H_{c_2} exceeds the Pauli paramagnetic limit*? Indeed, one of the most intriguing feature of a spin triplet pairing superconductor, as opposed to its spin singlet counterpart, is the absence of a paramagnetic pair-breaking effect, and consequently a survival of superconductivity at magnetic field exceeding the so-called Pauli limit (or Clogston-Chandrasekhar limit) at which the Zeeman energy becomes comparable to the condensation energy [11, 12]. Although early studies on resistively derived upper critical field values show little evidence for exceeding the Pauli limit [3, 9, 83, 153], subsequent experiments with better samples and magnetic field alignment find that H_{c_2} can significantly go beyond the Pauli limit in both $(TMTSF)_2ClO_4$ [21, 75] and $(TMTSF)_2PF_6$ [23]. Specifically, we have discussed in Chap. 3 the experiment by Lee *et al.*, showing that $H_{c_2}^{b'}$ and $H_{c_2}^{a}$ can exceed the Pauli limit by a factor of about 3, and demonstrate upward curvature at low temperatures [23]. Due to the experimental complication of high pressure, the upper critical field in PF_6 is derived from resistivity measurement only.

In this section, we focus on two recent experiments on the upper critical field in $(TMTSF)_2ClO_4$, where thermodynamic measurement [138] and angle-resolved tunability [139] are available at ambient pressure. Taking advantage of these extraordinary controllabilities, Oh and Naughton [138] confirm $H_{c_2}^{b'}$ indeed exceeds the Pauli limit from a simultaneous resistivity and magnetic torque measurement, while Yonezawa *et al.* [139] observe the same behavior within the whole x-y plane.

However, we would like to stress that the excess of Pauli paramagnetic limit can not set an unambiguous distinction between singlet and triplet pairing states. In quasi-1D superconductors, even singlet superconductivity can be stable far above the Pauli limit by forming a Larkin-Ovchinnikov-Fulde-Ferrell (LOFF) state [16, 29], which is characterized by a spatially modulating superconducting order parameter [30, 31]. In fact, Yonezawa *et al.* [139] rely on a model of LOFF state with rotating wave vector to explain their results on the angular dependence of in-plane H_{c_2}, while theoretical efforts have denoted along this line [154, 155, 156]. The possibility of a LOFF state in Bechgaard salts will be discussed with more details in Chap. 6.

The key improvement of the experiment by Oh and Naughton [138] is that they manage to obtain H_{c_2} not only from resistivity, but also from a simultaneous measurement of magnetic torque, which is expected to vanish as the sample enters a superconducting state due to a rotation of symmetry axis from $\boldsymbol{b}^*$ to $\boldsymbol{b}'$. In order to achieve this, the $(TMTSF)_2ClO_4$ sample is mounted onto a micromachined catilever magnetometer [157], and wired for resistivity measurements along the z axis. This type of design allows simultaneous and independent resistivity and magnetization measurements. The magnetic signal is calibrated using integrated planar coils placed on the cantilever through which a current produces a gauge torque. The deflection of this cantilever is detected via capacity measurement with a bandwidth of 1 Hz. The sample is cooled with a slow cooling rate of $\sim 1\,\mathrm{K/hour}$, and is tested to be in the relaxed state by a high residual resistivity ratio $RRR = \rho_z(300\mathrm{K})/\rho_{T_0}$, which is 450 for $T_0 = 4.2\,\mathrm{K}$ and 1400 for $T_0 = 0\,\mathrm{K}$. All the magnetic data are obtained on the second run of cooling down. The resistivity results, however, are recorded in both the first and second runs, and dramatic difference is observed for temperature below $\sim 0.3\,\mathrm{K}$. This difference is accounted to the creation of micro cracks during thermal cycle.

In Fig. 4.3, the simultaneously measured torque and resistivity results are shown for magnetic field aligned along the y direction at a temperature of 25 mK. The field is chosen to be along this direction because it is predicted to be the most favorable direction for field induced dimensional crossover (FIDC), and in turn the most robust direction for superconductivity. Indeed, this direction is where anomalously large critical fields have been reported in transport measurements [21, 23, 75, 158]. The resistivity data show a rather broad transition from the superconducting state to the normal state with increasing magnetic field. This broad transition, starting at about 2.5 T and completing

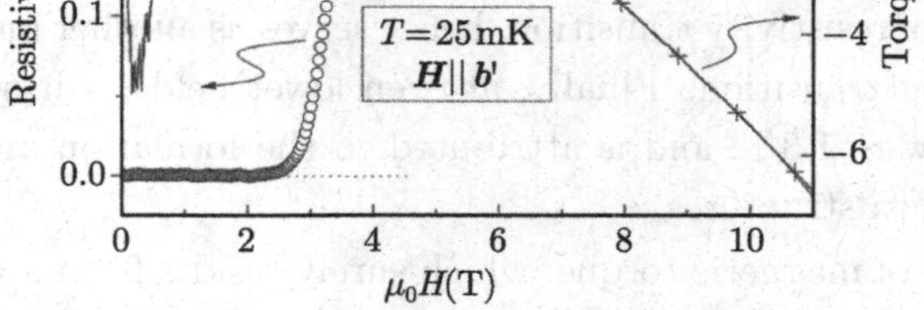

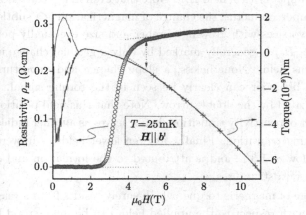

Figure 4.3 Resistivity and torque magnetization in $(TMTSF)_2ClO_4$ at $T = 25\,\mathrm{mK}$ with a magnetic field applied along the $\boldsymbol{b}'$ axis. The dotted line and + symbols associated with the torque data represent a temperature-independent normal state contribution. The onsets of diamagnetism and decreasing resistivity are indicated by the double arrow near $H_{c_2} \sim 5\,\mathrm{T}$. Arrows in the low field hysteretic loop indicate field sweep directions. From Ref. [138].

at about 5.0 T, brings inherent ambiguity in defining H_{c_2}. By extending the resistivity behavior from the high field side (dashed line in Fig. 4.3), one can in principle determine the transition point at which the curve starts to bend down. However, where exactly to place H_{c_2} is still not obvious.

In assistance to determine the superconducting transition, the torque data is considered and plotted in Fig. 4.3. From this result, three regimes can be clearly identified showing different field dependence. First, a high-field regime is observed and characterized by a well-behaved, T-independent, quadratic behavior. Such a signal in the normal state is consistent with expectations for a clean metal, since both Pauli paramagnetic and Landau diamagnetic susceptibilities are generally T and H independent. A fit to this normal state signal using data above 7 T is plotted atop the raw torque signal as cross (+) symbols and also as a dotted line in the vicinity of 5 T. Near this field, the torque signal starts to deviate from the normal state background, and enters the second regime which is attributed to the appearance of superconductivity with decreased magnetization. Hence, the point at which the deviation starts indicates the upper critical field H_{c_2}. Note that even in the conventional model of a type-II superconductor, the change of magnetization is a subtle feature at H_{c_2}. As the vortices with growing number and size eventually penetrate the whole sample, H_{c_2} is generally marked by only a gradual change in magnetic moment versus field. Nonetheless, a slope change and departure from the normal state behavior can clearly be seen in the torque signal, starting at the field indicated by the double arrow. Note that this field position coincides with the onset of resistivity transition, hence serves as another indicator for the superconducting transition. Finally, at even lower fields, a hysteretic regime is evident below ~ 1.3 T, and is attributed to the formation and destruction of complex vortex structures.

The change of magnetic torque, which surely results from a magnetization vector tilted with respect to the applied field, can be understood by noticing that the symmetry axis rotates from $\boldsymbol{b}^*$ direction (the normal to the $\boldsymbol{a}$-$\boldsymbol{c}^*$ plane) to $\boldsymbol{b}'$ axis upon entering the superconducting phase. In the normal state, the symmetry axis $\boldsymbol{b}^*$ is $\sim 5.5°$ away from the field direction, $\boldsymbol{H} \parallel \boldsymbol{b}'$. This explains the non-zero background torque signal. For the superconducting state, the symmetry axis becomes $\boldsymbol{b}'$, and the magnetic torque would vanish if there is no other effects like shape anisotropy or vortex pinning. However, in realistic examples, both terms are present and contribute to a finite torque signal, as shown in Fig. 4.3.

By taking the same resistivity and torque measurements at different temperatures, Oh and Naughton [138] determine $H_{c_2}(T)$ from both transport and thermodynamic properties, and find a good agreement between the two schemes. In Fig. 4.4, representative magnetic moment data derived from the torque are depicted, where $\Delta m = \Delta\tau/\mu_0 H$ and $\Delta\tau$ is the difference between the raw torque signal and the normal state background. Data from up and down magnetic field sweeps are both included to show the reversible nature of the magnetization. In this plot, the onset of the diamagnetic signal associated with the emergence of the superconducting phase becomes evident. At the lowest temperature obtained in this experiment, the magnetic torque data gives an $H_{c_2} = 4.92 \pm 0.05\,\mathrm{T}$ at $T = 25\,\mathrm{mK}$, while the resistivity measurement during the same field sweep provides $H_{c_2} = 5.02 \pm 0.15\,\mathrm{T}$. The uncertainty for resistivity data is much larger due to the rounded transition characteristic of such measurements (see Fig. 4.3).

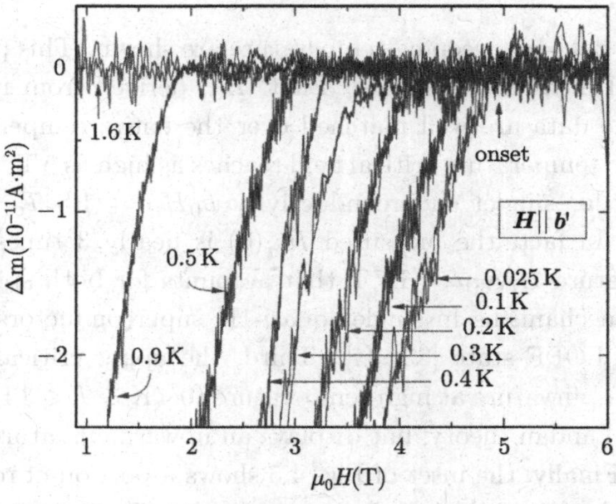

Figure 4.4 Contribution to the magnetization due to superconductivity at various temperatures. H_{c_2} is obtained at the onset of finite moment Δm, as indicated by the arrow for $T = 25\,\mathrm{mK}$. From Ref. [138].

The resulting phase diagram is shown in Fig. 4.5, where H_{c_2} positions determined from magnetic torque and resistivity data by sweeping field, and

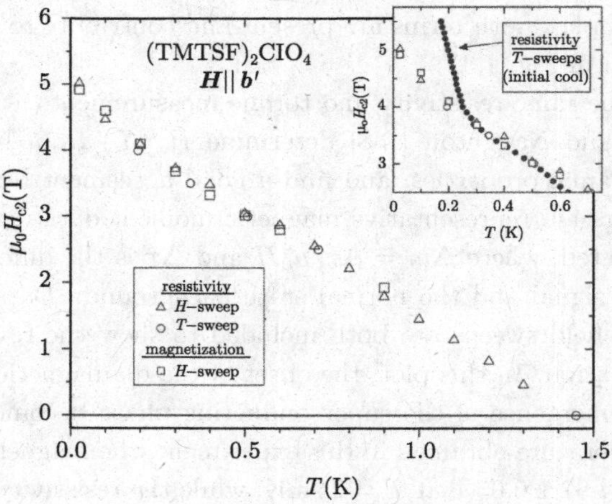

Figure 4.5 Upper critical field along $\boldsymbol{b}'$ axis in $(TMTSF)_2ClO_4$, extracted from both resistivity and magnetization measurements. The inset show resistively derived H_{c_2} from the initial cooldown of the same sample. From Ref. [138].

from resistivity data by sweeping temperature are shown. This phase diagram acquires several intriguing features. First, H_{c_2} derived from resistivity and magnetic torque data are well matched over the entire temperature regime. Second, the zero temperature critical field reaches as high as 5 T, which exceeds the Pauli limit for singlet superconductivity $\mu_0 H_P = 1.84 T_c = 2.6\,\mathrm{T}$ by a factor of ~ 2. In fact, the measured $H_{c_2}(0)$ is nearly 3 times a calculated critical temperature $H_{\mathrm{LOFF}} = 1.7\,\mathrm{T}$ that accounts for both spin and orbital pair breaking mechanisms in singlet quasi-1D superconductors allowing the possibility of a LOFF state [32, 34]. Third, the upper critical field $H_{c_2}(T)$ shows a negative curvature at high temperature ($0.4\,\mathrm{K} \leqslant T \leqslant 1\,\mathrm{K}$) as expected for a Ginzburg-Landau theory, but displays an upward curvature down at low temperatures. Finally, the inset of Fig. 4.5 shows a portion of resistivity data collected with the initial cool of the same sample. These data appear nearly identical to those from the second thermal circle at high temperatures, but shows a distinct upturn below $\sim 0.25\,\mathrm{K}$. This sharp raise of $H_{c_2}^{b'}$ is similar to the results reported for $(TMTSF)_2PF_6$ [23, 158], also for $\boldsymbol{H} \parallel \boldsymbol{b}'$. Oh and Naughton [138] do not report any magnetization data for the initial cool. They attribute the origin of the sharp difference in $H_{c_2}(T)$ below 0.25 K between the first and second cooldowns to the creation of microcracks during the first run. They further propose that these microcracks may be related to the mechanical kinks about the (210) dislocation plane [52], hence can cause large jumps in the in-lane resistance, with basically no impact on T_c and ρ_z. Thus, microcracks of this kind should not be considered as impurities in the usual sense, but rather as mesoscopic mechanical deformations that affect the connectivity of the sample, and in turn its conductivity. However, a thorough investigation is needed to understand the thermal cycle effect.

Putting the difference between the first and second cooldowns aside, we believe it is fair to say that the experimental results by Oh and Naughton [138] confirm the excess of upper critical field over the Pauli paramagnetic limit for fields along the $\boldsymbol{b}'$ direction. Besides, the positive curvature developing in $H_{c_2}(T)$ as H increases is compatible with the picture of FIDC, where the applied magnetic field along $\boldsymbol{b}'$ axis are strong enough to confine electronic states essentially in two dimensions (i.e., the x-y plane as discussed in Chap. 2). However, this is not the case for magnetic field pointing out of the y-z plane. Specifically, when $\boldsymbol{H} \parallel \boldsymbol{a}$, the electronic state remains anisotropic 3D, and the picture of FIDC can not be applied directly. Meanwhile, data for $H^a_{c_2}(T)$ of $(TMTSF)_2PF_6$ [23] do reveal quite interesting features as discussed in Chap. 3: it has a steep slope near $H = 0$, saturates at intermediate magnetic field $H \lesssim 1.8\,\mathrm{T}$, and increases again with an obvious upward curvature below ~ 0.5 K and well exceeds the Pauli paramagnetic limit. For $(TMTSF)_2ClO_4$, a through investigation on $H_{c_2}(T)$ for magnetic field pointing along arbitrary direction within the x-y plane is conducted by Yonezawa *et al.* [139]. The results provide evidence for the excess of Pauli limit when $\boldsymbol{H} \parallel \boldsymbol{a}$, and suggests some interesting angular-dependence of $H_{c_2}(T)$ within the x-y plane.

In the experiment by Yonezawa *et al.* [139], the upper critical field H_{c_2} is derived from resistance measurement along the z direction on single crystal samples of $(TMTSF)_2ClO_4$ using an ac four-probe method. Temperature is measured using a RuO_2 resistance thermometer with magnetoresistance correction and can reach as low as 80 mK. All samples are cooled at a slow rate of 2 mK/min across the anion ordering temperature T_{AO} to ensure that the whole sample is in the relaxed state. Magnetic fields are applied using the "Vector Magnet" system [159], with which the field direction can be tuned without mechanical heatings. The directions of the crystallographic axes of the sample are gauged from the anisotropy of H_{c_2} at 0.1 K. The accuracy of field alignment with respect to the x-y plane and of the x axis in the x-y plane are both within 0.1°. In the following discussion of this section, the azimuthal angle ϕ is used to denote the rotation of $\boldsymbol{H}$ within the x-y plane with respect to the x axis.

A typical set of resistance data $R_{c^*}(T)$ in zero field is shown in Fig. 4.6(a). The results show that $R_{c^*}(T)$ starts to drop with a fairly sharp transition at about 1.45 K, reaches zero value around 1.30 K, and increases again below 0.8 K. The drop and zero-resistivity behavior clearly indicate the superconducting transition, while the anomalous increase is probably attributed to the microcracks created during cooldown. For magnetic field $H = 50\,\text{kOe}$ applied along $\boldsymbol{b}'$, a decrease of resistance R_{c^*} can also be identified below 0.2 K, indicating a transition to the superconducting phase. This indeed can be confirmed by adding a small out-of-plane field $H_{c^*} = 0.5 \sim 1.0\,\text{T}$, with the effect of destroying superconductivity while bringing negligible magnetoresistance change. As plotted in Fig. 4.6(b), the decrease of resistance disappears with H_{c^*}, hence is

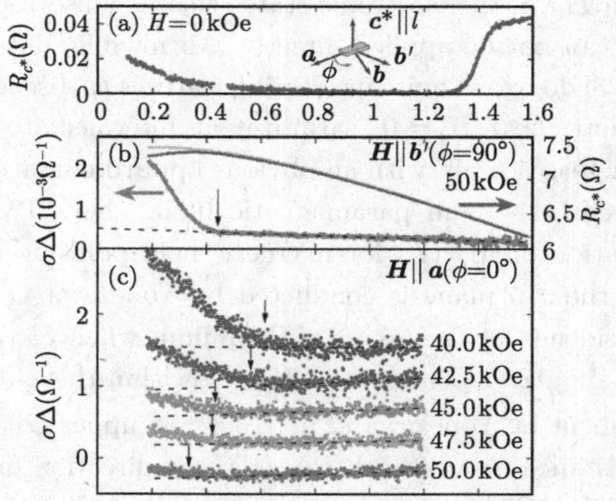

Figure 4.6 (a) Temperature dependence of R_{c^*} in the absence of magnetic field. (b) Conductance difference $\Delta\sigma$ under an in-plane magnetic field of 50 kOe applied along the $\boldsymbol{b}'$ axis. Resistances for zero field (blue) and $H_{c^*} = 0.5\,\text{kOe}$ (light blue) are plotted against the right vertical axis. (c) Temperature dependence of $\Delta\sigma$ with the in-plane field applied along the $\boldsymbol{a}$ axis and $H_{c^*} = 1.0\,\text{kOe}$. From Ref. [139].

validated to be a signature of superconducting transition. The precise position of the transition is then extracted from the conductance difference

$$\Delta\sigma(H) \equiv \frac{1}{R_{c^*}(H_{c^*}=0)} - \frac{1}{R_{c^*}(H_{c^*}>0)}. \tag{4.2}$$

The sharp increase of $\Delta\sigma$ shown in Fig. 4.6(b) gives a fairly satisfactory definition for the very onset temperature of superconductivity T_c, as indicated by arrows in Figs. 4.6(b) and 4.6(c). This scheme has the advantage that T_c is not affected by the small increase of R_{c^*} with decreasing temperature, for it is cancelled in the subtraction. For magnetic field along $\boldsymbol{c}^*$ direction, $T_c(H)$ is determined similarly from the conductance difference

$$\Delta\sigma(H) \equiv \frac{1}{R_{c^*}(H_{c^*}=H)} - \frac{1}{R_{c^*}(H_{c^*}=H+\Delta H)}. \tag{4.3}$$

The phase diagrams for magnetic field along the $\boldsymbol{a}$, $\boldsymbol{b}'$, and $\boldsymbol{c}^*$ axes are presented in Fig. 4.7. It is interesting that the overall features of all three curves are similar to those for $(TMTSF)_2PF_6$ shown in Fig. 3.2 [23]. At low magnetic fields, linear temperature dependence is observed for all field directions, as expected for a Ginzburg-Landau theory for a clean type-II superconductor. For $\boldsymbol{H} \parallel \boldsymbol{b}'$, $H_{c_2}^{b'}(T)$ keeps the linear T dependence up to 35 kOe and exhibits a rapid upturn in higher fields. This behavior is consistent with the first-cool data depicted in Fig. 4.5 observed by Oh and Naughton [138]. For $\boldsymbol{H} \parallel \boldsymbol{a}$, the curve shows an apparent saturating behavior at intermediate temperatures, which is consistent with the Pauli limiting effect with estimated value of $H_P = 1.84T_c = 26.7\,\text{kOe}$. At even lower temperature $T \lesssim 0.5\,\text{K}$, $H_{c_2}^a$ displays an upward curvature, and reaches as high as 50 kOe, exceeding the Pauli limit by nearly a factor of 2.

Yonezawa *et al.* also take advantage of their angle-resolved apparatus, and investigate the variation of T_c when magnetic fields are rotated within the x-y plane. The data are shown in Fig 4.8 using polar plots of $T_c(\phi)$, where the direction of each point corresponds to the field direction and the distance from the origin corresponds to the value of T_c. At low fields, $T_c(\phi)$ exhibits a sharp cusp at $\phi = 0°$ (i.e., $\boldsymbol{H} \parallel \boldsymbol{a}$) and a broad minimum around $\phi = \pm 90°$ $(\boldsymbol{H} \parallel \boldsymbol{a})$. As the field exceeds 20 kOe, dips of $T_c(\phi)$ emerge at $\phi \approx \pm 17.1°$,

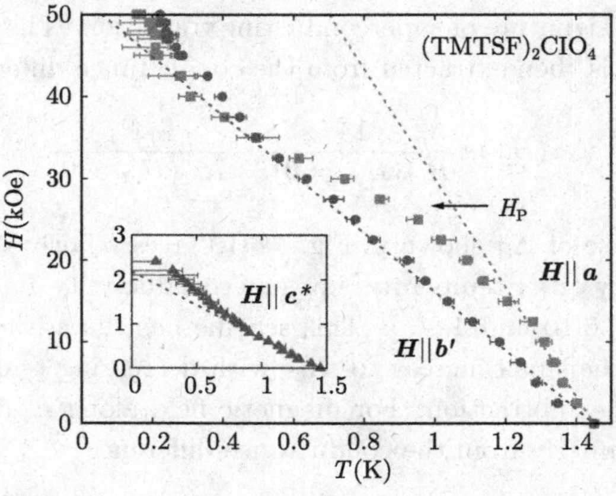

Figure 4.7 Phase diagrams for $\boldsymbol{H} \parallel \boldsymbol{a}$ (solid squares) and $\boldsymbol{H} \parallel \boldsymbol{b}'$ (solid dots). The phase diagram for $\boldsymbol{H} \parallel \boldsymbol{c}^*$ is shown in the inset. The dashed lines indicate the initial slopes of each curves. From Ref. [139].

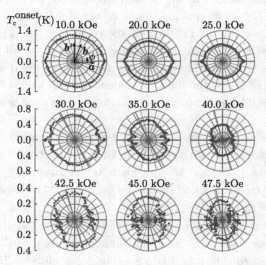

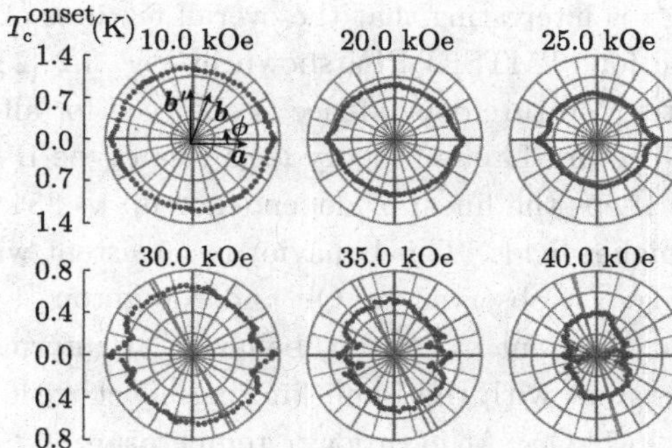

Figure 4.8 Polar plots of $T_c(\phi)$. The points for $|\phi| > 90°$ are duplicated from the same data as those for $|\phi| \leqslant 90°$. They are shown to give a symmetric figure. Error bars are not shown for the sake of clarity. Purple crosses indicate that no SC anomaly in $\Delta\sigma(T)$ was observed above 80 mK. Solid red lines indicate the new symmetry axis $\boldsymbol{X}$ (see text). From Ref. [139].

which may be attributed to the onset of FIDC. When H further increases above 30 kOe, the $\boldsymbol{b}'$ axis of $\phi = \pm 90°$ is no longer a symmetry axis of T_c, and a new symmetry axis $\boldsymbol{X}$ appears around $\phi \sim -70°$, as depicted by the solid line in Fig. 4.8. Besides, $T_c(\phi)$ exhibits an enhancement around the new axis $\boldsymbol{X}$, in clear contrast to the broad minimum around the old axis $\boldsymbol{b}'$. The $\boldsymbol{X}$ axis also tends to rotate toward the $\boldsymbol{b}'$ direction as the field increases. In fact, the deviation of $\boldsymbol{X}$ from the $\boldsymbol{b}'$ axis is reduced to about 10° at 47.5 kOe. This change of symmetry can not be explained by a field misalignment because it is only present at high fields. Yonezawa *et al.* [139] account the emergence of new symmetry axis to a rotation of wave vector in a LOFF state. Further discussion regarding this possibility can be found in Chap. 6.

4.3 Thermal conductivity and heat capacity

In the previous sections, we have discussed experiments on non-magnetic impurity effect and upper critical fields of superconducting $(TMTSF)_2ClO_4$. The combined results strongly suggest an unconventional superconductor featured by a drastic suppression of T_c upon adding non-magnetic disorders and an exceeding of Pauli paramagnetic limit. To characterize this unconventional superconducting phase, much experimental efforts have been endeavored to unveil the symmetry of the superconducting gap. In this section, we will discuss some recent investigations on thermodynamic quantities of superconducting $(TMTSF)_2ClO_4$. These measurements are not available for $(TMTSF)_2PF_6$ due to experimental difficulties imposed by high pressure. In particular, Belin and Behnia [39] report results on the thermal conductivity and suggest a fully gapped quasiparticle spectrum. On the other hand, Yonezawa *et al.* [40] draw a rather contradicting conclusion of a superconducting gap with lines of nodes, based on their field-angle-resolved calorimetry results.

Measurements of the thermal conductivity tensor is a powerful probe of the order parameter structure in unconventional superconductors. For instance, in the particular case of the heavy fermion superconductor UPt_3, thermal conductivity measurements are a major source of the current understanding of the angular distribution of nodes in the gap function [160, 161, 162]. In addition, for the high-T_c superconductor $YBa_2Cu_3O_{6.9}$, there are several important signatures of d-wave superconductivity in a variety of heat transport measurements [163, 164, 165]. For Bechgaard salts, however, measurements of thermal conductivity used to be restricted to temperatures well above the transition to the superconducting state [74]. In the experiment by Belin and Behnia (BB) [39], technical difficulties were surmounted and thermal, as well as electrical, conductivities were measured at low temperatures in $(TMTSF)_2ClO_4$.

The temperature dependence of thermal conductivity κ and of electrical resistance R of one single crystal sample are shown in Fig. 4.9, at low temperatures. This sample is fairly clean since the ratio of room temperature resistance to residual ($T \to 0$) resistance (RRR) is about 440. The superconducting transition is depicted by the jump of resistivity around $T = 1\,\text{K}$

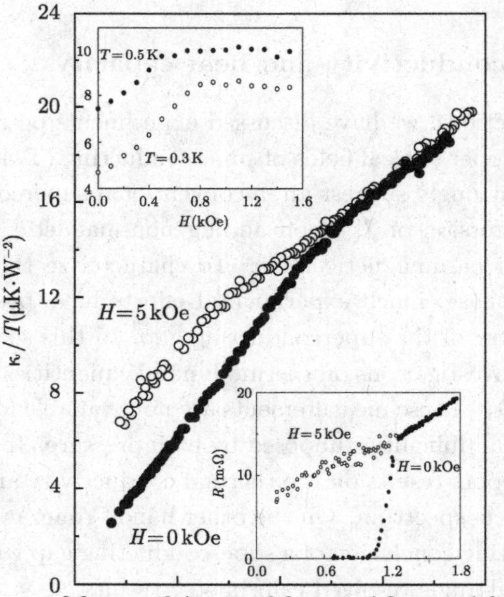

Figure 4.9 Plots of thermal conductivity divided by temperature (κ/T) for a relaxed sample of $(TMTSF)_2ClO_4$. The superconducting state is suppressed by applying a small magnetic field along the c^* axis. The lower insert shows the temperature dependence of the electrical resistance for the same sample. The upper insert shows the field dependence of κ/T for two different temperatures: $T = 0.5\,\text{K}$ and $T = 0.3\,\text{K}$. From Ref. [39].

in zero magnetic field. Corresponding changes in κ between normal ($H =$ 5 kOe along the c^* axis) and superconducting ($H = 0$) states can also be observed for $T < 1$ K. The ratio of thermal and electrical conductances at the onset of superconductivity indicates that heat transport is dominated by phonons and that the electronic contribution accounts only for a small fraction of total thermal conductivity for $T > 1$ K.

In general, the separation between lattice and quasiparticle components of the thermal conductivity is not a straightforward procedure. In the particular case of superconductors, the condensation of electrons into Cooper pairs can alter lattice contributions if there is a substantial electron-phonon coupling. To unveil the effect of the superconducting instability on heat carriers, BB analyzed the ratio $\Delta\kappa/T$ corresponding to the difference between the two experimental curves of κ/T (at $H = 5$ kOe and at $H = 0$), as a function of temperature. As shown in Fig. 4.10, $\Delta\kappa/T$ increases monotonically with decreasing temperature before saturating at about $T = 0.4$ K. The saturation value $(\Delta\kappa/T)|_{\text{sat}}$ is 3.7 ± 0.4 μW$\cdot$K^{-2} in this sample, which reflects the contribution comes from normal electrons at low temperatures, and indicates the absence of low energy quasiparticle heat carriers. Consider the Sommerfeld

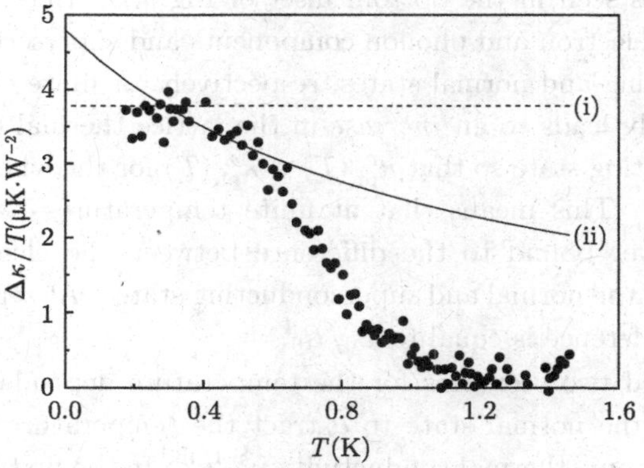

Figure 4.10 The difference between the thermal conductivity of $(TMTSF)_2ClO_4$ at $H = 5$ kOe and $H = 0$ kOe. Note the saturation at $T = 0.4$ K. The horizontal line (i) represents the ratio L_0/R_0, while line (ii) represents the behavior of $\Delta\kappa/T$ when the normal state thermal conductivity follows the behavior imposed by the electrical resistance $R(T)$ and by the Wiedemann-Franz (WF) law. From Ref. [39].

value of Lorenz number $L_0 = 2.45 \times 10^{-8}\,\Omega \cdot \mathrm{m} \cdot \mathrm{W} \cdot \mathrm{K}^{-2}$ and the residual resistance R_0 of the same crystal, this saturation value is very close to the expected maximum electronic contribution to heat transport according to the Wiedemann–Franz (WF) law ($L_0/R_0 = 3.8 \pm 0.5\,\mu\mathrm{W} \cdot \mathrm{K}^{-2}$). This result is not in disagreement with the theoretical prediction [166] that the separation of spin and charge degrees of freedom in an interacting 1D electron gas leads to the violation of WF law and to a divergence of Lorenz number at zero temperature. In fact, since the interplane tunneling energy t_z is estimated to be a few kelvins, the quasi-1D system $(TMTSF)_2ClO_4$ is expected to behave like an anisotropic 3D Fermi liquid at temperatures much below this scale.

In order to refine the argument that $\Delta\kappa/T$ is a measure of the difference between normal and superconducting electronic thermal conductivities, BB separated the electron and phonon contributions to κ and formally expressed $\Delta\kappa/T$ as

$$\frac{\Delta\kappa(T)}{T} = \left(\frac{\kappa^n_{\mathrm{el}}(T)}{T} - \frac{\kappa^s_{\mathrm{el}}(T)}{T}\right) + \left(\frac{\kappa^n_{\mathrm{ph}}(T)}{T} - \frac{\kappa^s_{\mathrm{ph}}(T)}{T}\right), \tag{4.4}$$

where the magneto-resistance are completely neglected since it is very small at $H = 5\,\mathrm{kOe}$ as seen in the bottom inset of Fig. 4.9. Here, the subscripts (el, ph) refer to electron and phonon components and superscripts (s, n) refer to superconducting and normal states, respectively. A finite electron-phonon coupling typically leads to an *increase* in the lattice thermal conductivity in the superconducting state so that $\kappa^n_{\mathrm{ph}}(T) \leqslant \kappa^s_{\mathrm{ph}}(T)$ for the whole temperature range below T_c. This means that at finite temperatures below T_c, $\Delta\kappa/T$ represents a lower bound to the difference between the electronic thermal conductivities of the normal and superconducting states. At zero temperature, however, this difference is equal to L_0/R_0.

BB then used two scenarios for the temperature dependence of thermal conductivity in the normal state to extract the temperature dependence of normalized electronic thermal conductivity $\kappa^s_{\mathrm{el}}/\kappa^n_{\mathrm{el}}$. In the first scenario, κ^n_{el}/T is the constant L_0/R_0 (see curve (i) in Fig. 4.10). In the second scenario, the thermal conductivity in the normal state follows the behavior imposed by the temperature dependence of electrical resistivity and the WF law (see curve (ii) in Fig. 4.10). The two scenarios are both over simplified. However, as seen in Fig. 4.11, the normalized $\kappa^s_{\mathrm{el}}/\kappa^n_{\mathrm{el}}$ curves are not very different for the two possible scenarios. In addition, the data from the unconventional superconductor UPt_3 [160], and the theoretical predictions from the conventional Bardeen–Richaysen–Tewordt (BRT) theory [167] with $\Delta(0)/T_c = 1.9$ and 2.3, are shown for comparison.

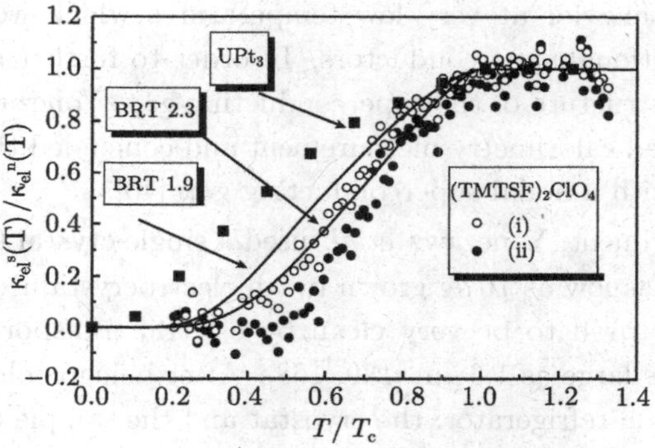

Figure 4.11 Normalized electronic thermal conductivity $\kappa_{\rm el}^s/\kappa_{\rm el}^n$ versus normalized temperature T/T_c for $(TMTSF)_2ClO_4$ in two different scenarios. The open circles correspond to the extracted data within the assumption that $\kappa_{\rm el}/T = L_0/R_0$. The filled circles correspond to the extracted data within the assumption that the normal κ follows the temperature dependence of $R(T)$ and the WF law. The experimental results are compared with the predictions of Bardeen–Richaysen–Tewordt [167] theory for a fully gapped s-wave superconductor for $\Delta(0)/T_c = 1.9$, 2.3, and with the published data on UPt_3 for a heat current along the $\boldsymbol{b}'$ axis. From Ref. [39].

As indicated in Fig. 4.11, the decrease in $\kappa_{\rm el}^s/\kappa_{\rm el}^n$ is much faster in $(TMTSF)_2$ ClO_4 than in UPt_3. For instance, at $T/T_c = 0.4$, $\kappa_{\rm el}^s/\kappa_{\rm el}^n$ drops virtually to zero in $(TMTSF)_2ClO_4$, but remains a sizable value (~0.38) in UPt_3. It is expected that a finite electron-phonon coupling would lead to an even sharper decrease in $\kappa_{\rm el}^s/\kappa_{\rm el}^n$ below T_c. Thus, in spite of several simplifications to obtain the plots of Fig. 4.11, the main conclusion from BB's work is clear: their data can not be made consistent with a gap function vanishing at any place in the Fermi surface. Their analysis is directly based on the saturation presented in Fig. 4.10, which is also observed in all other samples they have studied.

This result seems be in contradiction with the outcome of an earlier experimental investigation of the order parameter structure of $(TMTSF)_2ClO_4$ using

NMR techniques [13]. However, it is important to emphasize that the temperature range where nuclear relaxation rate $1/T_1$ showed a T^3 dependence was limited only to a rather high temperature ($T_c/2 < T < T_c$). This T^3 behavior at high temperatures can not be considered as convincing evidence for nodes in the order parameter, because the system could still crossover to an exponential behavior at very low temperatures which would be characteristic of conventional superconductors. In order to further clarify the issue regarding nodal structure of the superconducting gap, Yonezawa performed a field-angle-resolved calorimetry measurement and concluded that their results are compatible with a nodal superconducting gap [40].

In their experiment, Yonezawa *et al.* used a single crystal $(TMTSF)_2ClO_4$ sample weighing as low as 76 μg grown by an electrocrystallization technique. This sample is proved to be very clean by electric transport study, with a mean free path as large as 1.6 μm [139, 168]. After being cooled to 4.2 K with a ^{3}He$-^4$He dilution refrigerator, the cryostat and the sample are heated back up to 26 K, and cooled it very slowly across the anion ordering temperature T_{AO} at 4 mK/min. This procedure is required to make sure that the anions order well to be in the relaxed state. The calorimetry measurement is done with a high-resolution ($\sim$ 100 pJ/K at 1 K) calorimeter shown in Fig. 4.12(b). This apparatus is developed based on a modification of the "bath-modulating method", with an advantage that a heater on the sample holder is not necessary such that the background heat capacity of the sample holder can be minimized. Although this background contribution is not deducted from the data shown, it has been checked that the background is nearly field independent, hence brings little influence on the net results. The magnetic field is applied along the crystallographic axes with an accuracy of alignment of approximately 0.1°.

The results for heat capacity as functions of applied field along all three crystallographic axes are shown in Fig. 4.12. For field perpendicular to the conducting x-y ($\boldsymbol{a}$-$\boldsymbol{b}'$) plane, i.e., $\boldsymbol{H} \parallel \boldsymbol{c}^*$ or $\boldsymbol{H} \parallel \boldsymbol{z}$, the function $C(H)/T$ exhibits an $H^{0.5}$ dependence at low temperatures [see Fig. 4.12(b)]. Considering the fact the electronic heat capacity $C_e(T) \propto NT$ at low temperatures, where N is the quasiparticle density of states (QDOS), this dependence is compatible with a superconducting state with lines of nodes, where $N \propto N_F \sqrt{H/H_{c2}}$ with N_F the density of states of the normal state at Fermi surface [169]. As a comparison, when the magnetic field is aligned along the x (i.e., the $\boldsymbol{a}$) axis, the function $C(H)/T$ shows a upward curvature as H approaches H_{c_2} from below.

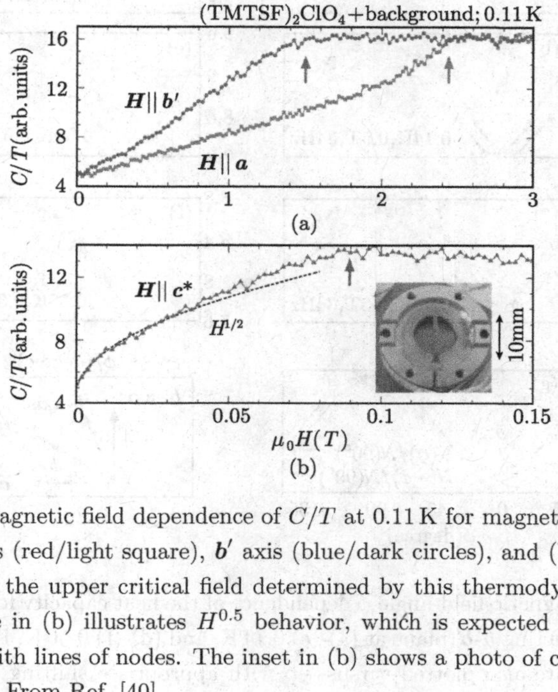

Figure 4.12 Magnetic field dependence of C/T at 0.11 K for magnetic fields applied along (a) $\boldsymbol{a}$ axis (red/light square), $\boldsymbol{b}'$ axis (blue/dark circles), and (b) $\boldsymbol{c}^*$ axis. The arrows indicate the upper critical field determined by this thermodynamic method. The dotted line in (b) illustrates $H^{0.5}$ behavior, which is expected for a superconducting state with lines of nodes. The inset in (b) shows a photo of calorimeter used in experiments. From Ref. [40].

To investigate the nodal structure of the superconducting gap, Yonezawa *et al.* further measured the heat capacity with magnetic field pointing along different directions within the x-y plane, and presented their results of C/T as functions of the azimuthal angle ϕ of the field measured from the x axis. From Fig. 4.13(a)–(f), we can observe a clear angular dependence of C/T, indicating a variation of the QDOS upon rotation of magnetic field.

In the low-temperature and low-field limit, the variation of QDOS can be approximately described in a semiclassical manner, where the energy of quasiparticles are "Doppler shifted" by an amount of

$$\delta\omega(\boldsymbol{r},\boldsymbol{k}) \propto \boldsymbol{v}_{\rm s}(\boldsymbol{r}) \cdot \boldsymbol{v}_{\rm F}(\boldsymbol{k}). \tag{4.5}$$

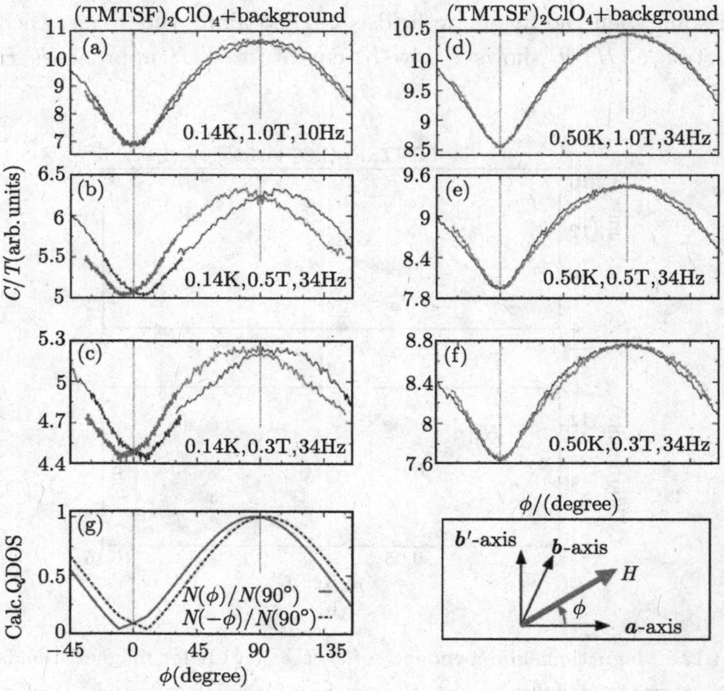

Figure 4.13 Magnetic-field-angle ϕ dependence of the heat capacity for fields rotated within the conducting $\boldsymbol{a}$-$\boldsymbol{b}'$ plane at (a)-(c) 0.14 K, and (d)-(f) 0.50 K. For comparison, the same data are also plotted versus $-\phi$ with appropriate shifting (black curves). The deviation of the two curves indicates the asymmetry in the $C(\phi)/T$ curve. (g) Quasiparticle density of states (QDOS) $N(\phi)/N(90°)$ (blue solid) and $N(-\phi)/N(90°)$ (black dotted) calculated by Eq. (4.9) assuming two nodes (n1 and n2) with parameters $\phi_{\mathrm{n1}} = -10°$ and $\phi_{\mathrm{n2}} = +10°$, $A_{\mathrm{n2}}/A_{\mathrm{n1}} = 0.3$, and $\gamma = H_{\mathrm{c_2}}(0°)/H_{\mathrm{c_2}}(90°) = 3.5$. From Ref. [40].

Here, $\boldsymbol{v}_{\mathrm{s}}(\boldsymbol{r})$ is the velocity of the supercurrent around a vortex and $\boldsymbol{v}_{\mathrm{F}}(\boldsymbol{k})$ is the Fermi velocity [169, 170]. When $\delta\omega$ exceeds the pairing gap $\Delta(\boldsymbol{k})$, quasiparticles with wave vector $\boldsymbol{k}$ will be significantly excited. Thus, when the superconducting gap acquires nodes at some wave vectors $\boldsymbol{k}_{\mathrm{node}}$, most of the excitation would occur in the vicinity of these nodes and the most important contribution to QDOS is originated from the Doppler shift around the nodes

$$\delta\omega^{\text{node}} \propto \boldsymbol{v}_{\text{s}} \cdot \boldsymbol{v}_{\text{F}}^{\text{node}}, \tag{4.6}$$

where $\boldsymbol{v}_{\text{F}}^{\text{node}}$ is used to denote $\boldsymbol{v}_{\text{F}}(\boldsymbol{k}_{\text{node}})$. When the applied magnetic field $\boldsymbol{H}$ is parallel to $\boldsymbol{v}_{\text{F}}^{\text{node}}$, the supercurrent velocity is perpendicular to $\boldsymbol{H}$ and to $\boldsymbol{v}_{\text{F}}^{\text{node}}$, leading to a *zero* Doppler shift $\delta\omega^{\text{node}} = 0$. In this circumstance, the QDOS induced by this shift becomes small. This mechanism can be easily understood from a schematic drawing in Fig. 4.14. When the applied field is parallel to $\boldsymbol{v}_{\text{F}}$ at a node, the quasiparticle excitation at this node is reduced and so does its contribution to electronic heat capacity.

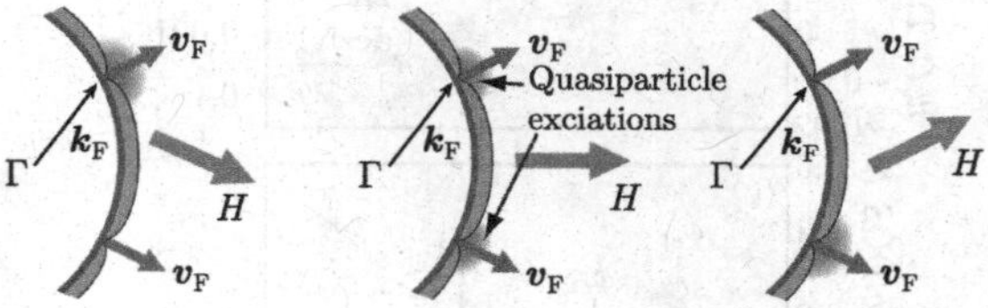

Figure 4.14 Illustration of quasiparticle excitation due to the Doppler shift. When the field is parallel to $\boldsymbol{v}_{\text{F}}$ at a node, the quasiparticle excitation at this node is reduced. From Ref. [40].

Adopting this scheme, we can read immediately from Fig. 4.13 the directions of $\boldsymbol{v}_{\text{F}}$ at the nodal positions. Indeed, the central feature of results shown in Fig. 4.13 is that $C(\phi)/T$ becomes asymmetric with respect to the x axis at low temperatures and low fields, as depicted in Figs. 4.13(b) and 4.13(c). As a comparison, the curves are nearly symmetric at hight temperatures [Figs. 4.13(d–f)] or in high fields [Fig. 4.13(a)]. Thus, the asymmetry is not due to a misalignment of the field. For instead, it is originated from two separated minimal angles of $C(\phi)/T$ which correspond to directions of Fermi velocity and at least two nodes. Besides, a close inspection of the data at $T = 0.14\,\text{K}$ and $H = 0.3\,\text{T}$ reveals two small kinks at around $\phi \approx \pm 10°$, in consistent with the observed minimal angles [see Fig. 4.15(a)]. The kink signatures are more evident in the first and second derivatives, where step-like behavior and peaks are observed in Fig. 4.15(b), respectively. We also notice that the heat capacity anomalies at $\phi \approx \pm 10°$ disappear at high temperatures or in a high magnetic field. This observation is qualitatively compatible with the explanation attributing the anomalies to the gap anisotropy, considering the fact that high temperature and field can populate more high energy excitations and eventually blur out any low-energy features.

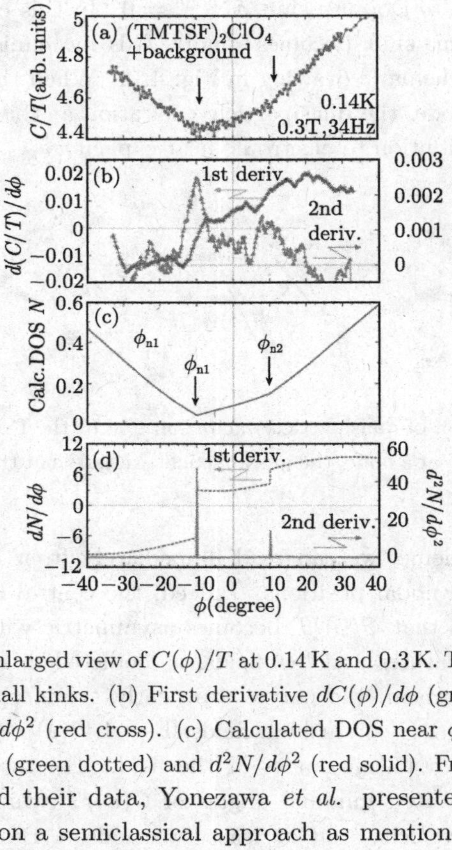

Figure 4.15 (a) Enlarged view of $C(\phi)/T$ at 0.14 K and 0.3 K. The arrows indicate the positions of the small kinks. (b) First derivative $dC(\phi)/d\phi$ (green cross) and second derivative $d^2C(\phi)/d\phi^2$ (red cross). (c) Calculated DOS near $\phi = 0$. (d) Qualitative behavior of $dN/d\phi$ (green dotted) and $d^2N/d\phi^2$ (red solid). From Ref. [40].

To understand their data, Yonezawa *et al.* presented a phenomenological model based on a semiclassical approach as mentioned above, where the excitations are described by Doppler shift. In order to incorporate the large in-plane H_{c_2} anisotropy, they assume that the QDOS for $H \ll H_{c_2}$ takes the following form

$$N(\phi) \propto \sqrt{\frac{H}{H_{c_2}(\phi)}} \sum_n A_n |\sin(\phi - \phi_n)|, \tag{4.7}$$

where the summation runs over all nodes of the superconducting gap, and ϕ_n

is the direction of Fermi velocity $\boldsymbol{v}_{\mathrm{F}}$ at the nth node. To take into account of the triclinic band structure, they introduce nodal-position-dependent amplitudes A_n. The upper critical field H_{c_2} is approximated by the effective mass model

$$H_{\mathrm{c}_2}(\phi) = \frac{H_{\mathrm{c}_2}(\phi = 0^\circ)}{\sqrt{\gamma^2 \sin^2\phi + \cos^2\phi}}, \tag{4.8}$$

with $\gamma = H_{\mathrm{c}_2}(\phi = 0^\circ)/H_{\mathrm{c}_2}(\phi = 90^\circ) = H_{\mathrm{c}2}^{a}/H_{\mathrm{c}2}^{b'}$. Thus, the QDOS, which is proportional to C_{e}/T, reads

$$N(\phi) \propto \sqrt{\frac{H}{H_{\mathrm{c}_2}(0^\circ)}}\left(\gamma^2 \sin^2\phi + \cos^2\phi\right)^{1/4} \sum_n A_n |\sin(\phi - \phi_n)|. \tag{4.9}$$

Using this approach, and assuming $\phi_{1,2} = \pm 10^\circ$ *a priori*, the calculated $N(\phi)$ and its derivatives are plotted in Figs. 4.13(g) and 4.15(c–d), respectively. As expected, the results can qualitatively reproduce some features of experimental data, but additional efforts are required to give a microscopic theory within a full quantum mechanical level. Within such a theory, contributions from quasiparticles, as well as from vortex cores and thermal excitations should be all taken into account with an equal footing. For discussion along this line, readers are referred to Chap. 5.

Knowing the direction of Fermi velocity at nodes is in principle not enough to determine the nodal structure in $\boldsymbol{k}$-space. For quasi-two-dimensional or three-dimensional systems, the assumption $\boldsymbol{k}_{\mathrm{F}} \parallel \boldsymbol{v}_{\mathrm{F}}$ is reasonable as a simple model. However, in quasi-one-dimensional system like Beckgaard salt, $\boldsymbol{v}_{\mathrm{F}}$ is dominated by a large x component $|\boldsymbol{v}_{\mathrm{F}}| \approx \boldsymbol{v}_{\mathrm{F},x}$, and is *not* parallel to $\boldsymbol{k}_{\mathrm{F}}$ even in the simplest model. This can be seen clearly from Fig. 2.4 in Chap. 2. Thus, further information about the band structure is required to extract the nodal positions from the heat capacity results. In Fig. 4.16(a), the angle of Fermi velocity as a function of k_y is plotted along the Fermi surface, based on a tight-binding band structure [171]. Notice that a two-fold Fermi surface structure is present, with outer (O-FS) and inner (I-FS) Fermi surfaces at $\pm k_{\mathrm{F}}$. This is because as cooled slowly through the anion order temperature T_{AO}, the orientation of anions is uniform along the stacking axis but alternate along the y direction, leading to the appearance of a $2b$ superlattice and the creation of two sheets of open Fermi surfaces. Fig. 4.16(a) implies that $\phi = 10^\circ$ or -10° on the outer Fermi surface (O-FS) at $k_y/b^* \sim \pm 0.25, +0.36$, and -0.06, where

$b^* = 2\pi/b$ is the size of the first Brillouin zone along k_y. Yonezawa *et al.* then propose a simple structure with lines of nodes running at $k_y/b^* = \pm 0.25$, which is depicted in Fig. 4.16(b), and suggest pairing states with d-wave or g-wave symmetries as shown in the same figure.

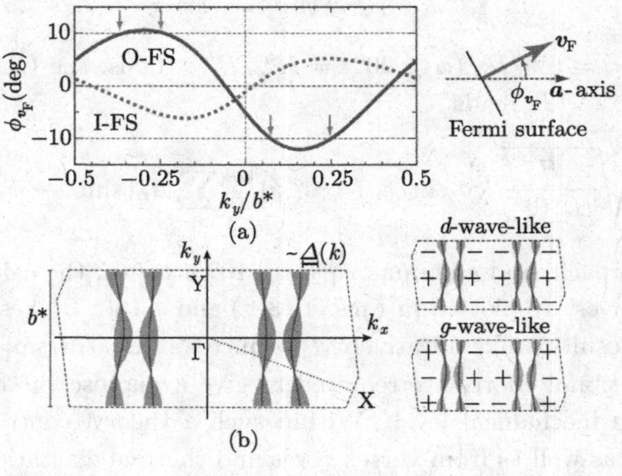

Figure 4.16 (a) Dependence of $\phi_{v_F} \equiv \arctan(v_y/v_x)$ on k_y for the outer (O-FS) and inner (I-FS) Fermi surfaces. The arrows indicate the points where $\phi_{v_F} = \pm 10°$. (b) Plausible gap structure with nodes at $k_y \sim \pm 0.25b^*$. Examples of the superconducting states that satisfy the observed nodal structure are shown: the d-wave-like state and the g-wave-like state. From Ref. [40].

In this section, we have introduced two experiments both focus on the nodal structure of the superconducting gap in $(TMTSF)_2ClO_4$. Separated by 15 years, the two works suggest more or less contradicting results, where Belin and Behnia [39] concluded a fully gapped phase from thermal conductivity measurement, and Yonezawa *et al.* [40] favored a superconducting state with lines of nodes based on their heat capacity data. That far, we believe it is fair to say that the controversy regarding the superconducting nodal structure is not fully settled yet.

4.4 Muon and nuclear spin resonances

In the previous section, we briefly go over the controversy regarding the nodal structure of the superconducting gap in $(TMTSF)_2ClO_4$. In this section, we will discuss some spectroscopic measurements, which can provide us further information regarding the spin symmetry of the superconducting phase. Among them, Luke *et al.* [140] performed zero-field muon spin resonance (μSR) measurements to search for spontaneous magnetic fields below T_c in $(TMTSF)_2ClO_4$, and observed no evidence for such fields within experimental resolution. Therefore, they suggested that time reversal symmetry is likely not broken in superconducting $(TMTSF)_2ClO_4$. Besides, Shinagawa *et al.* [38] conducted ^{77}Se NMR Knight shift and relaxation rate measurements at low magnetic fields, and found a decrease in spin susceptibility which is consistent with singlet spin pairing.

Zero-field-μSR is the most effective method to detect the existence of local magnetic fields. In the case of unconventional superconductors characterized by broken time reversal symmetry, Cooper pairs carry either spin and/or orbital angular momentum and may produce non-zero static magnetic fields. These local magnetic fields were identified by zero-field-μSR in the cases of heavy fermion systems [172] and strontium ruthenate [173], and have indicated the possibility of time reversal symmetry breaking in these systems. However, in zero-field-μSR experiments for $(TMTSF)_2ClO_4$ within the resolution of $\approx 0.25\,$G, Luke *et al.* [140] did not observe any increase in the relaxation rate of deuterated $(TMTSF)_2ClO_4$ samples below the superconducting transition temperature T_c (see Fig. 4.17). Therefore, they suggested that time reversal symmetry is not broken for superconducting $(TMTSF)_2ClO_4$. This absence

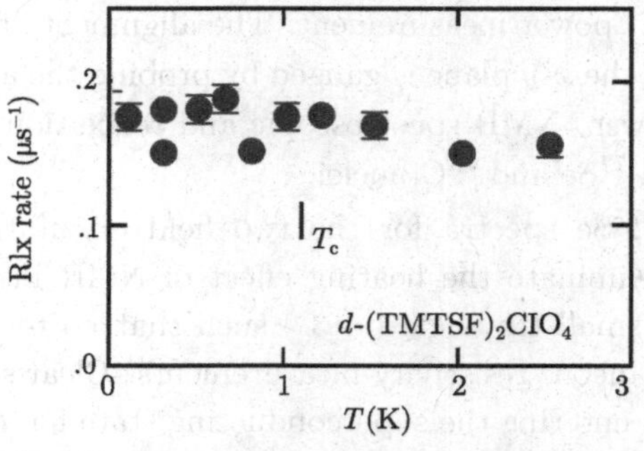

Figure 4.17 Zero-field-μSR relaxation rate in deuterated $(TMTSF)_2ClO_4$ showing no increase below $T_c \approx 1.1\,$K. From Ref. [140].

of time reversal symmetry breaking does not rule out either singlet or triplet pairing states. However, it restricts the number of possibilities for the corresponding order parameter. For instance, all pairing states listed in Tab. 5.3 in Sec. 5.7 have time reversal symmetry broken, thus are not compatible with the μSR experiments.

Compared to μSR method, the NMR measurement is most discriminating in the spin symmetry of the superconducting order parameter. As noted in Sec. 3.2, the electronic spin susceptibility for a singlet superconductor should be strongly suppressed at low temperatures below the superconducting transition, leading to a decrease of the NMR Knight shift. In contrast, for an equal spin triplet pairing (ESTP) superconductor, the Knight shift will remain unchanged upon crossing T_c. In fact, it is indeed the ^{77}Se Knight shift measurement provides the strongest indication that the superconducting state in $(TMTSF)_2PF_6$ is a triplet pairing state [33, 35]. For $(TMTSF)_2ClO_4$, however, recent results on ^{77}Se Knight shift reported by Shinagawa *et al.* [38] are more compatible with a singlet paring scenario, at least in the low-magnetic-field regime.

In their experiment, Shinagawa *et al.* conducted measurements on two single crystal samples of $(TMTSF)_2ClO_4$, with one configured for magnetic field alignment along the $\boldsymbol{a}$ direction and the other for field along the $\boldsymbol{b}'$ direction. By slowly cooling down at the rate of 7 mK/min across the anion ordering temperature T_{AO}, the samples are prepared in the relaxed state with an onset of superconductivity at $T_c = 1.4$ K. The superconducting transition is observed in the resistivity measurement along the z direction in reflected radio-frequency (rf) power measurement. The alignment of the magnetic field direction to within the x-y plane is gauged by probing the angular dependence of the reflected power. NMR spectroscopic and relaxation measurements are performed on both ^{77}Se and ^{13}C nuclei.

The resulting ^{77}Se spectra for the two field orientations are shown in Fig. 4.18(a). To eliminate the heating effect of NMR measurement, Shinagawa *et al.* use a small tip angle ($< 3°$) such that no temperature rises are detected by simultaneous resistivity measurements. Clear shifts of the spectra are identified upon entering the superconducting state for $\boldsymbol{H} \parallel \boldsymbol{b}'$ (top) and for $\boldsymbol{H} \parallel \boldsymbol{a}$ (bottom), while the relative change is much smaller for the former case. The applied NMR fields are calibrated to better than 10 ppm by measurements of the ^{63}Cu (in the coil) and ^{3}He (in the mixture in the vicinity of the coil) resonances. The onset of superconductivity also generates screening currents, and in turn a demagnetization field. However, this field is determined to be less than 100 ppm from ^{13}C spectroscopy for $\boldsymbol{H} \parallel \boldsymbol{b}'$. The temperature and field effects on the spectroscopic shifts are shown in Figs. 4.18(b) and (c). Notice that no deviation from the normal state shift is observed for $H = 40$ kOe, although the resistivity data suggest that the lowest measured temperature (100 mK) is well below the onset temperature of superconductivity (see discussion below).

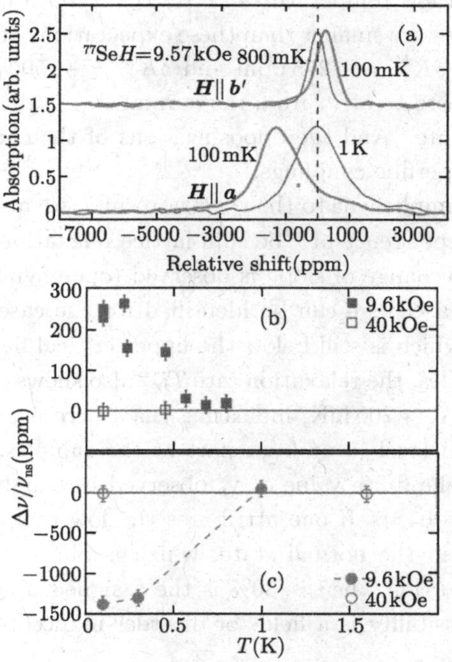

Figure 4.18 (a) ^{77}Se spectra in the normal and superconducting states of $(TMTSF)_2ClO_4$ for two orthogonal field orientations within the crystallographic layers. (b-c) Relative shifts versus temperature for $\boldsymbol{H} \parallel \boldsymbol{b}'$ and $\boldsymbol{H} \parallel \boldsymbol{a}$ axes, respectively. At $H =$ 40 kOe, there is no sizable change from the normal state shifts at the lowest measured temperature (100 mK). Zero shift is arbitrarily set to the normal state first moment. From Ref. [38].

The hyperfine coupling is nearly uniaxial: the dominant contribution is a p_z orbital originating from selenium [174, 175]. The normal state paramagnetic shift is given by

$$\begin{aligned} K &= K_{\rm iso} + K_{\rm ax}\left(3\cos^2\theta - 1\right), \\ K_{\rm iso} &= 3.9 \times 10^{-4}, \\ K_{\rm ax} &= 10.5 \times 10^{-4}. \end{aligned} \tag{4.10}$$

Thus, for a singlet superconductor, a Knight shift change for $\delta K_{\rm s}^{b'} \sim -700\,\text{ppm}$ ($\theta = \pi/2$, $\boldsymbol{H} \parallel \boldsymbol{b}'$) and $\delta K_{\rm s}^{a} \sim +2500\,\text{ppm}$ ($\theta = 0$, $\boldsymbol{H} \parallel \boldsymbol{a}'$) is expected in low fields ($H \to 0$) and low temperatures ($T \ll T_{\rm c}$). From Figs. 4.18(b) and (c), the observed changes are smaller than these expectations at magnetic field just below 10kOe with $\delta K_{\rm s}^{b'} = -275\,\text{ppm}$ and $\delta K_{\rm s}^{a} = +1500\,\text{ppm}$. Nonetheless, there is indeed a decrease of χ_x from the normal-state value upon entering the superconducting state. And the opposing signs of the change are consistent with the known hyperfine couplings.

To serve as a complement to the spectra results, we also show in Fig. 4.19 the temperature dependence of ^{77}Se spin-lattice relaxation rate. At low magnetic fields, a clear change of slope is observed (open symbols). However, no signature for such a change can be identified for the case with $H = 40\,\text{kOe}$ (closed symbols), which is still below the upper critical field $H_{\rm c_2}$ discussed in Sec. 4.2 [138]. Besides, the relaxation rate T_1^{-1} also shows a weak temperature dependence below $T \approx 200\,\text{mK}$, indicating that there is a nonzero density of states at the Fermi level in at least part of the samples. This observation is consistent with the finite value of $\chi_{\rm s}$ observed in Knight shift at low temperatures (see Fig. 4.18). If one attributes the low temperature features to a volume fraction in the normal state, which is phase segregated from the superconducting portion, then $\sim 30\%$ is the assigned fraction in the normal state. Another possibility is a field- or disorder-induced density of states at the Fermi energy.

To further investigate the magnetic field effect, the ^{77}Se relaxation rate T_1^{-1} data for varying magnetic fields at $T = 100\,\text{mK}$ are shown in Fig. 4.20, for both cases of $\boldsymbol{H} \parallel \boldsymbol{a}$ and $\boldsymbol{H} \parallel \boldsymbol{b}'$. The results for z axis resistance R_{zz} at the same temperature is also displayed with $\boldsymbol{H} \parallel \boldsymbol{a}$. Most notable is the presence of a high-field regime ($H \gtrsim 20\,\text{kOe}$), within which the resistance already shows signature of superconductivity while the slope of relaxation rate still maintains the normal-state value. In particular, the zero-resistance state is well established for field below 30 kOe, but the change of the slope of relaxation rate only starts at $H < H_s \sim 20$ kOe. The different behaviors of T_1^{-1} and resistance suggest the existence of two regimes including a low-field superconducting regime ($H < H_s$, LSC) and a high-field superconducting regime ($H > H_s$, HSC). The LSC regime exhibits a drop in χ_s, hence strongly suggests a singlet superconducting state.

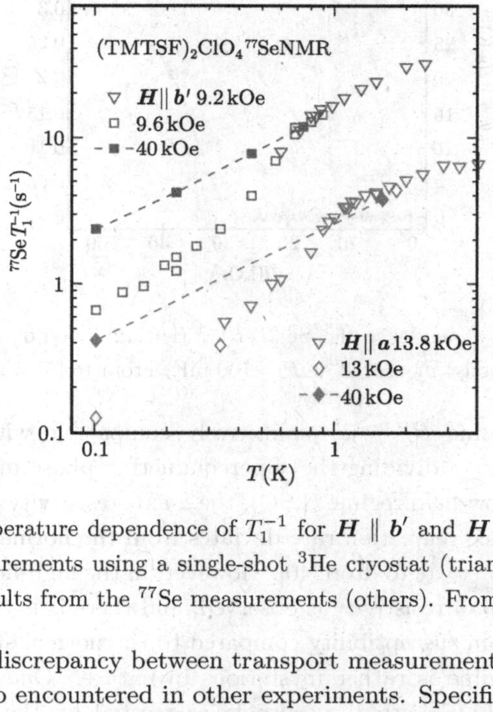

Figure 4.19 Temperature dependence of T_1^{-1} for $\boldsymbol{H} \parallel \boldsymbol{b}'$ and $\boldsymbol{H} \parallel \boldsymbol{a}$. Included are results from measurements using a single-shot ^{3}He cryostat (triangle, adapted from Ref. [176]) and results from the ^{77}Se measurements (others). From Ref. [38].

The obvious discrepancy between transport measurement and other types of methods is also encountered in other experiments. Specifically, in the heat capacity measurement we discussed in Sec. 4.3, Yonezawa *et al.* also map out the thermodynamic superconducting phase diagrams with magnetic field pointing along all three crystallographic axes, and compare with transport phase diagrams obtained via resistivity measurement along the z axis [139, 168] (see Fig. 4.21). Their thermodynamic upper critical field $H_{c_2}^{\text{thm}}$ is determined by the onset of the decrease of C/T, as indicated by arrows in Fig. 4.12. They find that $\mu_0 H_{c_2}^{\text{thm}}(0) \approx 2.5\,\text{T}$ for $\boldsymbol{H} \parallel \boldsymbol{a}$, which agrees with the Pauli paramagnetic limiting field $\mu_0 H_P \sim 2.3 - 2.6\,\mathrm{T}$, but is much smaller than the expectation value for the orbit pairing breaking,

$$\mu_0 H_{c_2}^{\rm orb}(0) = -0.73 T_c \mu_0 \frac{dH_{c_2}}{dT}\bigg|_{T=T_c} \sim 7.7\,\mathrm{T}. \tag{4.11}$$

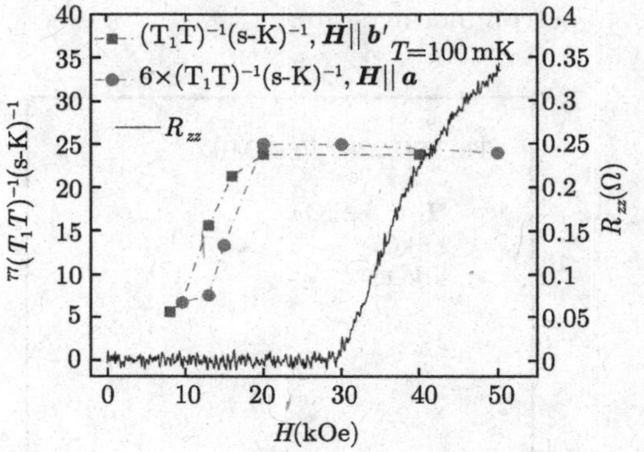

Figure 4.20 Field dependence of ^{77}Se $(T_1T)^{-1}$ (left axis) and of interlayer resistance $R_{zz}(\boldsymbol{H} \parallel \boldsymbol{a})$ (right axis) measured at $T = 100$ mK. From Ref. [38].

The thermodynamic $H_{c_2}^{\rm thm}$ is qualitatively compatible with the relaxation rate measurement in dividing the superconducting phase into LSC and HSC regimes. In the low-field regime (LSC), the z axis resistivity tends to zero, the nuclear spin-lattice relaxation rate deviates from the normal-state value [38], and heat capacity starts to drop [40]. However, in the high-field regime (HSC), only a sharp drop of resistivity is observed, and there is no noticeable change of entropy nor spin susceptibility compared to the normal state.

The HSC regime is rather mysterious in nature. One possibility is the emergence of a LOFF state, as also be suggested by the in-plane angular dependence of H_{c_2} within the x-y plane (see, e.g., discussions in Sec. 4.2 and in Chap. 6). However, the absolute value $H_{c_2}(T = 0)$ exceeds estimates for the paramagnetic limit of a LOFF state [32]. An alternative choice is a transition to a triplet paring state [44]. Nonetheless, the true nature of this HSC regime is one of the most intriguing problems to be clarified in the future.

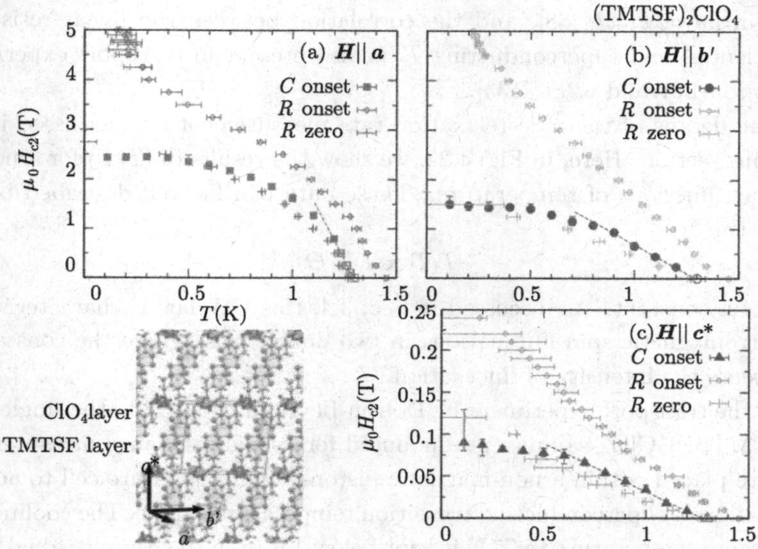

Figure 4.21 Thermodynamic superconducting phase diagrams of $(TMTSF)_2ClO_4$ compared with the transport phase diagrams for fields (a)$\|$ $\boldsymbol{a}$ axis, (b)$\|$ $\boldsymbol{b}'$ axis, and (c)$\|$ $\boldsymbol{c}^*$ axis. The solid symbols are obtained from field sweeps and the open symbols from temperature sweeps. For comparison, the onset temperature of the $\boldsymbol{c}^*$ axis resistance R_{c^*} (open diamonds) and the temperature where R_{c^*} becomes zero (crosses) are also shown (from Ref. [139]). The dotted lines indicate the slope $\mu_0 dH_{c_2}(T)/dT$ near T_c, with values of (a) -8.1 T/K, (b) -2.3 T/K, and (c) -0.11 T/K. From Ref. [40].

4.5 Non-Fermi liquid behavior of the normal state

In Sec. 3.4, we have learned that the analysis of normal state properties in PF_6 can shine lights on the understanding of the neighboring superconducting phase. The Curie-Weiss (CW) behavior of the spin relaxation rate, as well as the linear temperature dependence and the relation between linear coefficient A and superconducting T_c, imply that the antiferromagnetic spin fluctuation has dominant effect in the normal state, or even serves as a pairing mechanism in the superconducting phase. In $(TMTSF)_2ClO_4$, the same experiments have also been conducted and helped to draw somewhat similar conclusions. In particular, a CW behavior is also observed in ^{77}Se relaxation rate measurement by Shinagawa *et al.* [38], and the correlation between the linear resistivity coefficient A and superconducting T_c is also present in transport experiment by Doiron-Leyraud *et al.* [133].

The details of the ^{77}Se relaxation rate measurement are discussed in the previous section. Here, in Fig. 4.22, we show the results of (T_1T) for a normal state as functions of temperature. These data can be well described by the CW form

$$T_1T \propto T + \Theta \tag{4.12}$$

with Θ a constant. As discussed in Sec. 3.4, this CW law is characteristic of antiferromagnetic spin fluctuations in two dimensions, where the constant Θ represents the intensity of fluctuation.

In the transport experiment by Doiron-Leyraud *et al.* [133], the single crystal $(TMTSF)_2ClO_4$ samples are mounted for x-axis resistance measurement, and are placed within a non-magnetic piston-cylinder pressure cell to achieve a variation of superconducting transition temperature $T_c(P)$. The cooling rate from room temperature to 77 K is kept below 1 K/min to ensure gradual freezing of the pressure medium and a satisfactory level of pressure homogeneity, and to avoid generating cracks in the samples. Below 77 K, the cooling rate is kept below 5 K/hour such that the systems is prepared in the relaxed state below the anion ordering temperature $T_{AO} \approx 25$ K at low pressures.

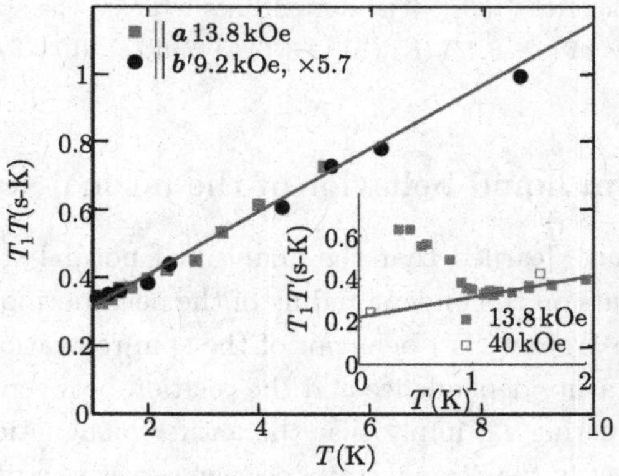

Figure 4.22 Temperature dependence of ^{77}Se T_1T, demonstrating that the normal state behavior follows the CW law. Inset: an expanded scale for $T < 2$ K and with $\boldsymbol{H} \parallel \boldsymbol{a}$. From Ref. [38].

A typical set of resistivity data $\rho_a(T)$ in $(TMTSF)_2ClO_4$ is shown in Fig. 4.23 with pressure $P = 4.9$ kbar. At this pressure, the onset of superconductivity is starting at $T_c = 0.82$ K. Compared to the similar results for $(TMTSF)_2PF_6$ (see, e.g., Fig. 3.18), the difference between low and high temperature regimes for ClO_4 is not so transparent. A significant quadratic contribution remains at low temperatures and no additional saturation could be detected at very low temperatures. Effectively, the data in Fig. 4.23 reveal a resistivity which retains a finite temperature dependence approaching zero temperature, but follows a quadratic dependence above ~ 12 K. Taking into account these two contributions, Doiron-Leyraud *et al.* fit their experimental data in the normal state by a second order polynomial form

$$\rho_a(T) = \rho_{0a} + AT + BT^2. \tag{4.13}$$

Here, both A and B are temperature-dependent prefactors, while the residual resistivity ρ_{0a} is only determined by pressure. The value of ρ_{0a} is first extracted from the fit between $0.1 - 4.0$ K. For the determination of A and B, a fit of the experimental data is performed at the temperature T over a temperature window of 4 K centered on T, with the fixed value of ρ_{0a}. The analysis is only restricted to the temperature domain below 16 K since the actual temperature dependence above 16 K is affected by anion ordering occurring at around 25 K. The results of this sliding fit procedure is displayed in the inset of Fig. 4.23. The results of six consecutive pressure funs (1.5, 4.9, 5.8, 7.3, 10.4, and 17 kbar) performed on the same sample are shown in Fig. 4.24, where only the inelastic contribution $(\rho_a - \rho_{0a})$ is shown. A monotonic increase of residual resistivity ρ_{0a} upon increasing pressure is also observed and depicted in the inset of Fig. 4.24, and is attributed to the anion ordering becoming less complete at T_{AO} under pressure.

The correlation between the linear coefficient A and the superconducting T_c can be established in Fig. 4.25. Here, A_{LT}, the maximum value of A determined at low temperatures and T_c given by the onset of superconductivity are used for demonstration. Notice that A_{LT} and T_c are quite parallel for pressures $P \lesssim 10$ kbar, indicating a linear dependence between the two quantities as revealed in $(TMTSF)_2PF_6$ (see discussion in Sec. 3.4). For the experimental run at $P = 17$ kbar, no signature of superconductivity can be observed to the lowest measured temperature 100 mK, while a finite value of A_{LT} can still be extracted. However, the complicated data fitting scheme used by Doiron-

Leyraud *et al.* makes the determination of such a small $A_{LT} \sim 0.07$ rather ambiguous.

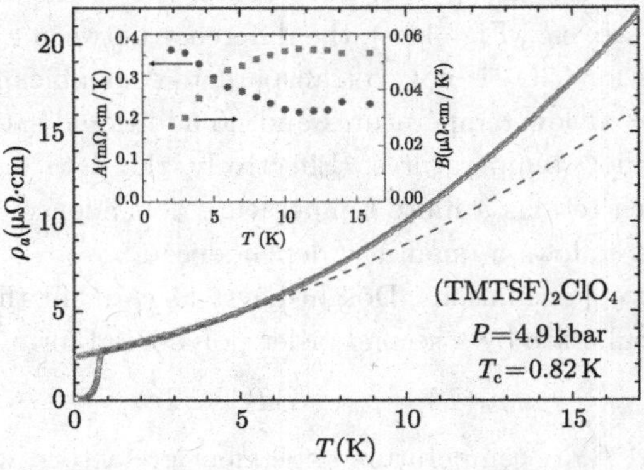

Figure 4.23 Temperature dependence of the longitudinal resistivity R_a of $(TMTSF)_2ClO_4$ measured at $P = 4.9$ kbar below 17 K, both at zero field and under $H = 0.05$ T along the $\boldsymbol{c}^*$ axis in order to suppress superconductivity. The second order polynomial fit of Eq. (4.13) using the sliding fit procedure is shown for the temperature intervals 2–6 K and 13–17 K in blue and red respectively. The top inset provides the temperature dependence of the A and B coefficients. From Ref. [133].

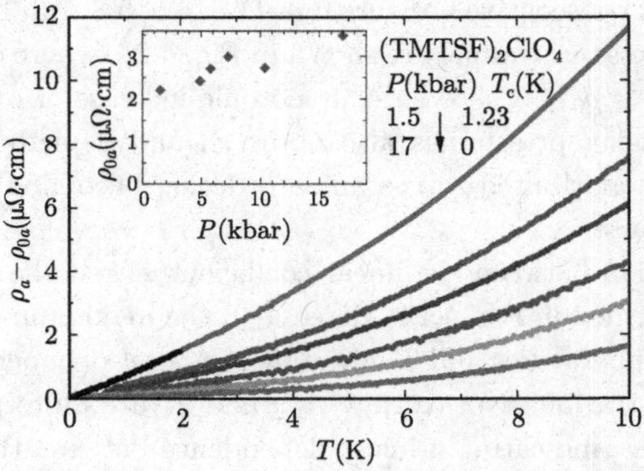

Figure 4.24 Inelastic contribution $(\rho_a - \rho_{0a})$ of the normal state resistivity at different pressures for $(TMTSF)_2ClO_4$. Inset shows the pressure dependence of the residual resistivity ρ_{0a} deduced from low temperature fit. From Ref. [133].

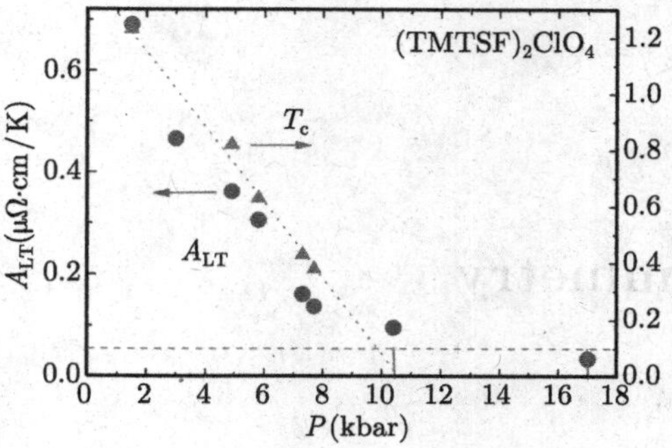

Figure 4.25 Pressure dependence of T_c onset and A_{LT} coefficient for $(TMTSF)_2ClO_4$. A_{LT} is obtained from the second order polynomial fit described in the text in the temperature window 0.1–4 K. The horizontal dashed line indicates the lowest reached temperature of 0.1 K. The dotted line is a linear fit of T_c onset data points. From Ref. [133].

5
Pairing Symmetry

One of the central topics in the research on quasi-one-dimensional organic superconductors is to unveil the nature of the superconducting state. In the previous two chapters, we have discussed some recent experimental efforts which have been devoted toward this ultimate goal. There is no doubt that a considerable amount of valuable information has been gathered from these excellent works. We now know that the upper critical fields for both $(TMTSF)_2PF_6$ and $(TMTSF)_2ClO_4$ exceed the Pauli paramagnetic limit. $(TMTSF)_2PF_6$ shows no change in spin susceptibility across the superconducting transition at a fairly high magnetic field, while $(TMTSF)_2ClO_4$ presents an obvious decrease at low fields. Both of the two compounds display similar non-Fermi liquid behavior in the normal state close to the superconducting transition, regardless of the different phase diagrams where $(TMTSF)_2PF_6$ has a neighboring spin-density wave state but $(TMTSF)_2ClO_4$ has not. However, there are still many unsolved questions circulating around the superconducting phase of Bechgaard salts.

One of the most important and natural questions would be the symmetry of the superconducting order parameter. Is it singlet or triplet? In either case, what is the orbital symmetry? These two questions catch a tremendous amount of attention since the discovery of superconductivity in Bachgaard salts, and are still hanging around after more than three decades. In this chapter, we introduce some theoretical proposals aiming to solve this puzzle. In order to give an un-biased overview of this field, we include discussions on both singlet and triplet pairing states, as well as their corresponding orbital symmetries.

We organize our discussion in this chapter in order to emphasize the accommodation of theoretical works to experimental evidences discussed in the previous two chapters. In Secs. 5.1 to 5.4, we analyze in detail the behavior of superconductivity in the presence of high magnetic fields. We will introduce the origin of the excess of Pauli paramagnetic limit, the intriguing possibility of a reentrant superconducting phase and its anomalous paramagnetic nature, and some attempts to achieve quantitative comparison with the experimental data of H_{c_2}. In Sec. 5.5, we discuss the theoretical work regarding the angular dependence of the heat capacity measurement [40]. Sec. 5.6 is devoted to the study on vortex effects upon NMR results [35]. Finally, we discuss in Sec. 5.7 some group theory analysis, which systematically enumerate and categorize all possible symmetries of the superconducting phase.

5.1 Superconductivity in high magnetic fields

The equilibrium state of type-II superconductors in a magnetic field was first described by Abrikosov using a phenomenological Ginzburg–Landau (GL) theory [177], which was later justified by Gorkov using a microscopic model [178]. The Ginzburg–Landau–Abrikosov–Gorkov (GLAG) theory treats the magnetic field in the semiclassical phase (Eikonal) approximation and neglects all quantum effects of the magnetic field. Hence, this approximation is valid only when $\omega_c \ll \min[T, 1/\tau]$, where ω_c is the characteristic magnetic energy and τ is the elastic scattering time. This assumption is expected to breakdown at high magnetic fields and low temperatures for sufficiently clean materials. In this limiting case, the determination of $H_{c_2}(T)$ and of the vortex lattice structure requires studies of quantum effects of the external magnetic field.

The influence of Landau level quantization in isotropic superconductors was first investigated in 1960s [69, 70] and were discussed more recently by Tesanovic *et al.* [64]. When only the lowest Landau level (LLL) is occupied in isotropic superconductors, the system becomes one-dimensional, and pairing occurs only between states with opposite momenta along the direction of the field. In this case, the orbital frustration associated with transverse motion vanishes and superconductivity is only limited by impurity scattering and the Pauli pair breaking effect. However, in Bechgaard salts $(TMTSF)_2X$, the semiclassical orbits are open in the presence of an external field. As a result, there is no Landau level quantization but there is a magnetic field induced dimensional crossover (FIDC) from anisotropic three-dimensional to anisotropic two-dimensional [179]. This FIDC mechanism can also suppress orbital frustration, and lead to the existence of superconductivity in strong magnetic fields.

The initial idea that the semiclassical H_{c_2} of quasi-1D superconductors can be largely exceeded in strong magnetic fields parallel to the $\boldsymbol{b}'$ axis was first proposed by Lebed [16]. The main purpose of Lebed's work was to call attention to the important role played by the electronic orbits of quasi-1D systems in a magnetic field. The incorporation of these effects lead theoretically to the preservation of superconductivity at fields exceeding the semiclassical upper critical field; and to a change in the sign of the derivative dH_{c_2}/dT at $H^*_{c_2} \gg H^0_{c_2}$, where $H^*_{c_2}$ corresponds to the minimum T_c. In this case, the upper critical field is not limited by the Pauli paramagnetic limit, as was first pointed out by Buzdin and Tugushev [180].

A year after Lebed's initial effort, Burlachkov, Gorkov and Lebed (BGL) [17] published a paper where the upper critical field was calculated numerically for both singlet and triplet superconductors. BGL further emphasized the strong reentrant phase for triplet quasi-1D superconductors. After the work of Lebed [16] and BGL [17], Dupuis, Montambaux and Sá de Melo (DMS) [18] revisited the magnetic field versus temperature phase diagram of quasi-1D superconductors and calculated the corresponding vortex lattice structure. DMS found a more general order parameter equation than that discussed by Lebed and BGL, and proposed that the usual GL regime at low fields was followed by a cascade of superconducting phases at higher fields separated by first order transition lines. These new phases exhibit a vortex lattice symmetry of the laminar type. DMS also showed that the Zeeman splitting did not completely suppress the high field reentrant phase in the case of a singlet superconductor, where the ground state resembled the LOFF state [30, 31].

These theoretical predictions can be illustrated by considering a quasi-1D system with the linearized electronic dispersion spectrum Eq. (2.2)

$$\varepsilon_\alpha(\boldsymbol{k}) = v_{\rm F}(\alpha k_x - k_{\rm F}) - t_y \cos(k_y b) - t_z \cos(k_z c)$$

with a superconducting critical temperature $T_c \ll t_z$. This latter condition ensures that the smallest coherence length in the system is always much larger than the spacing between chains $[\xi_z(T) > \xi_z(0) \gg c]$ at zero magnetic field. Discussion about the Josephson regime $T_c > t_z$ can be found in a publication by Dupuis and Montambaux [29]. The quantum effects of the magnetic field are taken into account by writing down the Hamiltonian

$$\mathscr{H} = \mathscr{H}_0 + \mathscr{H}_{\text{int}}, \tag{5.1}$$

where $\mathscr{H}_0 = \varepsilon(\boldsymbol{k} \to -i\nabla - e\boldsymbol{A})$ is the standard non-interacting Hamiltonian obtained via the Peierls substitution.

It is worthy noticing that the dispersion relation $\varepsilon(\boldsymbol{k})$ allows for regions of perfect electron-hole nesting with the vector $(2k_{\text{F}}, \pi/b, \pi/c)$, so that it is necessary to consider simultaneously the logarithmic divergences appearing in the electron-hole and electron-electron channels. However, a slightly more realistic dispersion eliminates electron-hole nesting and suppresses quickly any electron-hole instabilities. In Bechgaard salts, a magnetic field along the z direction could restore the logarithmic divergence in the electron-hole channel, a situation known as quantized nesting and responsible for the field induced spin density wave phases (FISDW) [54, 181]. However, for a field along the y axis, the FISDW phases do not occur because the deviation from perfect nesting in the y direction is too large to be restored by accessible magnetic fields. Furthermore, these minor corrections to $\varepsilon(\boldsymbol{k})$ do not affect the superconducting instability in the electron-electron channel, hence can be neglected in the calculation of the upper critical along the y direction.

For a uniform magnetic field $\boldsymbol{H} \parallel \boldsymbol{b}'$ and the Landau gauge $\boldsymbol{A}(0, 0, -Hx)$, the non-interacting Hamiltonian $\mathscr{H}_0$ can be easily diagonalized and the critical temperature $T_{\text{c}}(H)$ is given by the linearized order parameter equation

$$\Delta(x, k_z) = \lambda_\eta \int dx' K_{\sigma,\sigma'}(x, x', k_z) \Delta(x', k_z). \tag{5.2}$$

Here $\lambda_{\eta=\text{s}}$ $(\lambda_{\eta=\text{t}})$ is the effective coupling constant of singlet (triplet) superconductivity, and the Kernel has the form

$$\begin{aligned} K_{\sigma,\sigma'}(x, x', k_z) = & \frac{T}{bcv_{\text{F}}^2} \frac{\cos\left[(\sigma - \sigma')\mu_{\text{B}} H(x - x')/v_{\text{F}}\right]}{\sinh\left[|x - x'| 2\pi T/v_{\text{F}}\right]} \\ & \times J_0\left(\frac{4t_z}{\omega_{\text{c}}} \sin\left[\frac{G}{2}(x - x')\right] \sin\left[k_z \frac{c}{2} - \frac{G}{2}(x + x')\right]\right). \end{aligned} \tag{5.3}$$

The order parameter $\Delta(x, k_z)$ appearing above is the Fourier transform of the order parameter $\Delta(x, z)$ with respect to z. In the high magnetic field limit, only equal spin triplet pairing (ESTP) and singlet states are possible due to the strong Zeeman splitting effect, then it is proper to choose $\sigma = \sigma' = 1$ and $\sigma = -\sigma' = 1$ for ESTP and singlet superconductivity, respectively. The critical temperature $T_c(H)$ is determined from Eq. (5.2) when the right hand side integral operator has eigenvalue 1. Notice that $T_c(H)$ is independent of k_z since the wavevector component only shifts the origin of the x axis by $k_z c/2G$, hence it is allowed to set $k_z = 0$.

There is an obvious periodic solution of the order parameter equation above, as was discussed by Lebed [16]. The proposed solution for the triplet case is $\Delta_0(x + \pi/G) = \Delta_0(x)$, while for the singlet case is $\Delta_0(x + \pi/G) = \cos(2\mu_B H x)\Delta_0(x)$. These choices lead to logarithmic divergences in Eq. (5.2) as $T \to 0$, and this instability of the metallic state in an arbitrary magnetic field determines $H_{c_2}(T)$.

However, this solution of $\Delta_0(x)$ is only a special form of the general solutions of Eq. (5.2). In fact, since the Kernel $K(x, x')$ has the property

$$K(x, x', 0) = K(x + \pi/G, x' + \pi/G, 0), \tag{5.4}$$

the solution of Eq. (5.2) with $k_z = 0$ can be written without any loss of generality as [18]

$$\Delta_Q(x) = e^{iQx}\tilde{\Delta}_Q(x), \tag{5.5}$$

where $\tilde{\Delta}_Q(x + \pi/G) = \tilde{\Delta}_Q(x)$ is a periodic function and $-G < Q \leqslant G$. Notice that Lebed's choice [16, 17] corresponds to the case where $Q = 0$.

In the ESTP case, all possible values of Q correspond to the same $T_c(H)$ in the low field GL regime ($\omega_c \ll T$). However, in the quantum regime ($\omega_c \gg T$), this degeneracy is lifted, and the highest $T_c(H)$ is always obtained for either $Q = 0$ (Lebed's solution) or $Q = G$. The phase diagram is shown in Fig. 5.1. Notice that the GL regime is followed by a different regime where the solutions $Q = 0$ and $Q = G$ alternate for increasing magnetic field. The last phase is reentrant for very strong field and corresponds to the $Q = 0$ solution. These two solutions are characterized by a different structure of the order parameter, which implies that the magnetic field induces a cascade of superconducting phases separated by first order transitions. For singlet pairing, the best values of Q are $\pm 2\mu_B H/v_F$ and $\pm(G - 2\mu_B H/v_F)$. A shift $\pm 2\mu_B H/v_F$ of the value of Q displaces the Fermi surfaces of spin $\uparrow$ and spin $\downarrow$ relative to each other and compensates partially the effect of Zeeman splitting as it is the case in the

LOFF state [30, 31]. The order parameter, however, remains uniform along the magnetic field direction. As a result, one half of the phase space is again available for pairing so that the reentrant phase in high magnetic field is not completely suppressed as shown in Fig. 5.1 (a) (long-dashed line). The cascade of first order phase transitions will also exist but at very low temperatures, i.e., $T_c(H)/T_c \sim 10^{-3}$.

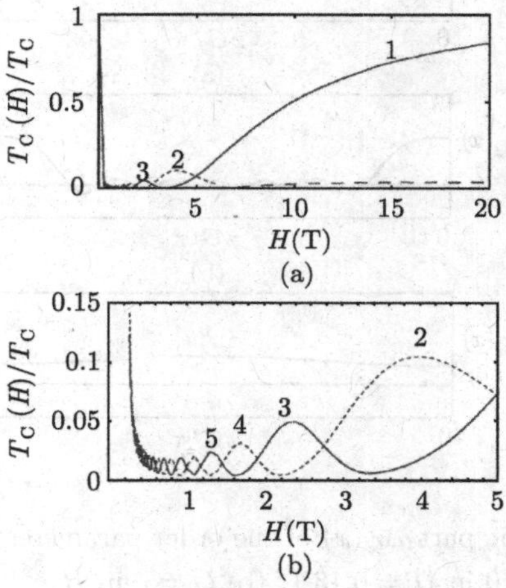

Figure 5.1 Scaled critical temperature $T_c(H)/T_c$, where $T_c = T_c(0)$ versus magnetic field for parameters $t_z = 20\,\mathrm{K}$ and $t_z/T_c(0) \sim 13$. In (a) the solution for $Q = 0$ (solid line) and $Q = G$ (dashed line) alternate for increasing magnetic field in the triplet case. Each phase is labeled by an integer corresponding to the periodicity of the rectangular vortex lattice along the z direction. In the singlet case, $T_c(H)$ is strongly reduced, but the reentrant phase is still not fully suppressed at high magnetic fields (long-dashed line) due to the formation of LOFF-type state. In (b) the low temperature region of the triplet case is shown. The cascade of first order phase transitions is easily seen via the discontinuous change of the periodicity of the vortex lattice along the z direction. From Ref. [18].

The reentrant phases in high fields are significantly different from the superconducting states in low fields. A numerical calculation [18] of the order parameter for temperatures slightly below $T_c(H)$ provides a picture of vortex lattice evolution from GL to the quantum regime. The periodic part $\tilde{\Delta}_Q(x)$ of the order parameter is shown in Fig. 5.2 for the case of $Q = 0$ (one of the triplet states). In the GL regime, $\tilde{\Delta}_Q(x)$ is localized around the points $x_n = n\pi/G$. The Abrikosov vortex lattice for triplet superconductivity is represented as a linear combination of the gaussian solutions $f(x - x_n - k_z c/2G)$ with the full

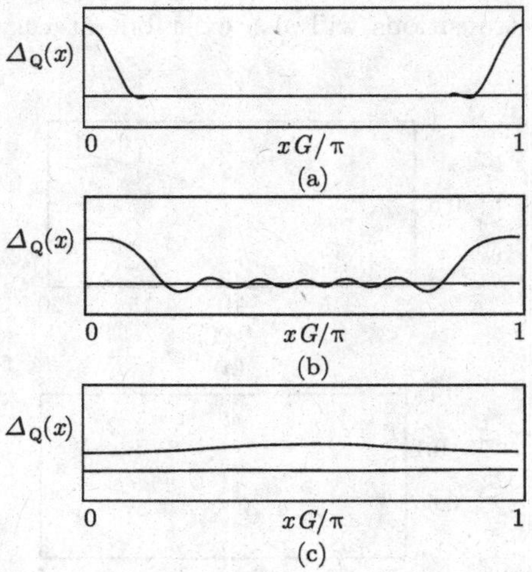

Figure 5.2 The periodic part $\tilde{\Delta}_Q(x)$ of the order parameter is shown. The solution corresponds to (a) $Q = 0$ in $H = 0.13$ T, (b) $Q = 0$ in $H = 0.48$ T, and (c) $Q = 0$ in $H = 5.8$ T. Note that $\tilde{\Delta}_Q(x)$ is real for $Q = 0$. From Ref. [18].

k_z dependence. However, since the vortex lattice rests on a crystal structure with lattice constant c along the z direction, a periodicity $a_z = Nc$ with N an integer must be imposed into the vortex lattice. Therefore, the vortex cores will lie between two planes to minimize the free energy. In this state, only the $Q = 0$ or $Q = G$ solution is necessary to describe the Abrikosov vortex lattice:

$$\Delta_{Q,N}(x, z = mc) = C_0 \sum_{l=m \bmod N} \Delta_{2l}^{Q} e^{i(Q+2lG)x}, \tag{5.6}$$

where $Q = 0$ (G) if N is odd (even). For simplicity, only the rectangular lattice is considered here, and the comparisons between triangular and rectangular lattices can be found elsewhere [20, 29]. In the quantum regime where the degeneracy with respect to Q is lifted, $\tilde{\Delta}_Q(x)$ becomes extended, which implies that the vortex lattice is strongly modified or even suppressed when $\omega_c \gg T$. The order parameter $\Delta_{Q,N}(x, z = mc)$ in Eq. (5.6) can be naturally extended into the quantum regime, where the best solution corresponds to either $Q = 0$ or $Q = G$ depending on the value of the field. Notice that $|\Delta_{Q,N}(x, m)|$ has periodicity $a_x = \pi/NG$ and $a_z = Nc$ for every value of the field. As a result the usual flux relation $Ha_x a_z = \phi_0$ is satisfied, showing that there is one flux quanta in a unit cell (a_x, a_z). Therefore, if the value of N is known for a given phase, the order parameter $\Delta_{Q,N}(x, z = mc)$ in Eq. (5.6) can be obtained from the numerical solution of $\tilde{\Delta}_Q(x)$ defined in Eq. (5.5). For more details, the reader is referred to Refs. [18, 20, 29].

5.2 Paramagnetism of reentrant phases

The theoretical works by Lebed [16], BGL [17] and DMS [18] predict anomalous ground states and suggest exotic phase diagrams in quasi-1D systems at high magnetic fields. Among the several phases corresponding to various N with increasing magnetic field, the most interesting one is the reentrant phase with $N = 1$ (for rectangular lattice of Josephson vortices). For a triplet state, this reentrant phase can survive in arbitrarily high magnetic fields, and $T_c(H)$ is a monotonically increasing function of H, saturating at $T_c(\infty) = T_c^{(2D)}(0)$. This anomalous $T_c(H)$ indicates the last reentrant phase has a paramagnetic excess magnetization $\Delta M > 0$, which is a rather surprising result at first glance when compared with the usual diamagnetism in weak fields.

This conclusion can be reached through a general thermodynamic consideration [20]. Within the Ginzburg–Landau theory, the mean field free energy difference $\Delta\mathscr{F} = \mathscr{F}_s - \mathscr{F}_n$ between superconducting and normal states can be written as

$$\Delta\mathscr{F} \approx -A\,[T - T_c(H)]^2 + \mathscr{F}_b, \tag{5.7}$$

where $A = 8\pi^2 N(E_F)/[7\zeta(3)]$ is a positive number, and $\mathscr{F}_b = \int d\boldsymbol{r}\,[\boldsymbol{h}_s(\boldsymbol{r})]^2/8\pi$ is the magnetic energy. In the case of magnetic field pointing along the $\boldsymbol{b}'$ direction, the total excess magnetization can be calculated from the thermodynamic relation $\Delta\boldsymbol{M} = \boldsymbol{B}/4\pi - \partial\Delta\mathscr{F}/\partial\boldsymbol{B}$, which leads to the result

$$\Delta\boldsymbol{M} \approx 2A\,[T_c(H) - T]\,\frac{dT_c(H)}{dH}\hat{\boldsymbol{y}}. \tag{5.8}$$

Notice that the sign of the excess magnetization is controlled by the slope of $T_c(H)$, as proposed by Dupuis [19]. In addition, the excess entropy is negative as expected, $\Delta S \approx -2A\left[T_c(H) - T\right]$, since the superconducting state is more ordered than the normal state. A direct calculation of specific heat jump ΔC from $\Delta\mathscr{F}$ leads to a result proportional to $T_c(H)$, which deviates from the BCS expression only through $T_c(H)$. Notice that the relation between the sign of ΔM and the slope $dT_c(H)/dH$ or $dH_{c_2}(T)/dT$ is neither a result of the mean field approximation, nor a consequence of the microscopic model applied to the specific quasi-1D systems. This relation has its origin on pure thermodynamic considerations.

More generally, in the vicinity of a second order phase transition, the specific heat change at constant H obeys the general thermodynamic relation

$$\Delta C = T\left(\frac{dH_{c_2}(T)}{dT}\right)^2\left(\frac{\partial \Delta M}{\partial H}\right)_T, \tag{5.9}$$

while the entropy change obeys

$$\left(\frac{\partial \Delta S}{\partial H}\right)_T = -\frac{dH_{c_2}(T)}{dT}\left(\frac{\partial \Delta M}{\partial H}\right)_T. \tag{5.10}$$

However, in a second order (continuous) phase transition ΔM and ΔS must vanish at H_{c_2} from below. Thus, in the vicinity of H_{c_2}, the changes in magnetization and entropy have the power law forms

$$\Delta M = M_0(1 - H/H_{c_2})^{\alpha_m}, \tag{5.11}$$

$$\Delta S = S_0(1 - H/H_{c_2})^{\alpha_s}. \tag{5.12}$$

As a result, there exists a relation connecting ΔS with ΔM near $T_c(H)$:

$$\Delta S = -\frac{\alpha_m}{\alpha_s}\frac{dH_{c_2}(T)}{dT}\Delta M, \tag{5.13}$$

where $\alpha_s > 0$ and $\alpha_m > 0$. Eq. (5.13) shows that when $\Delta S < 0$ the product $[dH_{c_2}(T)/dT]\Delta M > 0$, i.e., a diamagnetic (paramagnetic) excess magnetization $\Delta M < 0$ ($\Delta M > 0$) is always linked to a decrease (increase) in $H_{c_2}(T)$, i.e., $dH_{c_2}/dT < 0$ ($dH_{c_2}/dT > 0$). Conversely, when $T_c(H)$ decreases (increases) with H there is a diamagnetic (paramagnetic) excess magnetization. The GL results discussed in Eq. (5.8) satisfy the relation above with $\alpha_m = \alpha_s = 1$, since it is a mean field theory. Relation (5.13) is always valid provided that the system is in thermodynamic equilibrium and the phase transition is second order. Fluctuation effects only affect the ratio α_m/α_s between critical exponents and do not change the qualitative aspects contained in Eq. (5.13). For quasi-1D Bechgaard salts, the relation between the sign of ΔM and the slope of $T_c(H)$ should be valid over the *entire* phase diagram and, in particular, to all intermediate quantum phases discussed in the previous section. In these predicted intermediate phases the sign of ΔM has an oscillatory behavior governed by $dH_{c_2}(T)/dH$ as magnetic field increases. For instance, in a given phase the excess magnetization ΔM of the superconducting state near $T_c(H)$ will be first paramagnetic $[dH_{c_2}(T)/dT] > 0$, vanish at $[dH_{c_2}(T)/dT] = 0$, and then be diamagnetic $[dH_{c_2}(T)/dT] < 0$. A similar scenario is also applicable to the isotropic case in high fields, where the competition between paramagnetic and diamagnetic currents in the Abrikosov lattice should also produce an excess ΔM whose sign would follow the slope of $H_{c_2}(T)$.

Microscopic descriptions are not needed to reach the qualitative statements discussed above. However, specific solutions of the order parameter equation (5.2) are required to find quantitative results. Searching for a general numerical or analytical solution is a very hard task. But in the last reentrant phase of triplet pairing, the static order parameter is expected to be non-uniform and has the form

$$\Delta(\boldsymbol{r}) = [\Delta_0 + 2\Delta_2 \cos(2Gx)] \exp[i\phi(x,y)] . \tag{5.14}$$

This form corresponds to an ESTP state with $N = 1$, $Q = 0$, which has periodicity $a_x = \pi/G$, $a_z = c$, and holds one flux quanta inside the plaquette (a_x, a_z), i.e., $Ha_xa_z = \phi_0$. Within this static approximation, the free energy $\mathscr{F}$ can be calculated near $T_c(H)$ up to order $(t_z/\omega_c)^2$ [20]:

$$\mathscr{F} = \int d\boldsymbol{r}\, [\mathscr{F}_n + \mathscr{F}_1 + \mathscr{F}_2 + \mathscr{F}_3 + \mathscr{F}_4 + \mathscr{F}_b] , \tag{5.15}$$

where $\mathscr{F}_n$ is the normal state contribution, and $\mathscr{F}_b = |\boldsymbol{B}(\boldsymbol{r})|^2$ is the magnetic energy density. The other four contributions have the forms

$$\mathscr{F}_1 \approx -(1/\lambda_t)(|\Delta_0|^2 + 2|\Delta_2|^2); \tag{5.16}$$
$$\mathscr{F}_2 \approx B_{20}|\Delta_0|^2 + (B_{11}\Delta_0\Delta_2^* + \text{C.C.}) + B_{02}|\Delta_2|^2; \tag{5.17}$$
$$\mathscr{F}_3 \approx C_{20}|\Delta_0|^2 + (C_{11}\Delta_0\Delta_2^* + \text{C.C.}) + C_{02}|\Delta_2|^2; \tag{5.18}$$
$$\mathscr{F}_4 \approx (1/2)\left[D_{40}|\Delta_0|^4 + (D_{31}|\Delta_0|^2\Delta_0\Delta_2^* + \text{C.C.}) + D_{22}|\Delta_0|^2|\Delta_2|^2\right] . \tag{5.19}$$

Here, $\mathscr{F}_1$ is a local quadratic term, $\mathscr{F}_2$ and $\mathscr{F}_3$ are two non-local quadratic terms, $\mathscr{F}_4$ is a non-local fourth order term, and C.C. stands for complex conjugate.

The saddle point equations $\delta\mathscr{F}/\delta\Delta_0 = 0$, $\delta\mathscr{F}/\delta\Delta_2 = 0$, and $\delta\mathscr{F}/\delta\boldsymbol{A} = 0$, where $\boldsymbol{A}$ is the vector potential, can be solved to give

$$|\Delta_0|^2 \approx \left[8\pi^2/7\zeta(3)\right] T_c(H) \left[T_c(H) - T\right] \left[1 + (t_z/\omega_c)^2\right], \qquad (5.20)$$

and the prefactor of the second term

$$|\Delta_2| \approx (t_z/\omega_c)^2 |\Delta_0|/2 \qquad (5.21)$$

for $E_F \gg \omega_c$. The phase of the order parameter is

$$\phi_n^{(0)}(x) = 2Gnx - (8\pi C_{20}\Delta_0^2/H_y^2) n \sin(2Gx). \qquad (5.22)$$

This solution corresponds to a rectangular Josephson vortex lattice with periodicity $a_x = \pi/G$ and $a_z = c$, as expected for the last reentrant phase.

These results lead to the final expression for the amplitude of the order parameter

$$|\Delta(\boldsymbol{r})| = |\Delta_0| \left[1 + (t_z/\omega_c)^2 \cos(2Gx)\right]. \qquad (5.23)$$

The circulating currents hence are given by $j_x(\boldsymbol{r}) \approx 0$, $j_y(\boldsymbol{r}) = 0$ and $j_z(\boldsymbol{r}) \approx j_0(T,H)\sin(2Gx)$, with the prefactor for j_z given by the expression

$$j_0(T,H) = 8|e|(t_z/\omega_c)^2 c N(E_F) \ln\left[\gamma_0\omega_c/\pi T_c(H)\right] |\Delta_0|^2, \qquad (5.24)$$

where $\gamma_0 = e^\gamma \approx 1.7811$ is the exponential of the Euler constant $\gamma \approx 0.57722$. Notice that the magnetic field induced Josephson coupling $j_0(T,H)$ is a consequence of the localization ($\omega_c \gg t_z$) and quantization effects ($\omega_c \gg 2\pi T$) of the magnetic field.

For weak fields in the regime $t_z \gg T_c(0)$ the planes are very strongly coupled, and the system is highly anisotropic but three-dimensional. However, in high magnetic fields, the normal state becomes unstable towards a weakly coupled quasi-two-dimensional superconductor with a magnetic field induced Josephson coupling between different x-y planes. This situation should be contrasted to the usual Josephson coupling present in weakly coupled layered superconductors $[t_z \ll T_c(0)]$ even in the absence of a magnetic field [182]. Notice that the gauge invariant form of $j_z(\boldsymbol{r})$ involves the phase difference

$\Delta\phi = \phi_{N+1}(\boldsymbol{r}) - \phi_N(\boldsymbol{r}) + 2|e| \int \boldsymbol{A} \cdot d\boldsymbol{s}$. In the last reentrant phase, $\phi_{N+1}(\boldsymbol{r}) = \phi_N(\boldsymbol{r})$ and the expression for $j_z(\boldsymbol{r})$ is gauge invariant as expected.

The local magnetic induction $\boldsymbol{B}(\boldsymbol{r}) = \boldsymbol{H} + \boldsymbol{h}_s(\boldsymbol{r})$ can be calculated from $\boldsymbol{j}(\boldsymbol{r})$ via the Maxwell's equation $\nabla \times \boldsymbol{h}_s(\boldsymbol{r}) = 4\pi\boldsymbol{j}(\boldsymbol{r})$ with specific boundary conditions [20]. Consider the length of the system along the x direction to be L_x. By imposing that $\boldsymbol{B}(\boldsymbol{r}) = \boldsymbol{H}$, i.e., $\boldsymbol{h}_s(0) = \boldsymbol{h}_s(L_x) = 0$ at the sample boundaries, the induced field has the form $\boldsymbol{h}_s(\boldsymbol{r}) = h_s(x)\hat{\boldsymbol{y}}$, where

$$h_s(x) = 4\pi[j_0(T,H)/2G]\,[1 - \cos(2Gx)]\,. \tag{5.25}$$

The number of flux quanta N_x enclosed between two consecutive x-y planes has to be $N_x = L_x G/\pi$ in order to ensure the appropriate boundary conditions for $\boldsymbol{h}_s(\boldsymbol{r})$. From Eq.(5.25), the local excess magnetization per plaquette $\Delta\boldsymbol{M}(\boldsymbol{r}) = \Delta M(\boldsymbol{r})\hat{\boldsymbol{y}}$ can be obtained as $\Delta\boldsymbol{M}(\boldsymbol{r}) = \boldsymbol{h}_s(\boldsymbol{r})/4\pi$. Hence the net excess magnetization per plaquette

$$\Delta M = j_0(T,H)/2G \tag{5.26}$$

is paramagnetic, as we concluded from the general thermodynamic consideration. In fact, it can be shown that the diamagnetic contributions to ΔM are of order $(t_z/\omega_c)^4$ in the limit of $(t_z/\omega_c \ll 1)$, thus being negligible in comparison with the paramagnetic contribution which is of the order $(t_z/\omega_c)^2$.

5.3 Upper critical fields in singlet and triplet superconductors

Theoretical efforts discussed above unveil exotic behaviors of quasi-1D superconductors in the presence of magnetic fields. For the triplet pairing case, the superconducting states can survive the detrimental effects of magnetic fields exceeding the Pauli limit, and can be paramagnetic instead of diamagnetic in the reentrant phase. Lee *et al.* [23] have observed that the Pauli limit is exceeded in $(TMTSF)_2PF_6$ for both $\boldsymbol{H} \parallel \boldsymbol{b}'$ and $\boldsymbol{H} \parallel \boldsymbol{a}$ (see Sec. 3.1). However, this experiment alone is not sufficient to rule out the possibility of singlet pairing since quasi-1D singlet superconductors can also exceed the Pauli limit [with much lower $T_c(H)$]. In particular, Miyazaki, Kishigi and Hasegawa (MKH) [183] considered a singlet state and concluded that the suppression of $T_c(H)$ is not very severe in high fields if one takes into account the optimal momentum of Cooper pairs and the effect of higher harmonic terms along the $\boldsymbol{b}'$ axis. Meanwhile, Lebed, Machida and Ozaki (LMO) [34] also calculated $H_{c_2}(T)$ for singlet and triplet states, and argued that $(TMTSF)_2PF_6$ under pressure was likely a triplet superconductor with order parameter $\boldsymbol{d}(\boldsymbol{k}) \equiv [d_x(\boldsymbol{k}) \neq 0,\ d_y(\boldsymbol{k}) = 0,\ d_z(\boldsymbol{k}) = ?]$, where $|d_x(\boldsymbol{k})| > |d_z(\boldsymbol{k})|$. Therefore, the experimental fact that the Pauli limit is exceeded in $(TMTSF)_2PF_6$ [23] can be understood qualitatively with either the singlet or triplet possibilities.

In MKH's singlet theory, quasi-1D systems are described by the non-interacting Hamiltonian in a magnetic field $\boldsymbol{H} \parallel \boldsymbol{b}'$ as

$$\begin{aligned}\mathscr{H}_0 = \sum_{\boldsymbol{k},\sigma} \big[& -t_x \cos(k_x a) - t_y \cos(k_y b) - t_z \cos(k_z c) \\ & + F(k_y) - \sigma\mu_B H - \mu \big] c^\dagger_{\boldsymbol{k}\sigma} c_{\boldsymbol{k}\sigma},\end{aligned} \tag{5.27}$$

where $\sigma = \pm$ represents spin up or down. Here,

$$F(k_y) = -t_{2y}\cos(2k_y b) - t_{3y}\cos(3k_y b) - t_{4y}\cos(4k_y b) \tag{5.28}$$

contains higher harmonic terms which represent next, third and fourth nearest neighbour hoppings along the $\boldsymbol{b}'$ direction. A similar model has been used to understand the FISDW state in the same kind of materials [184, 185, 186].

This modified energy dispersion can be linearized along k_x for each k_y, with eigenvalues

$$\varepsilon^{\alpha}_{k_x,k_y,\sigma} = v_F(k_y,\sigma)\,[\alpha k_x - k_F(k_y,\sigma)] \tag{5.29}$$

for electrons on the left ($\alpha = -$) or right ($\alpha = +$) sheet of the Fermi surface. Notice that in MKH's theory, the Fermi wave number k_F and Fermi velocity v_F along the x axis are both functions of k_y. For instance,

$$k_F(k_y,\sigma) = \frac{1}{a}\cos^{-1}\left[\frac{-t_y\cos(k_y b) + F(k_y) - \sigma\mu_B H - \mu}{t_x}\right]. \tag{5.30}$$

This k_y-dependence will change the Cooper pair momentum from its k_y-independent values ($= 0$ or G, as discussed in Sec. 5.1), and hence modify the upper critical field results. The eigenstates corresponding to eigenvalues (5.29) are

$$\psi^{\alpha}_{\boldsymbol{k},\sigma}(\boldsymbol{r}) = e^{i\boldsymbol{k}\cdot\boldsymbol{r}} \sum_n e^{in(k_z c - Gx)} J_n(\alpha t_z/\omega_c(k_y,\sigma)), \tag{5.31}$$

where $G = |e|Hc$, and $\omega_c(k_y, \sigma) = v_F(k_y, \sigma)G$ is also k_y- and σ-dependent.

Therefore, the one-particle Green's function in this mixed representation is

$$\begin{aligned} G^{\alpha}_{\sigma}(x, x', k_y, k_z, \omega_l) &= -\int_0^{\beta} d\tau e^{i\omega_l \tau} \left\langle \mathrm{T}_{\tau} c_{x,k_y,k_z,\sigma}(\tau) c^{\dagger}_{x',k_y,k_z,\sigma}(0) \right\rangle \\ &= \sum_{n,n'} \exp\left[in(k_z c - Gx) + in'(k_z c - Gx')\right] J_n(\bar{\alpha} t_z/\omega_c(k_y, \sigma)) \\ &\quad \times J_{n'}(\alpha t_z/\omega_c(k_y, \sigma)) \sum_{k_x} e^{ik_x(x-x')} \frac{1}{i\omega_l - \varepsilon^{\alpha}_{k_x,k_y,\sigma}}, \end{aligned} \tag{5.32}$$

where ω_l is the fermionic Matsubara frequency. The linearized order parameter equation for a singlet superconductor with isotropic s-wave pairing is given by

$$\begin{aligned} \Delta(x) = \lambda_s T \sum_{k_y,k_z} \sum_{\alpha,\omega_l} \int_{|x-x'|>d} dx' \Delta(x') \\ \times G^{\alpha}_{\sigma}(x, x', k_y, k_z, \omega_l) G^{\bar{\alpha}}_{\bar{\sigma}}(x, x', -k_y, -k_z, -\omega_l), \end{aligned} \tag{5.33}$$

where λ_s is the coupling constant, and d is the cutoff distance.

The solution of the order parameter equation above can be written without any loss of generality as

$$\Delta_Q(x) = e^{iQx} \tilde{\Delta}_Q(x), \tag{5.34}$$

where Bloch wave vector Q is taken as $-G < Q \leqslant G$, and $\tilde{\Delta}_Q(x) = \sum_m \Delta^Q_{2m} e^{2imGx}$ has period π/G. Then Eq.(5.33) becomes

$$\Delta^Q_{2m} = \lambda_s \sum_{m'} \Pi^Q_{2m,2m'} \Delta^Q_{2m'}, \tag{5.35}$$

where

$$\Pi^Q_{2m,2m'} = \sum_{k_y,N} S^N_{m,m'}(k_y) P_{k_y}(Q + NG). \tag{5.36}$$

The coefficient S is defined as

$$\begin{aligned} S^N_{m,m'}(k_y) = \sum_n J_{n+m}(\alpha t_z/\omega_c(k_y, \sigma)) J_{n+m'}(\alpha t_z/\omega_c(k_y, \sigma)) \\ \times J_{n-m+N}(\alpha t_z/\omega_c(k_y, \bar{\sigma})) J_{n-m'+N}(\alpha t_z/\omega_c(k_y, \bar{\sigma})), \end{aligned} \tag{5.37}$$

and P is given by

$$P_{k_y}(q_x) = T\sum_{\omega_l}\sum_{\alpha,k_x}\frac{1}{(i\omega_l - \varepsilon^{\alpha}_{k_x,k_y,\sigma})(-i\omega_l - \varepsilon^{\bar{\alpha}}_{q_x-k_x,-k_y,\bar{\sigma}})}. \tag{5.38}$$

A numerical solution for $T_c(H)$ can be obtained from Eq.(5.35), as plotted in Fig. 5.3 for $t_x = 1950\,\mathrm{K}$, $t_y = 195\,\mathrm{K}$, $t_{4y} = 0\,\mathrm{K}$, and $T_c(0) = 1.35\,\mathrm{K}$. The optimized $T_c(H)$ is obtained by tuning Q to give the maximum value of T_c. For a strictly two-dimensional case ($t_z/t_x = 0$), the optimized $T_c(H)$ in the absence of higher harmonic terms (solid dots) is higher than $T_c(H)$ with fixed momentum $Q = \pm 2\mu_B H/v_F$ (open squares), which was studied by Lebed [16] and DMS[18]. By adding the higher harmonic terms $t_{2y}/t_y = -0.1$ and $t_{3y}/t_y = -0.07$ (plotted by open dots), $T_c(H)$ increases further, especially at high magnetic fields ($H > 6\,\mathrm{T}$). If the interplane transfer integral t_z is taken into account by setting $t_z = 3.0\,\mathrm{K}$, the optimized $T_c(H)$ with higher harmonic terms is obtained (thick solid line). Notice that the upper critical field at zero temperature for this case reaches almost $9\,\mathrm{T}$, which is more than three times larger than the Pauli paramagnetic limit $H_P \approx 1.84 T_c(0) \approx 2.5\,\mathrm{T}$ for this isotropic s-wave pairing state [11, 12].

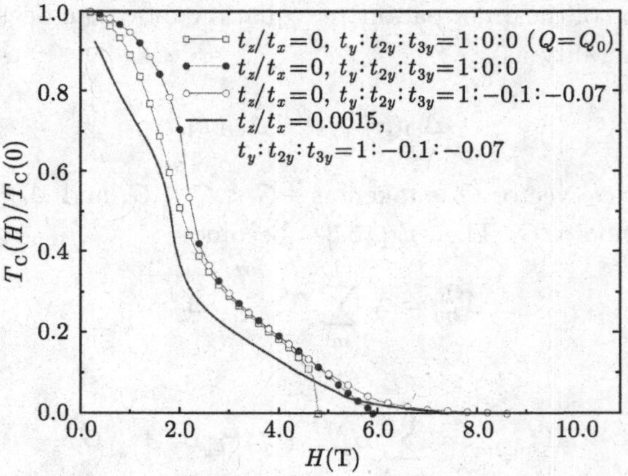

Figure 5.3 Transition temperature $T_c(H)$ for an isotropic s-wave pairing as a function of magnetic field in the case of $t_x = 1950$ K, $t_y = 195$ K, $t_{4y} = 0$ K, $T_c(0) = 1.35$ K for quarter filled electrons. The lines with solid dots and open dots and thick solid line are obtained from the optimal Q giving the maximum value of $T_c(H)$. If Q is fixed to $Q_0 = \pm 2\mu_B H/v_F$, the line with open squares is obtained. From Ref. [183].

MKH [183] also considered an anisotropic s-wave pairing state, which was discussed in Ref. [187]. The numerical results for this state (figure not shown) are very similar to those with isotropic s-wave pairing, except for a small difference of the $t_z \neq 0$ curve. These results, as obtained from a proposal of singlet pairing, can explain qualitatively the experimental results (upward curvature and excess of Pauli limit) of $H_{c_2}^a$ and $H_{c_2}^{b'}$ in $(TMTSF)_2 PF_6$ [23].

In addition to the singlet pairing states, either s- or d-wave symmetries as discussed above, the qualitative behavior of H_{c_2} can also be understood in a triplet pairing scheme, as discussed in previous sections. However, a general calculation for various triplet states is more complicated. In general, triplet pairing is characterized by pair wavefunctions of the type [188, 189]

$$\begin{aligned}\psi_t(\boldsymbol{k},\boldsymbol{R}) = &[-d_x(\boldsymbol{k},\boldsymbol{R}) + id_z(\boldsymbol{k},\boldsymbol{R})]|\uparrow\uparrow\rangle + d_y(\boldsymbol{k},\boldsymbol{R})(|\uparrow\downarrow\rangle + |\downarrow\uparrow\rangle)\\ &+[d_x(\boldsymbol{k},\boldsymbol{R}) + id_z(\boldsymbol{k},\boldsymbol{R})]|\downarrow\downarrow\rangle,\end{aligned} \tag{5.39}$$

where $\boldsymbol{R}$ corresponds to the center-of-mass coordinate of the pair. Due to the Pauli exclusion principle, the order parameter has odd parity, i.e., $\boldsymbol{d}(\boldsymbol{k},\boldsymbol{R}) = -\boldsymbol{d}(-\boldsymbol{k},\boldsymbol{R})$. Notice that the $\boldsymbol{d}$-vector defined above is different from the usual form (see, e.g., Refs. [50, 188]), since the quantization axis is chosen to be the y axis to incorporate the case $\boldsymbol{H} \parallel \boldsymbol{b}'$. By setting different values to d_x, d_y, and d_z, one can obtain several triplet states analogous to the A-phase, B-phase, and planar phase of ^{3}He [50, 188].

Among these states, Lebed, Machida and Ozaki (LMO) [34] considered the particular case

$$\boldsymbol{d}(\boldsymbol{k},\boldsymbol{R}) = \Big(d_x(\boldsymbol{k},\boldsymbol{R}) \neq 0;\ d_y(\boldsymbol{k},\boldsymbol{R}) = 0;\ d_z(\boldsymbol{k},\boldsymbol{R}) = ?\Big) \tag{5.40}$$

with $|d_x(\boldsymbol{k},\boldsymbol{R})| > |d_z(\boldsymbol{k},\boldsymbol{R})|$. This $\boldsymbol{d}$-vector corresponds to ESTP states with the pair wavefunction

$$\psi_t(\boldsymbol{k},\boldsymbol{R}) = [-d_x(\boldsymbol{k},\boldsymbol{R}) + id_z(\boldsymbol{k},\boldsymbol{R})]\,|\uparrow\uparrow\rangle + [d_x(\boldsymbol{k},\boldsymbol{R}) + id_z(\boldsymbol{k},\boldsymbol{R})]\,|\downarrow\downarrow\rangle. \tag{5.41}$$

In the limit of $H \to 0$, $\psi_t(\boldsymbol{k},\boldsymbol{R})$ and $\boldsymbol{d}(\boldsymbol{k},\boldsymbol{R})$ are independent of $\boldsymbol{R}$, the spin susceptibility at $T = 0$ is anisotropic and satisfies

$$\chi_{b'} = \chi_{\mathrm{n}}, \quad \chi_a \ll \chi_{\mathrm{n}}. \tag{5.42}$$

According to LMO [34], this special triplet state is the most stable at zero field since it corresponds to Cooper pairs [see Eq.(5.41)] with $S_{b'} = \pm 1$, which is consistent with the fact that $\boldsymbol{b}'$ is the easy axis for the spin direction in the SDW phase of $(\mathrm{TMTSF})_2\mathrm{PF}_6$ at $H = 0$.

Assuming the clean limit and open electron orbits, the upper critical field can be obtained for the triplet state proposed in Eq.(5.41) at $\boldsymbol{H} \parallel \boldsymbol{b}'$. By using a method suggested by Lebed [32], LMO obtained a linearized order parameter equation that determines the critical temperature $T_{\mathrm{c}}(H)$ of a quasi-1D system with linearized dispersion Eq.(2.2):

$$\begin{aligned} \Delta(x) = \frac{\lambda_t}{2} \int_{|x-x_1|>d} & \frac{2\pi T dx_1}{v_{\mathrm{F}} \sinh\left(\frac{2\pi T|x-x_1|}{v_{\mathrm{F}}}\right)} J_0 \left[\frac{2\sqrt{2} t_y \mu_{\mathrm{B}} H(x-x_1)\beta_y}{t_x v_{\mathrm{F}}}\right] \\ & \times J_0\left(\frac{4t_z}{\omega} \sin\left[\frac{\omega_{\mathrm{c}}(x-x_1)}{2v_{\mathrm{F}}}\right] \sin\left[\frac{\omega_{\mathrm{c}}(x+x_1)}{2v_{\mathrm{F}}}\right]\right) \\ & \times \cos\left[\frac{2\mu_{\mathrm{B}} H(x-x_1)\beta_y}{v_{\mathrm{F}}}\right] \Delta(x_1), \end{aligned} \tag{5.43}$$

where λ_t is the effective interaction in the triplet channel, and d is the cutoff distance.

To be more specific, LMO considered a triplet phase with $\boldsymbol{d}_1(\boldsymbol{k}, \boldsymbol{r}) \equiv (d_x = 1, d_y = 0, d_z = 0) f(\boldsymbol{k}) \Delta(x)$, which is a special case described by Eq.(5.40). Notice that for the ESTP state, the Zeeman coefficient $\beta_y = 0$. In Fig. 5.4, a numerical solution of the order parameter equation (5.43) is shown for $|dH_{\mathrm{c}_2}^{b'}/dT|_{T_{\mathrm{c}}} \simeq 2\,\mathrm{T/K}$, with parameters $t_x \simeq 3200\,\mathrm{K}$, $t_y \simeq 400\,\mathrm{K}$, $t_z \simeq 6\,\mathrm{K}$, $v_{\mathrm{F}} = t_x a/(2\sqrt{2}) = 10^7\,\mathrm{cm/sec}$, $T_{\mathrm{c}}(0) = 1.14\,\mathrm{K}$, and $c = 13.6\,\mathrm{Å}$. For $\boldsymbol{H} \parallel \boldsymbol{b}'$, LMO found that superconductivity survives for fields as high as $H \simeq 6\,\mathrm{T}$ with $T_{\mathrm{c}} \simeq 0.2 - 0.25\,\mathrm{K}$. This result is in qualitative agreement with experimental observations [23, 57].

In the case of $\boldsymbol{H} \parallel \boldsymbol{a}$, the upper critical field $H_{\mathrm{c}_2}^a$ of the order parameter (5.40) will be paramagnetic limited due to its non-zero d_x component. For this state, the corresponding linearized order parameter equation can be obtained as

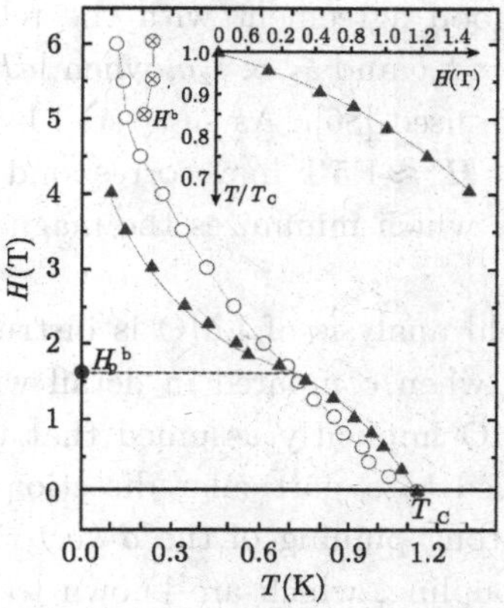

Figure 5.4 Upper critical fields along the $\boldsymbol{a}$ and $\boldsymbol{b}'$ directions. Circles stand for the critical magnetic fields along $\boldsymbol{b}'$ axis: open circles show experimental curve [57, 23], a full circle corresponds to the calculated paramagnetically limited value of $H_{\mathrm{P}}^{b'}$ at $T = 0$ in a singlet superconductor, and crossed circles show the calculated non-paramagnetically limited critical fields $H_{c_2}^{b'}(T)$ for a triplet order parameter (5.40). Triangles stand for the experimental critical fields [57, 23] along $\boldsymbol{a}$ axis, $H_{c_2}^{a}(T)$. The inset shows the experimental values of $H_{c_2}^{a}(T)$ (solid triangles) compared with the calculated paramagnetically limited field (full line) for the triplet order parameter of Eq.(5.39). From Ref. [34].

$$\Delta(x) = \frac{\lambda_t}{2}\int_0^{2\pi}\frac{d\phi}{2\pi}\int_{|x-x_1|>\sqrt{2}d|\sin\phi|/\eta}^{\infty}\Delta(x_1)\,\frac{\sqrt{2}\eta\pi T dx_1}{v_{\mathrm{F}}\sin\phi\sinh\left[\frac{\sqrt{2}\eta\pi T|x-x_1|}{v_{\mathrm{F}}\sin\phi}\right]}$$
$$\times J_0\left(\frac{4\sqrt{2}t_c\eta}{\omega_c\sin\phi}\sin\left[\frac{\omega_c(x-x_1)}{2v_{\mathrm{F}}}\right]\sin\left[\frac{\omega_c(x+x_1)}{2v_{\mathrm{F}}}\right]\right)$$
$$\times\cos\left[\frac{\sqrt{2}\eta\mu_{\mathrm{B}}H\beta_y(x-x_1)}{v_{\mathrm{F}}\sin\phi}\right], \qquad (5.44)$$

where $\eta = t_x a/(2t_y b)$. In Fig. 5.4, a numerical solution of the order parameter equation (5.44) is shown for the same parameters used in the solution of Eq.(5.43). In this figure, the best fit to experimental data [23, 57] in the interval $0 \leqslant H \leqslant 1.5\,\mathrm{T}$ occurs when $\beta_y \simeq 0.9$ and $|dH_{c_2}^{a}/dT|_{T_c} \simeq 8\,\mathrm{T/K}$. Based on this, LMO suggested that the value of β_y corresponded to $d_x \simeq 0.9$, and susceptibility $\chi_a \simeq 0.2\chi_{\mathrm{n}} \ll \chi_{\mathrm{n}}$. Furthermore, they argued that their fitted slope for $|dH^a_{c_2}/dT|_{T_c}$ was in good agreement with the relation $\xi_z|dH^a_{c_2}/dT|_{T_c} = \xi_y|dH^{b'}_{c_2}/dT|_{T_c}$, where $\xi_y \propto t_y b$ and $\xi_z \propto t_z c$, when $|dH^{b'}_{c_2}/dT|_{T_c} \simeq 2\,\mathrm{T/K}$ and the ratio $t_y/t_x \simeq 8.5$ were used [86]. As suggested by LMO [34], the drastic changes of $H^a_{c_2}(T)$ at $H \geqslant H^* \approx 1.5\,\mathrm{T}$ may correspond to the appearance of a triplet state with $d_x = 0$, which minimizes the magnetic contribution to the free energy.

The upper critical field analysis of LMO is instructive, however, further consideration is required when compared in detail with experiments due to three reasons. First, LMO implicitly assumed that the order parameter $\boldsymbol{d}$-vector was strongly pinned to a particular direction, with a very strong x component. However, strong pinning of the $\boldsymbol{d}$-vector is typically associated with strong spin-orbit coupling, which are known to be small in Bechgaard salts [23, 57, 86]. Furthermore, LMO argued that the H_{c_2} anisotropy inversion occurring at $H^* \approx 1.5\,\mathrm{T}$ (see Fig. 5.4 and Sec. 3.1) might be due to a $\boldsymbol{d}$-vector flop transition induced by the magnetic field applied along the $\boldsymbol{a}$ direction. However, this flop transition was not observed in NMR measurements [35], and a theoretical estimate [37] indicated that the magnetic field (along the $\boldsymbol{a}$ direction) required for such a flop transition is approximately 0.22 T (see Sec. 5.7), which is nearly an order of magnitude smaller than H^*. Second, the choice of a large component of the $\boldsymbol{d}$-vector along the x direction and the small susceptibility $\chi_a \approx 0.2\chi_n \ll \chi_n$ are in disagreement with NMR experiments performed by Lee *et al.* [35], which show no Knight shift change for magnetic fields parallel to the $\boldsymbol{a}$ direction at $H = 1.43\,\mathrm{T}$. Third, Eq.(5.44) is derived including only open electronic orbits. However, close orbits in momentum space are important at low magnetic fields parallel to the $\boldsymbol{a}$ axis, at least in the range of $0 < H < 2\,\mathrm{T}$. Therefore, the comparison between theoretical and experimental results of upper critical fields is qualitative at best.

5.4 Phase fluctuations and angular dependence of upper critical field

One of the most interesting properties of quasi-1D organic superconductors is the possible paramagnetic reentrant phase with $dH_{c_2}(T)/dT > 0$ at high magnetic fields, as predicted in Ref. [16, 17, 18] within mean field theories.

However, this unusual behavior has not been observed experimentally thus far. One possible reason for this discrepancy is the presence of fluctuations which were not taken into account in the mean field theory. In principle, the effects of thermal and quantum fluctuations tend to reduce the mean field critical temperature T_{MF} to a lower value T_c. This effect is most severe at or near the lower critical dimension, and thus, there is a range of temperatures below T_{MF}, but still above the true transition temperature T_c where an ordered state exists [190, 191]. In addition to fluctuation effects, the fact that the critical temperature $T_c(H)$ is very sensitive to magnetic field direction may be another reason. For instance, small deviations from $\boldsymbol{H} \parallel \boldsymbol{b}'$ can cause a dramatic reduction of $T_c(H)$. These two effects were discussed preliminarily by Vaccarella and Sá de Melo (VS) [49, 192]. In these calculations, VS considered only phase fluctuations [49] and magnetic fields applied in the y-z plane [192] for simplicity.

In field theory language, phase fluctuations can be considered as massless bosons with sine-Gordon Hamiltonian, as discussed in both classical XY [190] and in layered-superconductor [193, 194, 195] models. In the present problem of triplet superconductivity in quasi-1D systems, the order parameter has the form $\Delta_n = |\Delta_n| \exp(i\phi_n)$ in the nearly two-dimensional quantum regime, where n labels different layers. In this case the electronic motion is mostly confined within x-y planes by a large external magnetic field along y direction. The free energy describing phase fluctuations of this layered system takes the form

$$\mathscr{F}_{\mathrm{p}} = \sum_n \int dxdy \left[J_x \left| \frac{\partial}{\partial x}\phi_n - i2eA_{n_x} \right|^2 + J_y \left| \frac{\partial}{\partial y}\phi_n - i2eA_{n_y} \right|^2 \right.$$
$$\left. + J_z \cos\left(\phi_{n+1} - \phi_n - 2Gx - \frac{2e}{c}\int_{nc}^{(n+1)c} A_z dz \right) \right] + \mathscr{F}_{pm} \qquad (5.45)$$

where G is the inverse magnetic length, $\boldsymbol{A}$ is the fluctuation vector potential, and $\mathscr{F}_{pm} = \int d\boldsymbol{r}(\nabla\times\boldsymbol{A})^2/(8\pi)$ represents the magnetic energy contribution. The coupling coefficient within layers are

$$J_x = |C_0 N(E_{\mathrm{F}}) v_{\mathrm{F}}^2/T^3| \, |\Delta_n|^2, \qquad (5.46a)$$

$$J_y = |C_0 N(E_{\mathrm{F}}) t_y^2 b^2/(2T^2)| \, |\Delta_n|^2, \qquad (5.46b)$$

where $N(E_{\mathrm{F}})$ is the total density of states at the Fermi energy, and $C_0 = 7\zeta(3)/(16\pi^2)$. The second line of Eq. (5.45) represents the magnetic-field induced Josephson coupling between n and $(n+1)$ layers, with prefactor

$$J_z = N(E_{\mathrm{F}})\frac{t_z^2}{\omega_{\mathrm{c}}^2}\ln\left(\frac{\omega_{\mathrm{c}}\gamma_0}{2\pi T}\right)|\Delta_n|\,|\Delta_{n+1}|, \tag{5.47}$$

where ω_{c} is the cyclotron frequency and $\gamma_0 \approx 1.7811$ is the exponential of the Euler constant. The amplitude and phase saddle point solutions of order parameter are given by Eqs. (5.20) and (5.22), respectively. These solutions correspond to a rectangular Josephson vortex lattice with periodicity $a_x = \pi/G$ and $a_z = c$, and holds one flux quanta ϕ_0 per plaquette.

The phase solution beyond the saddle point approximation is a combination of $\phi_n^{(0)}$ and fluctuations $\delta_n(x,y)$, i.e., $\phi_n(x,y) = \phi_n^{(0)} + \delta_n(x,y)$. Following a similar method developed by Horovitz [193, 194] and Korshunov and Larkin [195] to study the vortex lattice melting in layered superconductors, the effective nonlocal sine-Gordon Hamiltonian [190] is obtained as $\mathscr{H} = \int dxdy(\mathscr{H}_x + \mathscr{H}_y + \mathscr{H}_z)$, where

$$\mathscr{H}_\mu = J_\mu \sum_{n,n'} \gamma_\mu(n,n')\frac{\partial\delta_n}{\partial\mu}\frac{\partial\delta_{n'}}{\partial\mu} \tag{5.48}$$

with μ represents either the x or y directions, and

$$\mathscr{H}_z = -\Gamma\sum_n \cos(\delta_{n+1} - 2\delta_n + \delta_{n-1}) \tag{5.49}$$

with the interlayer coupling constant Γ taking the form

$$\Gamma = \frac{J_z^2}{4\sqrt{J_xJ_y}\sqrt{\gamma_x(\pi)\gamma_y(\pi)h_z^2}}. \tag{5.50}$$

Here $h_z = (\nabla\times\boldsymbol{A})_z$ is the z-component of the fluctuation magnetic field, and

$$\gamma_\mu(n,n') = \gamma_\mu(n-n') = \int_{-\pi}^{\pi}\frac{dq}{2\pi}e^{i(n-n')q}\gamma_\mu(q) \tag{5.51}$$

is the Fourier transform of

$$\gamma_\mu(q) = \frac{1 - \cos(q)}{1 - \cos(q) + 16\pi^3 J_\mu c/\phi_0^2}. \tag{5.52}$$

When $\Gamma \to 0$, the effective Hamiltonian reduces to a layered anisotropic XY model, where phase fluctuations between layers n and n' are still coupled via the function $\gamma_\mu(n, n')$. Upon scaling the integration variables $x \to \tilde{x}\sqrt{J_x\gamma_x}$, $y \to \tilde{y}\sqrt{J_y\gamma_y}$, and defining the charges $q_{n_i} = \oint_{\partial \boldsymbol{r}_i} d\delta_n(\boldsymbol{r})/2\pi$, which correspond to the vorticities at position $\boldsymbol{r}_i$ in the n-th plane, the partition function is identical to the anisotropic quasi-two-dimensional Coulomb gas. When an expansion in powers of the small parameter Γ and Gaussian integration over δ_n are performed [195, 196], the partition function of this system becomes

$$Z = \sum_m \left[\prod_{i=0}^{2m}\left(\int d\boldsymbol{r}_i \sum_{n_i}\right)\right]\left(\frac{\Gamma}{2T}\right)^{2m} \exp\left[-\frac{1}{2}\sum_{i,j}^{2m} q_{n_i} M(\boldsymbol{r}_{ij}, n_{ij}) q_{n_j}\right], \tag{5.53}$$

where charges $q_{n_i} = +1$ for $i = 1, ..., m$ and $q_{n_i} = -1$ for $i = m+1, ..., 2m$. In addition, $\boldsymbol{r}_{ij} = \boldsymbol{r}_j - \boldsymbol{r}_i = (x_j - x_i, y_j - y_i)$ labels the separation between topological charges in the x-y plane, and $n_{ij} = n_i - n_j$ labels their separation in the z direction. In Eq.(5.53), $\Gamma/2T$ plays the role of the fugacity of the charges. The interaction

$$M(\boldsymbol{r}_{ij}, n_{ij}) = \int \frac{d^2\boldsymbol{k}}{(2\pi)^2}\frac{dq}{2\pi} M(\boldsymbol{k}, q) \exp\left[i(\boldsymbol{k}\cdot\boldsymbol{r}_{ij} + n_{ij}q)\right]$$

is the Fourier transform of the function $M(\boldsymbol{k}, q) = T/[k^2\sqrt{J_xJ_y}\sqrt{\gamma_x(q)\gamma_y(q)}]$. Upon integration over $\boldsymbol{k}$ and proper core regularization, the relevant part of the interaction term becomes

$$M(\boldsymbol{r}_{ij}, n_{ij}) \approx \frac{\alpha(n_{ij})T}{2\pi\sqrt{J_xJ_y}} \ln(r_{ij}/a_0), \tag{5.54}$$

where $\alpha(n_{ij}) = \int dq \exp(in_{ij}q)/[2\pi\sqrt{\gamma_x(q)\gamma_y(q)}]$. Notice that the charges of quasi-two-dimensional Coulomb gas interact logarithmically when they are in the same layer, and also logarithmically (but with a smaller strength) when they are in different layers. In the limiting case $\Gamma/2T \to 0$ (small fugacity), a RG analysis shows that the system has the same behavior as the ordinary two-dimensional Coulomb gas, as discussed by Kosterlitz [196]. Therefore, a phase transition belonging to the same universality class of the Berezinskii-Kosterlitz-Thouless (BKT) transition [197, 198] will be present in this system. In the present problem, this transition corresponds to the melting of the magnetic-field induced Josephson vortex lattice with transition temperature

$$T_m = \frac{\pi\sqrt{J_x(T_m)J_y(T_m)}}{1+\sqrt{J_x(T_m)J_y(T_m)}16\pi^3 c/\phi_0^2}. \tag{5.55}$$

Using the expressions for $J_\mu(T)$ [see Eq. (5.47) and the paragraph above it] and solving Eq. (5.55) for the melting temperature at infinite field $T_m(\infty)$, the following result is obtained:

$$T_m(\infty) = \frac{T_{\mathrm{MF}}(\infty)}{1+\eta}. \tag{5.56}$$

Here $\eta = 2\sqrt{2}\pi T_{\mathrm{MF}}(\infty)/t_y$, and since $\eta > 0$, the melting temperature $T_m(\infty)$ is smaller than the mean field critical temperature $T_{\mathrm{MF}}(\infty)$, indicating that classical phase fluctuations reduce the transition temperature from $T_{\mathrm{MF}}(\infty)$ to $T_m(\infty)$. However, this reduction is small for Bechgaard salts since $\eta \ll 1$. When the condition $H \to \infty$ is relaxed the first order correction to T_m can be computed using a perturbative renormalization group method [193, 194, 196] since J_z is much smaller than $T_{\mathrm{MF}}(\infty)$ and $T_m(\infty)$. This standard procedure leads to

$$T_m(H) = T_m(\infty)\left[1+\frac{\pi}{2}\left(\frac{\pi^2 Y}{2\omega_{\mathrm{c}}^2 T_m(\infty)}\right)^2\right], \tag{5.57}$$

where the magnetic field dependent coefficient

$$Y = N(E_{\mathrm{F}})T_{\mathrm{MF}}^2(\infty)\left(\frac{\sqrt{2}\pi^2 v_{\mathrm{F}}}{7\zeta(3)t_y b}\right)\left(\frac{\Delta_0^2 t_z^4}{\omega_{\mathrm{c}}^6}\right)\ln^2\left(\frac{\gamma_0\omega_{\mathrm{c}}}{2\pi T_{\mathrm{MF}}(\infty)}\right).$$

Plots of $T_m(H)$ are shown in Fig. 5.5, in comparison to the mean field $T_{\mathrm{MF}}(H)$. As shown in the figure, $T_m(H)$ is lower than $T_{\mathrm{MF}}(H)$, although only slightly even at high magnetic fields. A possible reason for this small reduction is that classical phase fluctuations cannot be considered alone. It might be necessary to include quantum (temporal) phase and amplitude fluctuations to obtain a more substantial reduction of $T_{\mathrm{c}}(H)$ from its mean field value.

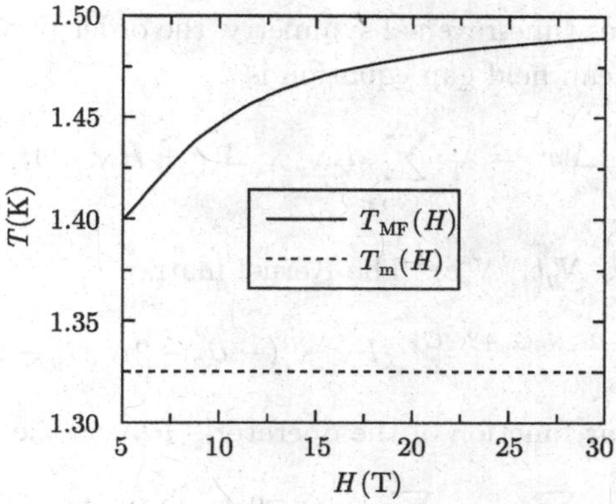

Figure 5.5 This figure shows the (mean field) critical temperature $T_{\mathrm{MF}}(H)$ and the melting temperature $T_m(H)$, in the regime $t_z/\omega_c \ll 1$ for applied fields $\boldsymbol{H}$ precisely aligned with the y ($\boldsymbol{b}'$) axis. The ratio $t_z/\omega_c \approx 1/3$ $(1/18)$ for fields $H = 5$ (30) T. The parameters used were $T_c(\infty) = 1.5\,\mathrm{K}$, $t_y = 100\,\mathrm{K}$, $t_z = 5\,\mathrm{K}$ with lattice constants characteristic of Bechgaard salts. From Ref. [49].

Another effect that can reduce $T_c(H)$ is the deviation from perfect alignment of $\boldsymbol{H} \parallel \boldsymbol{b}'$, as discussed by VS [192]. For simplicity, they considered only an external magnetic field in the y-z plane. Using the gauge $\boldsymbol{A} = (H_y z - H_z y, 0, 0)$, the non-interacting Hamiltonian

$$\mathscr{H}_0(\boldsymbol{k} - e\boldsymbol{A}) = \varepsilon_\alpha(\boldsymbol{k} - e\boldsymbol{A}) - \sigma\mu_{\mathrm{B}} H \tag{5.58}$$

with linearized dispersion Eq.(2.2) has eigenvalues

$$\varepsilon_{\mathrm{qn}} = v_{\mathrm{F}}(\alpha k_x - k_{\mathrm{F}}) - \sigma\mu_{\mathrm{B}} H + \alpha N_y \omega_{c_y} + \alpha N_z \omega_{c_z}. \tag{5.59}$$

The associated quantum numbers are $\mathrm{qn} = (\alpha, k_x, N_y, N_z, \sigma)$, with eigenfunctions

$$\psi_{\mathrm{qn}}(\boldsymbol{r}) = e^{ik_x x} J_{N_y - n_y}\left(\frac{\bar{\alpha} t_y}{\omega_{c_y}}\right) J_{N_z - n_z}\left(\frac{\bar{\alpha} t_z}{\omega_{c_z}}\right), \tag{5.60}$$

where $\boldsymbol{r} = (x, y, z)$, $n_y = y/b$ and $n_z = z/c$. The cyclotron frequencies ω_{c_μ} and inverse magnetic length G_μ with $\mu = y, z$ are defined as in Chap. 2. By introducing the interacting Hamiltonian Eq.(2.20) with $\lambda_s = 0$, the order parameter vector Δ_{t,m_s} is proportional to the expectation value $\langle O_{t,m_s}(\boldsymbol{r})\rangle$. If the system preserves time-reversal symmetry, the order parameter $\Delta_{t,\pm 1} = \Delta$. In this case, the mean field gap equation is

$$\Delta(\boldsymbol{r}) = \lambda_t \sum_{N_y, N_z} K_{N_y, N_z} \Delta(\boldsymbol{r} + \boldsymbol{R}_{N_y N_z}), \tag{5.61}$$

where $\boldsymbol{R}_{N_y N_z} = (0,\ N_y b,\ N_z c)$. The Kernel matrix

$$K_{N_y, N_z} = e^{-2i(N_y G_y + N_z G_z)x} F_{N_y, N_z}(-i\partial_x - 2N_y G_y - 2N_z G_z) \tag{5.62}$$

is a highly nonlinear function of the operator $-i\partial_x$ via the relation

$$\begin{aligned} F_{N_y, N_z}(\hat{q_x}) = \sum_{\alpha} \sum_{M_{1y}, M_{1z}, M_{2y}, M_{2z}} & P^{\alpha,\bar{\alpha}}_{N_y, N_z}(M_{1y}, M_{1z}, M_{2y}, M_{2z}) \\ & \times \Lambda^{\alpha,\bar{\alpha}} \left[\hat{q_x} + (M_{1y} + M_{2y}) G_y + (M_{1z} + M_{2z}) G_z \right]. \end{aligned} \tag{5.63}$$

Here, the weight functions

$$P^{\alpha,\bar{\alpha}}_{N_y, N_z} = W^{\alpha}_{N_y}(M_{2y}) W^{\alpha}_{N_z}(M_{2z}) W^{\bar{\alpha}}_{N_y}(M_{1y}) W^{\bar{\alpha}}_{N_z}(M_{1z}) \tag{5.64}$$

are defined with coefficients

$$W^{\alpha}_{N_\nu}(M_{i\nu}) = J_{M_{i\nu}}(\bar{\alpha} t_\nu / \omega_{c_\nu}) J_{M_{i\nu} - N_\nu}(\bar{\alpha} t_\nu / \omega_{c_\nu}) \tag{5.65}$$

for $i = 1, 2$ and $\nu = y, z$. In addition, the operator F_{N_y, N_z} contains the Cooper singularity contribution

$$\Lambda^{\alpha,\bar{\alpha}}(Q_x) = N(E_F) \left\{ \Psi\left(\frac{1}{2}\right) - \mathrm{Re}\left[\Psi\left(\frac{1}{2} + \frac{\alpha v_F Q_x}{4\pi i T} \right) \right] + \ln\left(\frac{2\omega_d \gamma_0}{\pi T} \right) \right\}, \tag{5.66}$$

where ω_d is the cut-off frequency, and $\Psi(x)$ is the Digamma Psi function.

In the nearly two-dimensional quantum regime, where the electronic motion along z direction is mostly confined by a magnetic field, the order parameter equation (5.61) can be simplified since the magnetic field along y direction can be treated semiclassically, when the angle θ between $\boldsymbol{H}$ and $\boldsymbol{b}'$ is changed to satisfy the conditions $t_y/\omega_{c_y} \gg 1$ and $t_z/\omega_{c_z} \ll 1$. Using the fact that q_y and q_z are good quantum numbers one can write

$$\Delta(x, y, z) = \exp[i(q_y y + q_z z)] u(x), \tag{5.67}$$

which substituted in Eq. (5.61) transforms it into the generalized Mathieu equation [49]

$$\left[-\beta_{2D}\frac{d^2}{d_x^2} + L_y(x) + A_{2D} + D_z\cos(q_z c - 2G_z x)\right]u(x) = 0, \tag{5.68}$$

where the prefactor of the gradient term is $\beta_{2D} = C_0(1 - t_z^2/\omega_{c_z}^2)v_F^2/T^2$, the second term is $L_y(x) = C_0(q_y b - 2G_y x)^2 t_y^2/2T^2$, the coefficient of the co-sine term is $D_z = (t_z^2/\omega_{c_z}^2)\ln|\omega_{c_z}\gamma_0/(2\pi T)|$, and the third term is

$$A_{2D} = \left(1 - \frac{t_z^2}{\omega_{c_z}^2}\right)\ln\left(\frac{2\omega_d\gamma_0}{\pi T}\right) + \frac{t_z^2}{\omega_{c_z}^2}\ln\left|\frac{2\omega_d}{\omega_{c_z}}\right| - \frac{1}{\lambda_t N(E_F)}. \tag{5.69}$$

The critical temperature resulting from this equation is

$$T_c = T_c^{(2D)}\left[1 - \frac{\sqrt{2}C_0 t_y \omega_{c_y}}{\left(T_c^{(2D)}\right)^2} - \left(\frac{t_z}{\omega_{c_z}}\right)^2 \ln\left|\frac{\omega_{c_z}\gamma_0}{\pi T_c^{(2D)}}\right|\right], \tag{5.70}$$

where $T_c^{(2D)} = (2\omega_d\gamma_0/\pi)\exp(-1/\lambda_t N(E_F))$. A plot of T_c as a function of magnetic field for various angles θ close to 90.0° ($\boldsymbol{H} \parallel \boldsymbol{b}'$) is shown in Fig. 5.6.

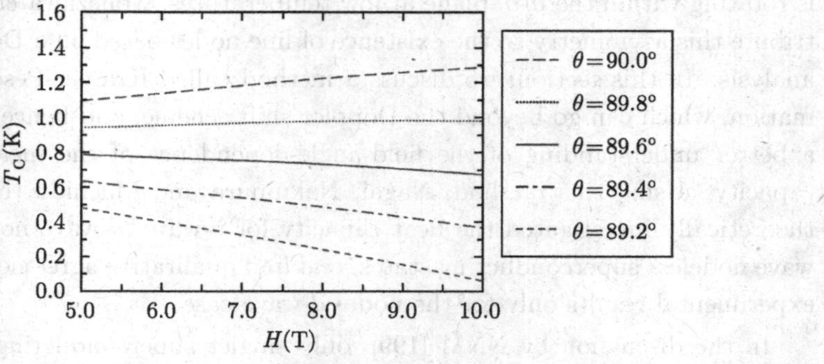

Figure 5.6 This figure shows the suppression of the critical temperature $T_c(H)$ for applied fields $\boldsymbol{H}$ along the yz plane in the two-dimensional regime. The direction $\theta = 90.0°$ corresponds to perfect alignment along the y axis. The parameters used were $T_c^{2D} = 1.5\,\mathrm{K}$, $t_y = 100\,\mathrm{K}$, $t_z = 10\,\mathrm{K}$, with lattice constants characteristic of Bechgaard salts. From Ref. [192].

Notice in Fig. 5.6 a rapid reduction in $T_c(H)$ for very small deviations from $\theta = 90.0°$. A similar drop in T_c is seen experimentally in the putative critical temperature of $(TMTSF)_2ClO_4$ at high magnetic fields for small angular deviations from the y axis [28]. However, since the alignment along $\boldsymbol{b}'$ direction in upper critical field measurements of $(TMTSF)_2PF_6$ by Lee *et. al.* [23] is accurate to within $\pm 0.015°$ with respect to rotations in the y-z plane towards the $\boldsymbol{c}^*$ axis, the calculation presented here is not sufficient to explain the absence of the reentrant superconducting state. A combined effect of this angular dependence and the fluctuation effect discussed above may be needed to fix this discrepancy. Furthermore, other facts such as triclinicity and SDW proximity effect may also be important.

5.5 Angle-dependent heat capacity

The very recent field-angle-resolved calorimetry measurement in $(TMTSF)_2$ ClO_4 [40] brings the nodal structure of the superconducting gap function into lively debate, as it was believed to be nodeless for nearly 15 years after the thermal conductivity experiments by Belin and Behnia [39]. In Sec. 4.3, we learn that the oscillation curve of the heat capacity C/T presents anomalous asymmetry with respect to the crystallographic $\boldsymbol{a}$ axis as a low magnetic field is rotating within the $\boldsymbol{a}$-$\boldsymbol{b}'$ plane at low temperatures. Yonezawa *et al.* [40] attribute this asymmetry to the existence of line nodes based on a Doppler shift analysis. In this section, we discuss a method called Kramer-Pesch approximation, which can go beyond the Doppler shift scheme, and hence can obtain a better understanding of the field-angle-dependence of the measured heat capacity. Using this method, Nagai, Nakumura and Machida (NNM) [199] theoretically investigated the heat capacity for s-wave, d-wave nodal, and d-wave nodeless superconducting states, and find qualitative agreement with the experimental results only for the nodal d-wave case.

In the discussion by NNM [199], only singlet superconducting states are considered, where the order parameter $\Delta(\boldsymbol{r}, \boldsymbol{k})$ is taken to be both position and momentum dependent within an inhomogeneous system. Adopting a quasi-classical approximation [200], the equation of motion for the Green's function can be written as

$$-i\boldsymbol{v}_{\mathrm{F}}(\boldsymbol{k}_{\mathrm{F}}) \cdot \nabla \begin{pmatrix} g & f \\ -f^* & -g \end{pmatrix} = \left[\begin{pmatrix} i\omega & -\Delta(\boldsymbol{r}, \boldsymbol{k}) \\ \Delta^*(\boldsymbol{r}, \boldsymbol{k}) & -i\omega \end{pmatrix}, \begin{pmatrix} g & f \\ -f^* & -g \end{pmatrix} \right], \tag{5.71}$$

where ω is the frequency, $\boldsymbol{k}_{\mathrm{F}}$ is the Fermi wave vector, and the square bracket $[\cdot,\cdot]$ stands for commutation. Here, the Green's function in the Nambu basis can be parametrized as

$$\begin{pmatrix} g & f \\ -f^* & -g \end{pmatrix} = \frac{-i\pi}{1+ab}\begin{pmatrix} 1-ab & 2ib \\ -2ib & -1+ab \end{pmatrix}, \tag{5.72}$$

where a and b are the solutions of the following Riccati differential equations

$$\boldsymbol{v}_{\mathrm{F}}\cdot\nabla a = \Delta - 2\omega a - \Delta^* a^2, \tag{5.73a}$$

$$\boldsymbol{v}_{\mathrm{F}}\cdot\nabla b = -\Delta^* + 2\omega a + \Delta^* b^2. \tag{5.73b}$$

Notice that the parametrization as in Eq.(5.72) automatically guarantees the normalization of the Green's function

$$\begin{pmatrix} g & f \\ -f^* & -g \end{pmatrix}^2 = -\pi^2\mathbf{1}. \tag{5.74}$$

The Riccati equations (5.73) can be reduced to a set of one-dimensional problems on a straight line by setting one coordinate axis to be along the direction of Fermi velocity $\boldsymbol{v}_{\mathrm{F}}$. Consider a single vortex along the $\hat{\boldsymbol{z}}_M$ axis. Since the pairing function Δ does not depend on z_M due to the translational symmetry along the z_M axis, the Riccati equations can be rewritten in the following form

$$v_{\mathrm{F}\perp}(\boldsymbol{k}_{\mathrm{F}})\frac{\partial a}{\partial s} = \Delta(s,y,\boldsymbol{k}_{\mathrm{F}}) - 2\omega a - \Delta^*(s,y,\boldsymbol{k}_{\mathrm{F}})a^2, \tag{5.75a}$$

$$v_{\mathrm{F}\perp}(\boldsymbol{k}_{\mathrm{F}})\frac{\partial b}{\partial s} = -\Delta^*(s,y,\boldsymbol{k}_{\mathrm{F}}) + 2\omega b + \Delta^*(s,y,\boldsymbol{k}_{\mathrm{F}})b^2, \tag{5.75b}$$

where $v_{\mathrm{F}\perp} = |\boldsymbol{v}_{\mathrm{F}} \times \hat{\boldsymbol{z}}_M|$ is the Fermi velocity component perpendicular to the $\hat{\boldsymbol{z}}_M$ axis, and the coordinates s and y are along the directions parallel and perpendicular to $\boldsymbol{v}_{\mathrm{F}}(\boldsymbol{k}_{\mathrm{F}})$, respectively.

With the solutions of the Riccati equations (5.75), the density of states can be obtained by integrating over the Fermi surface the imaginary part of the retarded Green's function and then averaging in real space around a vortex, leading to

$$\mathscr{N}(E) = \frac{1}{\pi R_2}\int_0^R rdr\int_0^{2\pi} d\theta\left[\frac{-1}{\pi}\int_{\mathrm{F.S.}}\frac{d^2\boldsymbol{S}}{2\pi^2|\boldsymbol{v}_{\mathrm{F}}|}\mathrm{Im}\left(g(i\omega\to E+i\eta)\right)\right], \tag{5.76}$$

where $R = \xi_0\sqrt{H_{c_2}/H}$ with ξ_0 the coherence length, $d^2\boldsymbol{S}$ is the Fermi-surface area element, and $i\omega \to E + i\eta$ is the conventional analytic continuation with $\eta \to 0^+$. The low temperature specific capacity is then given as

$$\frac{C(T)}{T} = \int_{-\infty}^{\infty} \frac{dE}{T}\frac{E^2}{T^2}\frac{\mathcal{N}(E)}{\cosh^2(E/2T)}. \tag{5.77}$$

Next, we discuss two approximation schemes which can be employed to solve the Riccati equations. We first briefly introduce the Doppler shift method adopted by Yonezawa *et al.* [40], and then the next level Kramer-Pesch approximation (KPA) [199]. The two methods bear roughly a same computational complexity, which is significantly simpler than that in direct numerical calculations. Meanwhile, the KPA scheme can give a quantitatively correct outcome for zero-energy density of states around a vortex as compared with the results of direct numerical methods [201]. Besides, KPA can also be used to calculate the density of states even in complicated Fermi surfaces, as has been demonstrated in other types of unconventional superconductors [201, 202, 203, 204].

Doppler shift scheme- The Doppler shift approximation can be adopted in the Riccati formalism by writing the pairing order parameter Δ in the form of separating amplitude and phase

$$\Delta(s, y, \boldsymbol{k}_{\mathrm{F}}) = |\Delta(s, y, \boldsymbol{k}_{\mathrm{F}})|e^{i\phi(s,y,\boldsymbol{k}_{\mathrm{F}})}. \tag{5.78}$$

Introducing the rotation $\tilde{a} = a\exp(-i\phi)$ and $\tilde{b} = b\exp(i\phi)$, the Riccati equations can be rewritten as

$$v_{\mathrm{F}\perp}(\boldsymbol{k}_{\mathrm{F}})\frac{\partial\tilde{a}}{\partial s} = |\Delta| + \left(-2\omega - iv_{\mathrm{F}\perp}(\boldsymbol{k}_{\mathrm{F}})\frac{\partial\phi}{\partial s}\right)\tilde{a} - |\Delta|\tilde{a}^2, \tag{5.79a}$$

$$v_{\mathrm{F}\perp}(\boldsymbol{k}_{\mathrm{F}})\frac{\partial\tilde{b}}{\partial s} = -|\Delta| + \left(2\omega + iv_{\mathrm{F}\perp}(\boldsymbol{k}_{\mathrm{F}})\frac{\partial\phi}{\partial s}\right)\tilde{b} + |\Delta|\tilde{b}^2. \tag{5.79b}$$

Assuming $\partial\tilde{a}/\partial s = \partial\tilde{b}/\partial s = 0$, the equations above can be exactly solved. The solution is equivalent to that in the bulk region by replacing the energy $i\omega$ with the Doppler shifted one

$$i\omega \to i\omega - \frac{v_{\mathrm{F}}(\boldsymbol{k}_{\mathrm{F}})}{2}\frac{\partial\phi}{\partial s}. \tag{5.80}$$

Thus, the Doppler shift method is an approximation neglecting the spatial variations of $|\tilde{a}| = |a|$ and $|\tilde{b}| = |b|$, and it is expected to break down as approaching a vortex core [201].

Kramer-Pesch scheme- Unlike Eq.(5.78), the order parameter around a vortex is taken as the following ansatz

$$\Delta(s,y,\boldsymbol{k}_{\mathrm{F}}) = f(x,y)\Delta_0 d(\boldsymbol{k}_{\mathrm{F}})e^{i\phi_r(\boldsymbol{k}_{\mathrm{F}})} = f(x,y)\Delta_0 d(\boldsymbol{k}_{\mathrm{F}})\frac{s+iy}{\sqrt{s^2+y^2}}e^{i\phi_v(\boldsymbol{k}_{\mathrm{F}})}, \tag{5.81}$$

where $f(s,y)$ describes the spatial variation of the pairing order parameter, Δ_0 is the amplitude of order parameter in the bulk, ϕ_r denotes the angle around the vortex core, and ϕ_v represents the directions of $\boldsymbol{v}_{\mathrm{F}\perp} \equiv \boldsymbol{v}_{\mathrm{F}} - (\boldsymbol{v}_{\mathrm{F}} \cdot \hat{z}_M)\hat{z}_M$. The function $f(s,y)$ thus takes the limiting value of 0 as $\sqrt{s^2+y^2} \to 0$ and approaches 1 as $\sqrt{s^2+y^2} \to \infty$. Introducing new variables in the rotating basis

$$\begin{aligned} \bar{a} &\equiv a e^{-i\phi_v}, \\ \bar{b} &\equiv b e^{i\phi_v}, \\ \bar{\Delta} &\equiv \Delta e^{-i\phi_v}, \end{aligned} \tag{5.82}$$

the Riccati equations (5.75) can be reformed as

$$v_{\mathrm{F}\perp}(\boldsymbol{k}_{\mathrm{F}})\frac{\partial \bar{a}}{\partial s} = \bar{\Delta} - 2\omega\bar{a} - \bar{\Delta}^*\bar{a}^2, \tag{5.83a}$$

$$v_{\mathrm{F}\perp}(\boldsymbol{k}_{\mathrm{F}})\frac{\partial \bar{b}}{\partial s} = -\bar{\Delta}^* + 2\omega\bar{b} + \bar{\Delta}\bar{b}^2. \tag{5.83b}$$

This set of equations can be further simplified by expanding the variables $\bar{a}$ and $\bar{b}$ with respect to the imaginary frequency and the imaginary part of the pair function $\bar{\Delta}$, as illustrated by Mel'nikov, Ryzhov, and Silaev [205]. By defining

$$\begin{aligned} \Delta_{\mathrm{R}} &= \mathrm{Re}\,\bar{\Delta} = f(s,y)\Delta_0 d(\vec{k}_{\mathrm{F}})\frac{s}{\sqrt{s^2+y^2}}, \\ \Delta_{\mathrm{I}} &= \mathrm{Im}\,\bar{\Delta} = f(s,y)\Delta_0 d(\vec{k}_{\mathrm{F}})\frac{y}{\sqrt{s^2+y^2}}, \end{aligned} \tag{5.84}$$

the expansion leads to the following solutions of the Riccati equations

$$\begin{aligned} \bar{a} &= \bar{a}_0 + \bar{a}_1 + \mathscr{O}(\omega^2, \Delta_{\mathrm{I}}^2, \omega\Delta_{\mathrm{I}}), \\ \bar{b} &= \bar{b}_0 + \bar{b}_1 + \mathscr{O}(\omega^2, \Delta_{\mathrm{I}}^2, \omega\Delta_{\mathrm{I}}). \end{aligned} \tag{5.85}$$

The corresponding terms in the expansion read

$$\bar{a}_0 = -\text{sign}\,(d(\vec{\boldsymbol{k}}_{\rm F})), \tag{5.86a}$$

$$\bar{b}_0 = \text{sign}\,(d(\vec{\boldsymbol{k}}_{\rm F})), \tag{5.86b}$$

$$\bar{a}_1(s) = \frac{e^{u(s)}}{v_{{\rm F}\perp}(\vec{\boldsymbol{k}}_{\rm F})}\int_{-\infty}^{s}(-2\bar{a}_0\omega + 2i\Delta_{\rm I}(s'))e^{-u(s')}ds' \tag{5.86c}$$

$$\bar{b}_1(s) = \frac{e^{u(s)}}{v_{{\rm F}\perp}(\vec{\boldsymbol{k}}_{\rm F})}\int_{\infty}^{s}(2\bar{b}_0\omega + 2i\Delta_{\rm I}(s'))e^{-u(s')}ds'. \tag{5.86d}$$

Using the relation

$$u(s) = 2\frac{|d(\vec{\boldsymbol{k}}_{\rm F})|}{v_{{\rm F}\perp}(\vec{\boldsymbol{k}}_{\rm F})}\int_0^s \Delta_0 f(s',y)\frac{s'}{\sqrt{s'^2+y^2}}ds', \tag{5.87}$$

the quasiclassical Green's function thus can be obtained in the following form

$$\begin{pmatrix} g & f \\ -f^* & g \end{pmatrix} = \frac{-2\pi i}{\bar{a}_1\bar{b}_0 + \bar{a}_0\bar{b}_1}\begin{pmatrix} 1 & i\bar{a}_0 \\ -i\bar{b}_0 & -1 \end{pmatrix}. \tag{5.88}$$

By substituting the equation above into Eqs.(5.76) and (5.77) and carrying the same analytic continuation $i\omega \to E + i\eta$, the density of states is given as

$$\mathscr{N}(E) = \frac{1}{\pi R_2}\int_0^R rdr\int_0^{2\pi}d\theta\left[\int_{\rm F.S.}\frac{d^2\boldsymbol{S}}{2\pi^2|\boldsymbol{v}_{\rm F}|}\frac{v_{{\rm F}\perp}(\boldsymbol{k}_{\rm F})}{C(y,\boldsymbol{k}_{\rm F})}e^{u(s)}\delta(E - D(y,\boldsymbol{k}_{\rm F}))\right], \tag{5.89}$$

where the functions are

$$C(y,\vec{\boldsymbol{k}}_{\rm F}) \equiv \int_{-\infty}^{\infty} e^{-u(s')}ds', \tag{5.90}$$

$$D(y,\vec{\boldsymbol{k}}_{\rm F}) \equiv \frac{|d(\vec{\boldsymbol{k}}_{\rm F})|\Delta_0}{C(y,\vec{\boldsymbol{k}}_{\rm F})}\int_{-\infty}^{\infty} f(s',y)\frac{y}{\sqrt{s'^2+y^2}}e^{-u(s')}ds'. \tag{5.91}$$

As a result, the heat capacity in the clean limit ($\eta \to 0^+$) within the KPA scheme reads

$$\frac{C(T)}{T} = \frac{1}{\pi R_2}\int_0^R rdr\int_0^{2\pi}d\theta\left[\int_{\rm F.S.}\frac{d^2\boldsymbol{S}v_{{\rm F}\perp}(\boldsymbol{k}_{\rm F})}{2\pi^2|\boldsymbol{v}_{\rm F}|T^3}\frac{D(y.\boldsymbol{k}_{\rm F})^2}{C(y,\boldsymbol{k}_{\rm F})}\frac{e^{-u(s)}}{\cosh^2\left(E(y,\vec{\boldsymbol{k}}_{\rm F})/2T\right)}\right]. \tag{5.92}$$

In their calculation, NNM then choose a special form of the spatial variation of the pairing order parameter [199]

$$f(s,y) = \frac{r}{\sqrt{r^2+\xi_0^2}} = \frac{\sqrt{s^2+y^2}}{\sqrt{s^2+y^2+\xi_0^2}}, \tag{5.93}$$

and integrate the functions $C(y,\boldsymbol{k}_{\rm F})$ and $D(y,\boldsymbol{k}_{\rm F})$ to obtain the analytic forms

$$\begin{aligned} C(y,\vec{\boldsymbol{k}}_{\rm F}) &= 2\sqrt{y^2+\xi_0^2}K_1(r_0(y,\vec{\boldsymbol{k}}_{\rm F})), \\ D(y,\vec{\boldsymbol{k}}_{\rm F}) &= |d(\vec{\boldsymbol{k}}_{\rm F})|\Delta_0 \frac{y}{\sqrt{y^2+\xi_0^2}} \frac{K_0(r_0(y,\vec{\boldsymbol{k}}_{\rm F}))}{K_1(r_0(y,\vec{\boldsymbol{k}}_{\rm F}))}, \end{aligned} \tag{5.94}$$

where

$$r_0(y,\vec{\boldsymbol{k}}_{\rm F}) \equiv \frac{2|d(\vec{\boldsymbol{k}}_{\rm F})|\Delta_0}{v_{\rm F\perp}(\vec{\boldsymbol{k}}_{\rm F})}\sqrt{y^2+\xi_0^2} \tag{5.95}$$

and $K_n(x)$ is the n-th modified Bessel function of the second kind.

Using this approach, NNM calculate the heat capacity for three singlet pairing states including isotropic s-wave, nodeless d-wave, and nodal d-wave symmetries. For simplicity, only a cylindrical symmetric vortex core is considered and the anisotropy of the upper critical field is neglected, i.e., $H_{c_2}(\phi) = H_{c_2}$. The size of a vortex core is chosen as $R = 5\xi_0$, which is comparable to the mutual vortices distance when $H \sim H_{c_2}/25$. In comparison, the upper critical field in Bechgaard salts acquires an obvious in-plane anisotropy and an anomalous anisotropy inversion at low temperatures, and the magnetic field used in heat capacity measurement is $H = 0.5\,\mathrm{T}$ with thermodynamic $H_{c2}^{a,\rm thm}(T=0) \sim 2.5\,\mathrm{T}$ and $H_{c2}^{b',\rm thm}(T=0) \sim 1.6\,\mathrm{T}$.

The results for the isotropic s-wave pairing state is shown in Fig. 5.7(a). In this case, the oscillatory behavior of the field-angle-resolved heat capacity originates exclusively from the anisotropy of the Fermi surface. As a consequence, the heat capacity oscillates with the angle ϕ and is symmetric with respect to the minimum at $\phi = 0$. In fact, as the magnetic field is applied along the Fermi velocity $\boldsymbol{v}_{\rm F}(\boldsymbol{k}_{\rm F})$, the perpendicular component $v_{\rm F\perp}(\boldsymbol{k}_{\rm F})$ becomes zero and leads to a vanishing value in the momentum $\boldsymbol{k}_{\rm F}$ dependent Kernel of the heat capacity in Eq.(5.92). Considering the fact that the Fermi velocity in Bechgaard salts acquires a dominant component along the $\boldsymbol{a}$ axis throughout the whole Fermi surface, a minimum of heat capacity at $\phi = 0$ (i.e., $\boldsymbol{H} \parallel \boldsymbol{a}$) is naturally expected.

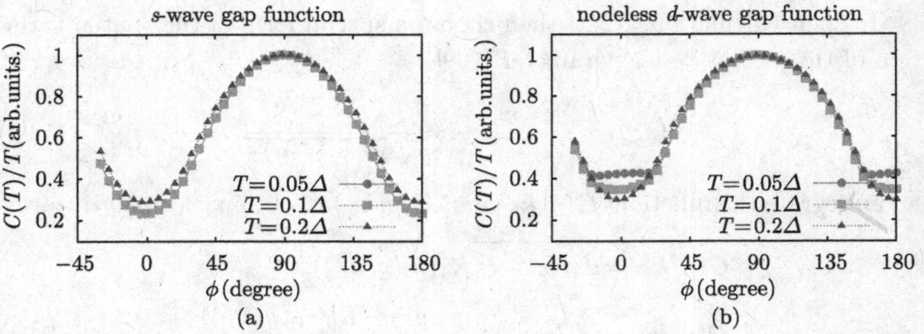

Figure 5.7 Angular dependence of the heat capacity with magnetic field applied within the x-y plane for (a) s-wave and (b) nodeless d-wave pairing scenarios. From Ref. [199].

For the nodeless d-wave case, NNM consider the following ansatz for the order parameter

$$\Delta^{\text{in}} = a_0 h^0(k_x, k_y) + b_0, \tag{5.96a}$$

$$\Delta^{\text{out}} = -a_1 h^1(k_x, k_y) - b_1, \tag{5.96b}$$

where $a_0 = 0.385$, $b_0 = -0.654$, $a_1 = 0.435$ and $b_1 = -0.130$ are chosen to be normalization factors. Having $(\text{TMTSF})_2\text{ClO}_4$ in mind, NNM [199] take into account the anion ordering and the resulting folding of the reciprocal lattice, and assign different order parameters along the inner (Δ^{in}) and the outer (Δ^{out}) Fermi surfaces. The functions are defined as

$$h^0(k_x, k_y) \equiv \cos\left((k_y - k_0)\frac{2\pi}{g_y}\right) - \frac{1}{2}\cos\left(2(k_y - k_0)\frac{2\pi}{g_y}\right), \tag{5.97a}$$

$$h^1(k_x, k_y) \equiv \cos\left((k_y - k_0)\frac{2\pi}{g_y}\right), \tag{5.97b}$$

where $k_0 \equiv -0.031$ sign (k_x) denotes the intersection of two Fermi surfaces, and $g_y = 2\pi/(b \sin\beta)$ with b the lattice constant and $\beta = 68.92°$. The angular dependence of the heat capacity for this nodeless d-wave superconducting state is shown in Fig. 5.7(b). A slightly concave but almost flat curve with a broad minimum at center lies around the angles $-20° < \phi < 20°$. This complex behavior is a combined effect of Fermi surface and order parameter anisotropies. Nonetheless, the heat capacity is symmetric respect to $\phi = 0$, which is inconsistent with the experimental observations [40].

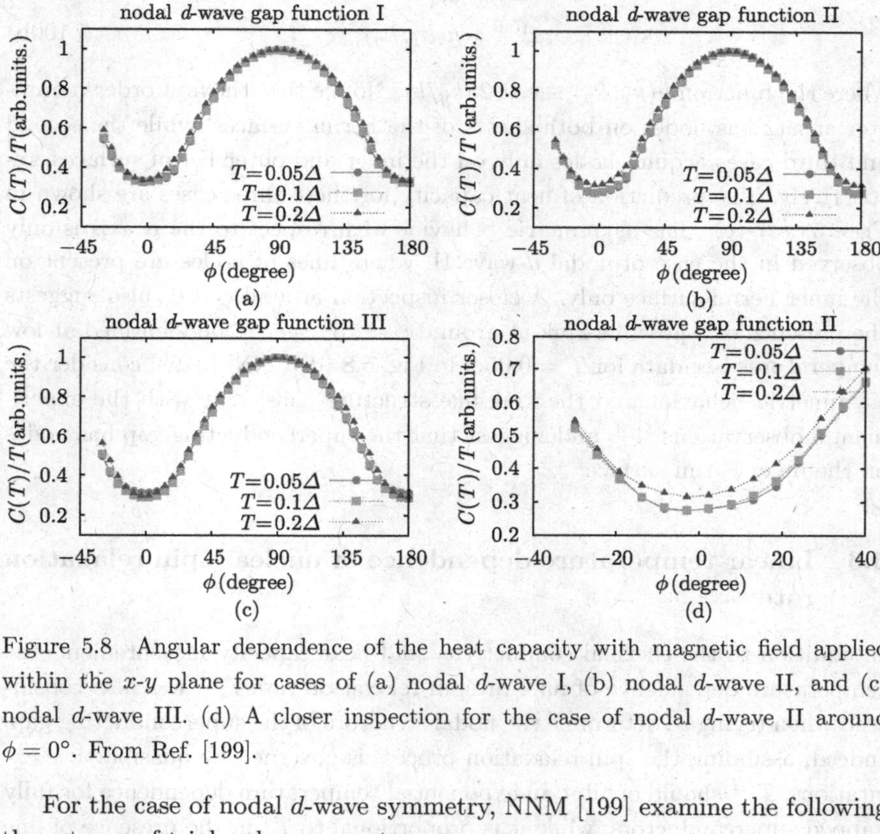

Figure 5.8 Angular dependence of the heat capacity with magnetic field applied within the x-y plane for cases of (a) nodal d-wave I, (b) nodal d-wave II, and (c) nodal d-wave III. (d) A closer inspection for the case of nodal d-wave II around $\phi = 0°$. From Ref. [199].

For the case of nodal d-wave symmetry, NNM [199] examine the following three types of trial order parameters:

nodal d-wave I-

$$\Delta^{\text{in}} = h(k_x, k_y), \tag{5.98a}$$

$$\Delta^{\text{out}} = h(k_x, k_y); \tag{5.98b}$$

nodal d-wave II-

$$\Delta^{\text{in}} = h(k_x, k_y), \tag{5.99a}$$

$$\Delta^{\text{out}} = 1; \tag{5.99b}$$

nodal d-wave III-

$$\Delta^{\text{in}} = 1, \tag{5.100a}$$

$$\Delta^{\text{out}} = h(k_x, k_y), \tag{5.100b}$$

where the function $h(k_x, k_y) \equiv \cos(2\pi k_y/b)$. Notice that the first order parameter ansatz has nodes on both sheets of the Fermi surfaces, while the second and third cases acquire nodes only on the inner and outer Fermi surfaces, respectively. The prediction of heat capacity for these three cases are shown in Figs. 5.8(a)–(c). The asymmetric behavior with respect to the $\boldsymbol{a}$ axis is only observed in the case of nodal d-wave II, where lines of nodes are present on the inner Fermi surface only. A closer inspection around $\phi = 0°$ also suggests the presence of a possible kink at around $\phi \approx 15°$, as can be identified at low temperatures [see data for $T = 0.05\Delta$ in Fig. 5.8 (d)]. NNM [199] consider the asymmetric behavior and the kink-like structure consistent with the experimental observations [40], and suggest that the superconducting gap has nodes on the inner Fermi surface.

5.6 Linear temperature dependence of nuclear spin relaxation rate

In addition to the thermal conductivity and heat capacity measurement, the temperature dependence of nuclear spin relaxation rate T_1^{-1} was also considered when trying to determine the nodal structure of the superconducting gap. Indeed, assuming the spin relaxation process is governed by quasiparticle excitations, T_1^{-1} should exhibit an exponential temperature dependence for fully gapped superconductors, while it is proportional to T^3 in the presence of line nodes in the gap. In the early investigation of $(\text{TMTSF})_2\text{ClO}_4$, the observation of $T_1^{-1} \propto T^3$ and the absence of Hebel-Slichter peak [13] at zero magnetic field indeed suggests a superconducting state with nodal lines [15]. However, recent experimental efforts reveal a linear temperature dependence $T_1^{-1} \propto T$ instead of a T^3 law in both $(\text{TMTSF})_2\text{PF}_6$ [33, 35] and $(\text{TMTSF})_2\text{ClO}_4$ [38], when a magnetic field is applied either along the $\boldsymbol{a}$ axis or the $\boldsymbol{b}'$ axis.

To understand the linear temperature dependence of T_1^{-1}, Takigawa *et al.* [206] study the effect of vortices generated by a magnetic field pointing along the $\boldsymbol{a}$ axis. By considering three pairing symmetries including fully gapped p-wave, gapless d-wave, and gapless f-wave states, Takigawa *et al.* find the linear behavior of $T_1^{-1}(T)$ is discriminating against nodeless superconducting states.

In order to extract the low energy excitations in the presence of vortices, Takigawa *et al.* [206] perform a real space Bogoliubov–de Gennes (BdG) calculation. Starting from the tight-binding spectrum Eq.(2.1), the pairing functions are assumed to be $\psi(k_x) = \sin(2k_x a)$ for p-wave, $\cos(2k_x a)$ for d-wave, and $\sin(4k_x a)$ for f-wave pairings. This choice of pairing functions thus leads to the following mean-field Hamiltonian

$$\mathscr{H} = \sum_{k_x, i, j, \sigma} K_{i,j}(k_x) a_{i\sigma}^\dagger a_{j\sigma} + \sum_{k_x, i} \left[\Delta_i \psi(k_x) a_{i,\uparrow}^\dagger a_{i,\downarrow}^\dagger + \text{H.C.} \right], \tag{5.101}$$

where i and j stand for site indices within the y-z plane $i = (i_y, i_z)$ and $j = (j_y, j_z)$, Δ_i is the order parameter amplitude at site i, and H.C. represents Hermitian conjugate. In the presence of a magnetic field along the x direction $\boldsymbol{H} = (H, 0, 0)$ and within the symmetric gauge $\boldsymbol{A}(\boldsymbol{r}) = (\boldsymbol{H} \times \boldsymbol{r})/2$, the dispersion $K_{i,j}(k_x)$ takes the form

$$K_{i,j}(k_x) = -\frac{t_\ell}{2} \exp\left[i \frac{\pi}{\phi_0} \int_{\boldsymbol{r}_j}^{\boldsymbol{r}_i} \boldsymbol{A}(\boldsymbol{r}) \cdot d^3\boldsymbol{r} \right] \tag{5.102}$$

for i and j are nearest neighbors along the $\ell = (y, z)$ direction, and becomes

$$K_{i,j}(k_x) = -t_x \cos(k_x a) - \mu \tag{5.103}$$

for $i = j$. Here, ϕ_0 is the flux quanta and μ is the chemical potential which is adjusted to keep the system at quarter filling. To describe the quasi-1D nature of the Bechgaard salts, Takigawa *et al.* [206] employ the ratio between hopping integrals as $t_x : t_y : t_z = 10 : 1 : 0.3$.

The BdG equation for each value of k_x can be derived from Eq.(5.101), leading to

$$\sum_j \begin{pmatrix} K_{i,j}(k_x) & \Delta_i \psi(k_x) \delta_{ij} \\ \Delta_i^\dagger \psi(k_x) \delta_{ij} & -K_{i,j}(k_x) \end{pmatrix} \begin{pmatrix} u_{\lambda,j} \\ v_{\lambda,j} \end{pmatrix} = E_\lambda \begin{pmatrix} u_{\lambda,i} \\ v_{\lambda,i} \end{pmatrix}, \tag{5.104}$$

where E_λ is the eigenenergy of the eigenstate labeled by λ, and u_λ and v_λ form the corresponding eigenfunction. The self-consistent equation for the order parameter is then given by

$$\Delta_i = g \sum_{\lambda} \frac{u_{\lambda,i} v^*_{\lambda,i} \psi(k_x)}{e^{\beta E_\lambda} + 1}, \tag{5.105}$$

where g is the strength of the pairing interaction and $\beta = 1/T$ is the inverse temperature. By solving Eqs. (5.104) and (5.105) iteratively until a self-consistent solution is obtained, the local density of states (LDOS) at a given site i can be obtained by

$$\mathcal{N}(E, \boldsymbol{r}_i) = \sum_{\lambda} \left[|u_{\lambda,i}|^2 \delta(E - E_\lambda) + |v_{\lambda,i}|^2 \delta(E + E_\lambda) \right]. \tag{5.106}$$

Takigawa *et al.* [206] consider a specific vortex configuration where two vortices accommodate in a unit cell of 20×6 lattice sites in the y-z plane and form a triangular lattice. They also take into account the anisotropy of coherence length with $\xi_y : \xi_z \propto \sqrt{t_y} : \sqrt{t_z}$. In the calculation, parameters are chosen as $g = -7.5t_y$ for p-wave pairing (giving bulk order parameter at zero magnetic field $\Delta_0 = 0.31t_y$ and $T_c = 0.175t_y$), $g = -17t_y$ for d-wave pairing (resulting $\Delta_0 = 1.2t_y$ and $T_c = 0.525t_y$), and $g = -13.75t_y$ for f-wave pairing ($\Delta_0 = 0.65t_y$ and $T_c = 0.275t_y$). Different values of g are employed to achieve similar order parameter amplitudes for different symmetries. Notice that the p-wave pairing is fully gapped throughout the whole Fermi surfaces, while the d-wave and f-wave states are gapless at around $k_x a = \pm\pi/4$, as illustrated in Fig. 5.9 for $T = 0$ and $H = 0$.

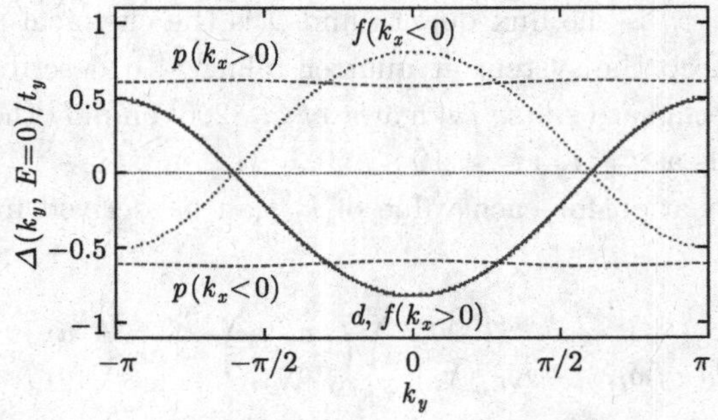

Figure 5.9 The gap function $\Delta_0\psi(k_x)$ along the Fermi surface for p-, d-, and f-wave pairing symmetries at $T = 0$ and $H = 0$. From Ref. [206].

In the absence of magnetic field, the calculated density of states for different pairing symmetries are shown by the dashed lines in Fig. 5.10. The dotted lines in the same figure denote the DOS in the corresponding normal states, where different results are rooted from the usage of different pairing intensities g. In the p-wave case [Fig. 5.10(a)], the full gap structure can be seen with two peaks at $E = \pm 0.325t_y$ reflecting the large enhancement of DOS at the gap edge. As a comparison, in the d-wave and f-wave symmetries, the order parameters have line nodes at $k_x a = \pm\pi/4$ on the Fermi surface, which lead to a V-shaped gap structure in the DOS as depicted by dashed lines in Figs. 5.10(b) and 5.10(c), respectively.

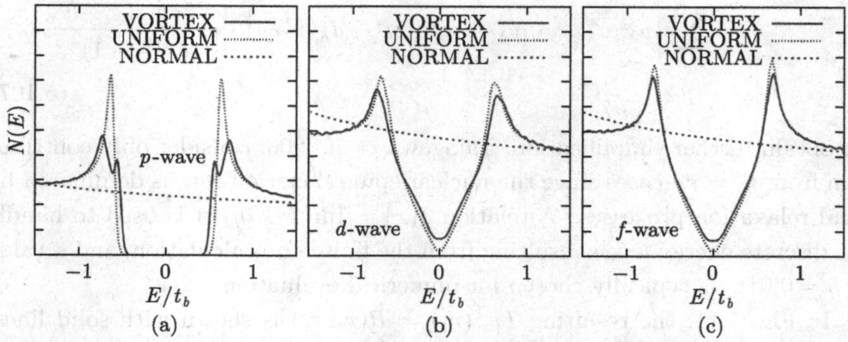

Figure 5.10 The LDOS at the site nearest to the vortex core is presented at $T = 0$ for the (a) p-wave, (b) d-wave, and (c) f-wave pairings. The LDOS in the vortex state is plotted with solid lines. Results in the uniform state at zero field (dashed) and in the normal state (dotted) are also shown as comparison. From Ref. [206].

When a magnetic field is applied along the $\boldsymbol{a}$ axis, the DOS becomes site dependent due to the inhomogeneity induced by the vortices. The LDOS in the vortex state is shown as solid lines in Fig. 5.10, where the lattice site nearest to the vortex is chosen for display. In the fully gapped p-wave case, quasiparticles have bound states around the Josephson-type vortex with energy near the full gap energy [207]. These bound states induce two smaller peaks in LDOS around $E \sim \pm 0.28t_y$. The larger peaks around $E \sim \pm 0.365t_y$ correspond to gap edges smeared by the magnetic field. Despite the double-peak structure at the gap edge, there is no quasiparticle states at the Fermi energy and a wide gap is still present even in the presence of magnetic field, as clearly depicted in Fig. 5.10(a). In contrast, in the d-wave [Fig. 5.10(b)] and f-

wave [Fig. 5.10(c)] symmetries, the V-shaped superconducting gap is partially filled with low-energy quasiparticle excitations induced by vortices. These low-energy quasiparticles can not induce a zero-energy peak of the vortex core state, since the suppression of order parameter is not complete near the vortex core.

With the solutions of low-energy excitations, the nuclear spin relaxation rate can be calculated from the imaginary part of the spin-spin correlation function $\chi(\boldsymbol{r}_i, \boldsymbol{r}_{i'}, i\omega_n)$ [208, 209]

$$\begin{aligned} R(\boldsymbol{r}_i, \boldsymbol{r}_{i'}) &= \mathrm{Im}\left[\frac{\chi\left(\boldsymbol{r}_i, \boldsymbol{r}_{i'}, i\Omega_n \to \Omega + i\eta\right)}{\beta\Omega}\right]_{\Omega\to 0} \\ &= \sum_{\lambda,\lambda'}\left[u_{\lambda,i}u^*_{\lambda',i'}v_{\lambda,i}v^*_{\lambda',i'} - v_{\lambda,i}u^*_{\lambda',i'}u_{\lambda,i}v^*_{\lambda',i'}\right]\frac{\pi e^{\beta E_\lambda}\delta(E_\lambda - E_{\lambda'})}{\left(e^{\beta E_\lambda} - 1\right)^2}. \end{aligned} \tag{5.107}$$

To obtain further simplification, Takigawa *et al.* [206] consider only contribution from $\boldsymbol{r}_i = \boldsymbol{r}_{i'}$ cases since the nuclear spin relaxation rate is dominated by local relaxation processes. A relation $\delta(x) = \mathrm{Im}(x - i\eta)/\pi$ is used to handle the discrete energy levels resulting from the finite size calculation, and a value of $\eta = 0.01t_y$ is typically chosen for numerical evaluation.

In Fig. 5.11, the resulting $T_1^{-1}(\boldsymbol{r}_i) = R(\boldsymbol{r}_i, \boldsymbol{r}_i)$ is shown with solid lines, where $\boldsymbol{r}_i$ is set to be the nearest site to the vortex core. In spite of an explicit $\boldsymbol{r}$

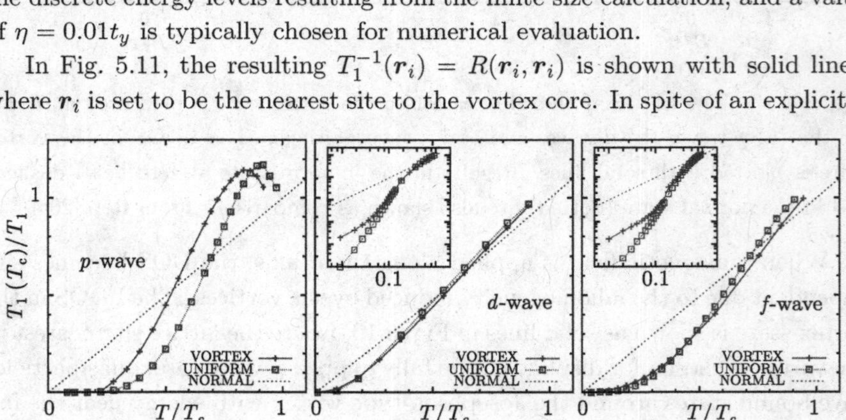

Figure 5.11 The nuclear spin relaxation rate T_1^{-1} is shown as a function of temperature for (a) *p*-wave, (b) *d*-wave, and (c) *f*-wave symmetries. Solid lines denote T_1^{-1} at the lattice site nearest to the vortex core. Results for the uniform state at zero field (dashed) and for the normal state (dotted) are also shown for comparison. Insets in (b) and (c) present the same data in a double logarithmic chart. From Ref. [206].

dependence in the definition, Takigawa *et al.* [206] find the spin relaxation rate T_1^{-1} is almost position independent. In the *p*-wave case [Fig. 5.11(a)], T_1^{-1} shows an exponential behavior with respect to temperature for both the vortex (with magnetic field) and the uniform (without magnetic field) states. Effects of the vortices on T_1^{-1} are almost negligible at low temperatures ($T \lesssim 0.2T_c$) since there are no low-energy quasiparticle excitations within the superconducting gap as illustrated in Fig. 5.10(a). On the other hand, in the *d*-wave [Fig. 5.11(b)] and *f*-wave [Fig. 5.11(c)] cases, T_1^{-1} acquires a linear T dependence instead of a T^3 behavior at low temperatures, as can be seen more clearly in the double logarithmic chart in the insets. This linear temperature dependence is compatible with the experimental observations in both $(TMTSF)_2PF_6$ [33, 35] and $(TMTSF)_2ClO_4$ [38]. Finally, the calculation reproduces the linear temperature dependence of the well-known Korringa law for the normal state as shown by dotted lines in Fig. 5.11.

From the results discussed above, Takigawa *et al.* [206] conclude that the presence of vortices have non-trivial effect on the spin relaxation rate in superconducting Bechgaard salts when a magnetic field is applied along the $\boldsymbol{a}$ axis, provided that the superconducting gap has lines of nodes. Since the $\boldsymbol{a}$ axis is the easiest axis for conducting, this vortex effect could be more evident when a magnetic field is applied along different directions within the $\boldsymbol{a}$-$\boldsymbol{b}'$ plane. Therefore, they attribute the linear temperature dependence of T_1^{-1} to this vortex effect, and infer that a superconducting state with line nodes is more compatible with existing experimental data. However, their analysis can not discriminate among different gapless superconducting symmetries.

5.7 Group theory and symmetry of distinct superconducting states

The orbital symmetry of superconducting order parameter is another important property that needs to be characterized. The previous singlet versus triplet discussion is helpful to refine the number of possible choices, since the wave function is antisymmetric due to Pauli principle. However, even if we clarify the spin symmetry, either singlet or triplet, there are still several candidates for the orbital symmetry. One way to investigate all possible symmetries systematically is through a group theoretical analysis, within the assumption that the normal state of quasi-1D superconductors do not break the full lattice symmetry. This group theoretical analysis was performed for the orthorhombic group by Duncan, Vaccarella and Sá de Melo (DVS) [36] and extended by Duncan, Cherng and Sá de Melo (DCS) [37], and for the triclinic structure by Powell [210].

In the case of weak spin-orbit coupling (SOC), rotations of the spin and spatial degrees of freedom are independent, hence the most general conventional symmetry group for the normal state is $\mathrm{G_n} = \mathrm{SO(3)} \times \mathrm{G_c} \times \mathrm{U(1)} \times \mathrm{T}$, where SO(3) is the group of rotations in spin space, $\mathrm{G_c}$ is the crystal space group, U(1) is the gauge group and T corresponds to time reversal symmetry (TRS). Since the superconducting state breaks U(1) symmetry, the possible superconducting order parameters correspond to different irreducible representations of $\mathrm{G_s} = \mathrm{SO(3)} \times \mathrm{G_c} \times \mathrm{T}$. In the case of strong SOC the normal state symmetry group is $\mathrm{G_n} = \mathrm{G_c^{(J)}} \times \mathrm{U(1)} \times \mathrm{T}$, where now $\mathrm{G_c^{(J)}}$ is identical to the space group $\mathrm{G_c}$ except when a rotation is involved. In this case, rotations of the spin and spatial degrees of freedom are not independent and the possible superconducting order parameters correspond to different irreducible representations of $\mathrm{G_s} = \mathrm{G_c^{(J)}} \times \mathrm{T}$. Notice that in the strong SOC case, spin is not a good quantum number and it is not proper to classify the states as singlet or triplet. However, by introducing the concept of pseudo-spin, the classification can be generalized to pseudo-spin space provided that the superconducting state does not break inversion symmetry. The singlet/triplet pseudo-spin classification was first introduced by Anderson [211] in the context of heavy fermions.

Considering quasi-1D superconductors of orthorhombic crystal structure, the relevant crystallographic point group is $\mathrm{D_{2h}}$, which has only one-dimensional representations [212] (see Tab. 5.1).

Table 5.1 Character table for the $\mathrm{D_{2h}}$ point group [37].

Representation	E	C_2^z	C_2^y	C_2^x	i	iC_2^z	iC_2^y	iC_2^x	Basis
A_{1g}	1	1	1	1	1	1	1	1	1
B_{1g}	1	1	−1	−1	1	1	−1	−1	XY
B_{2g}	1	−1	1	−1	1	−1	1	−1	XZ
B_{3g}	1	−1	−1	1	1	−1	−1	1	YZ
A_{1u}	1	1	1	1	−1	−1	−1	−1	XYZ
B_{1u}	1	1	−1	−1	−1	−1	1	1	Z
B_{2u}	1	−1	1	−1	−1	1	−1	1	Y
B_{3u}	1	−1	−1	1	−1	1	1	−1	X

The general form of the order parameter is a linear combination of singlet (s) and triplet (t) contributions,

$$\tilde{\Delta}(\boldsymbol{k}) = i\left(\Delta_{\rm s}(\boldsymbol{k})\tilde{1} + \Delta_{\rm t}(\boldsymbol{k})\boldsymbol{d}(\boldsymbol{k})\cdot\sigma\right)\sigma_y. \tag{5.108}$$

Here, $\tilde{\Delta}(\boldsymbol{k})$ must transform according to the one-dimensional representations of the orthorhombic point group D_{2h}, under the assumption that the order parameter does not break the crystal translational symmetry, i.e., the order parameter is invariant under all primitive lattice translations. The matrix $\tilde{1}$ is the identity and the function $\Delta_s(\boldsymbol{k})$ is symmetric (even) under the transformation $\boldsymbol{k} \to -\boldsymbol{k}$. The three-dimensional vector $\boldsymbol{d}(\boldsymbol{k})$ is antisymmetric (odd) under $\boldsymbol{k} \to -\boldsymbol{k}$, while the function $\Delta_t(\boldsymbol{k})$ is symmetric (even) under the same transformation. However, the total order parameter, in general, does not have fixed inversion parity.

Table 5.2 Time reversal invariant singlet states in an orthorhombic crystal, assuming weak SOC [37].

State	Residual Group	Order parameter $\Delta_s(\boldsymbol{k})$	$E_{\boldsymbol{k}} = 0$ ($\tilde{\mu} > 0$)	$E_{\boldsymbol{k}} = 0$ ($\tilde{\mu} < 0$)
$^1A_{1g}$	$SO_3 \times D_{2h} \times T$	1	none	none
$^1B_{1g}$	$SO_3 \times D_2(C_2^z) \times I \times T$	XY	lines	none
$^1B_{2g}$	$SO_3 \times D_2(C_2^y) \times I \times T$	XZ	lines	none
$^1B_{3g}$	$SO_3 \times D_2(C_2^y) \times I \times T$	YZ	lines	none

Table 5.3 Time reversal breaking triplet states in an orthorhombic crystal, assuming weak SOC [37].

State	Residual Group	Order parameter $\boldsymbol{d}(\boldsymbol{k})$	$E_{\boldsymbol{k}} = 0$ ($\tilde{\mu} > 0$)	$E_{\boldsymbol{k}} = 0$ ($\tilde{\mu} < 0$)
$^3A_{1u}(b)$	$D_\infty(E) \times D_2 \times I(E)$	$(1, i, 0)XYZ$	surface	none
$^3B_{1u}(b)$	$D_\infty(E) \times D_2(C_2^z) \times I(E)$	$(1, i, 0)Z$	surface	none
$^3B_{2u}(b)$	$D_\infty(E) \times D_2(C_2^y) \times I(E)$	$(1, i, 0)Y$	surface	none
$^3B_{3u}(b)$	$D_\infty(E) \times D_2(C_2^x) \times I(E)$	$(1, i, 0)X$	surface	none

Table 5.4 Time reversal invariant triplet states in an orthorhombic crystal, assuming weak SOC [37].

State	Residual Group	Order parameter $\boldsymbol{d}(\boldsymbol{k})$	$E_{\boldsymbol{k}} = 0$ ($\tilde{\mu} > 0$)	$E_{\boldsymbol{k}} = 0$ ($\tilde{\mu} < 0$)
$^3A_{1u}(a)$	$D_\infty(C_\infty) \times D_2 \times I(E) \times T$	$(0, 0, 1)XYZ$	lines	none
$^3B_{1u}(a)$	$D_\infty(C_\infty) \times D_2(C_2^z) \times I(E) \times T$	$(0, 0, 1)Z$	lines	none
$^3B_{2u}(a)$	$D_\infty(C_\infty) \times D_2(C_2^y) \times I(E) \times T$	$(0, 0, 1)Y$	lines	none
$^3B_{3u}(a)$	$D_\infty(C_\infty) \times D_2(C_2^x) \times I(E) \times T$	$(0, 0, 1)X$	none	none

Table 5.5 Time reversal invariant triplet states in an orthorhombic crystal, assuming strong SOC [37].

State	Residual Group	Order parameter $\boldsymbol{d}(\boldsymbol{k})$	$E_{\boldsymbol{k}} = 0$ ($\tilde{\mu} > 0$)	$E_{\boldsymbol{k}} = 0$ ($\tilde{\mu} < 0$)
A_{1u}	$D_2 \times I(E) \times T$	(AX, BY, CZ)	none, points, lines	none
B_{1u}	$D_2(C_2^z) \times I(E) \times T$	$(AY, BX, CXYZ)$	none, lines	none
B_{2u}	$D_2(C_2^y) \times I(E) \times T$	$(AZ, BXYZ, CX)$	none, lines	none
B_{3u}	$D_2(C_2^x) \times I(E) \times T$	$(AXYZ, BZ, CY)$	points, lines	none

In Tab. 5.2, a summary of some possible order parameter symmetries consistent with the D_{2h} point group is presented for singlet systems assuming weak spin-orbit coupling, while in Tab. 5.3, a summary of some possible order parameter symmetries for time-reversal-symmetry-breaking triplet states is shown. Tabs. 5.4 and 5.5 summarize the group theoretical analysis performed for the order parameter matrix $\tilde{\Delta}(\boldsymbol{k})$ for time-reversal-invariant triplet superconductors in the weak and strong SOC cases, respectively. All the tables mentioned above include the state nomenclature, the order parameter for singlet or triplet, and the type of zeros of the quasiparticle excitation spectrum $E_{\boldsymbol{k}}$, when $\tilde{\mu} = \mu - \min[\varepsilon_{\boldsymbol{k}}]$ is positive or negative. In Tab. 5.4, the vector (0,0,1) is indicated up to an arbitrary rotation in spin space. In Tab. 5.5, the numerical coefficients A, B and C can be determined through the joint solution of the order parameter and number equations.

It is important to emphasize that the basis functions $X(\boldsymbol{k}), Y(\boldsymbol{k})$ and $Z(\boldsymbol{k})$ transform like k_x, k_y and k_z under the crystallographic point group operations, but they cannot be chosen to be k_x, k_y and k_z. The reason is that the normal state dispersion $\varepsilon_{\boldsymbol{k}}$ intersects the boundary of the Brillouin zone along the y and z directions, thus it is necessary to take into account the periodicity of the order parameter matrix $\tilde{\Delta}(\boldsymbol{k})$ and of the order parameter vector $\boldsymbol{d}(\boldsymbol{k})$ in reciprocal (momentum) space. In the case of orthorhombic crystals with dispersion $\varepsilon_{\boldsymbol{k}}$ defined in Eq.(2.1) where $t_x \gg t_y \gg t_z$, the Fermi surface exists (for weak attractive interaction or high densities) and does intersect the Brillouin zones, therefore the minimal basis set must be periodic and may be chosen to be $X(\boldsymbol{k}) = \sin(k_x a)$, $Y(\boldsymbol{k}) = \sin(k_y b)$ and $Z(\boldsymbol{k}) = \sin(k_z c)$.

For the weak spin-orbit coupling (Tab. 5.4) case, the only candidate for weak attractive interaction ($\tilde{\mu} > 0$) is the state ${}^3B_{3u}(a)$, where the quasiparticle excitation spectrum $E_{\boldsymbol{k}}$ has no zeros and is fully gapped. In the case of strong SOC (Tab. 5.5) there are three candidates for weak attractive interaction ($\tilde{\mu} > 0$), i.e., the states A_{1u}, B_{1u} and B_{2u}, where the quasiparticle excitation spectrum $E_{\boldsymbol{k}}$ may have no zeros and be fully gapped. When $\tilde{\mu} > 0$, the state A_{1u} is fully gapped only for $A \neq 0$ and for any value of B and C; the state B_{1u} is fully gapped only for $B \neq 0$ and for any value of A and C; and the state B_{2u} is fully gapped only for $C \neq 0$ and for any value of A and B. Note in passing that in the case of strong attractive interactions where $\tilde{\mu} < 0$, the excitation spectrum $E_{\boldsymbol{k}}$ is fully gapped for all of the states involving Tabs. 5.4 and 5.5 for example.

A three-dimensional view of the Fermi surface and order parameter sign for possible weak SOC singlet states compatible with the orthorhombic (D_{2h}) group are shown in Fig. 5.12 for the case where $\tilde{\mu} > 0$. In Figs. 5.12(a), (b), (c) and (d) the signs of the orbital component of the order parameter are shown

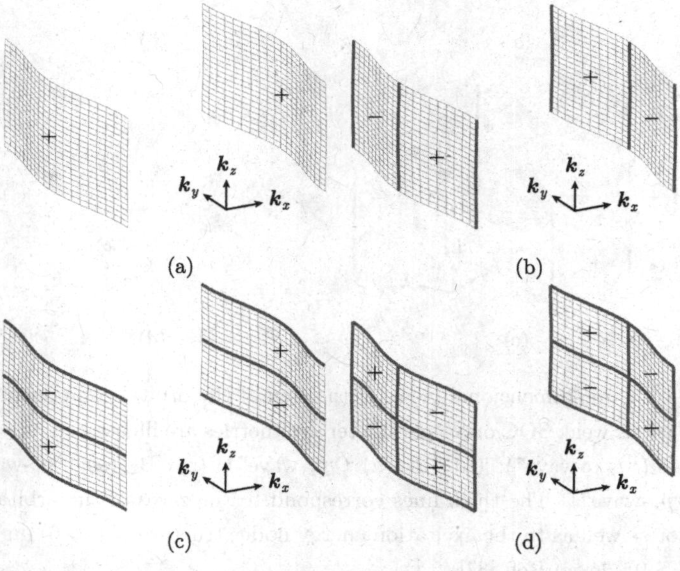

Figure 5.12 Three-dimensional views of the sign of the orbital component of the order parameter. Various weak SOC order parameter symmetries are illustrated (see Table 5.2): (a) $^1A_{1g}$ ("s-wave"); (b) $^1B_{1g}$ ("d_{xy}-wave"); (c) $^1B_{2g}$ ("d_{xz}-wave"); (d) $^1B_{3g}$ ("d_{yz}-wave"). The thick lines correspond to the zeros of the orbital parts of the order parameter as well as to the excitation energy node structure $E_{\boldsymbol{k}} = 0$ on the Fermi surface ($\tilde{\mu} > 0$). From Ref. [37].

for the $^1A_{1g}$ ("s-wave"), $^1B_{1g}$ ("d_{xy}-wave"), $^1B_{2g}$ ("d_{xz}-wave"), $^1B_{3g}$ ("d_{yz}-wave"), respectively. In Fig. 5.13 three-dimensional view of the Fermi surface and order parameter sign for possible weak SOC triplet states compatible with the orthorhombic group are shown for the case where $\tilde{\mu} > 0$. Figs. 5.13(a), (b), (c) and (d) correspond to the $^3A_{1u}(a)$ ("f_{xyz}-wave"), $^3B_{1u}(a)$ ("p_z-wave"), $^3B_{2u}(a)$ ("p_y-wave"), and $^3B_{3u}(a)$ ("p_x-wave"), respectively.

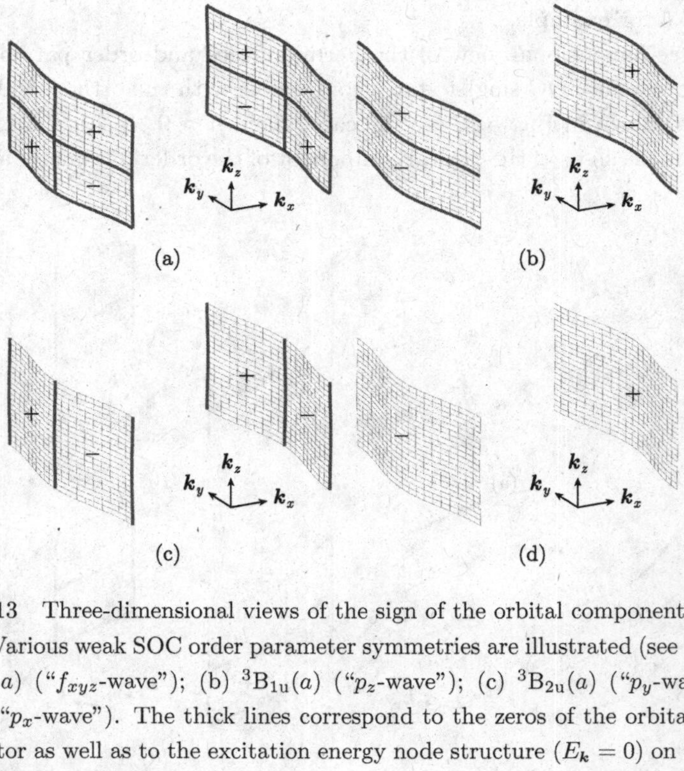

Figure 5.13 Three-dimensional views of the sign of the orbital component of the $\boldsymbol{d}$-vectors. Various weak SOC order parameter symmetries are illustrated (see Tab. 5.4): (a) $^3A_{1u}(a)$ ("f_{xyz}-wave"); (b) $^3B_{1u}(a)$ ("p_z-wave"); (c) $^3B_{2u}(a)$ ("p_y-wave"); (d) $^3B_{3u}(a)$ ("p_x-wave"). The thick lines correspond to the zeros of the orbital parts of the $\boldsymbol{d}$-vector as well as to the excitation energy node structure ($E_{\boldsymbol{k}} = 0$) on the Fermi surface ($\tilde{\mu} > 0$). From Ref. [37].

As discussed in Chap. 3, the combined experimental evidence of the upper critical field measurements [21, 23, 75] and the Knight shift and NMR relaxation experiments [33, 35] suggest that $(TMTSF)_2PF_6$ may be fully gapped triplet superconductors. Furthermore, the zero-field-μSR measurements performed by Luke *et al.* [140] suggest that the time reversal symmetry in likely not broken in $(TMTSF)_2ClO_4$. As listed in Tabs. 5.4 and 5.5, there are eight time reversal invariant triplet superconducting states comply with the orthorhombic lattice group. A quasi-1D superconductor in each of these states has unique quasiparticle excitation spectrum, and hence produces different thermodynamic properties which may be observable. Further inspired by the thermal conductivity measurement in $(TMTSF)_2ClO_4$ by Belin and Behnia [39], Duncan, Vaccarella and Sá de Melo (DVS) [36] and Duncan, Cherng and Sá de Melo (DCS) [37] focused on superconducting states with fully gapped order parameters, and discussed the quasiparticle density of states (QDOS) and uniform spin susceptibility at low temperatures for two particular examples ${}^3B_{3u}(a)$ and A_{1u} states.

When the zero center-of-mass momentum pairing approximation is used for weak or strong SOC, the anomalous Green's function can be written as

$$F_{\alpha\beta}(\boldsymbol{k}, i\omega_n) = \frac{\Delta_{\alpha\beta}(\boldsymbol{k})}{\omega_n^2 + E_{\boldsymbol{k}}^2}, \tag{5.109}$$

and the single particle Green's function has the form

$$G_{\alpha\beta}(\boldsymbol{k}, i\omega_n) = -\frac{i\omega_n + \xi_{\boldsymbol{k}}}{\omega_n^2 + E_{\boldsymbol{k}}^2}\delta_{\alpha\beta}. \tag{5.110}$$

These results can be derived by either using the equation of motion [188] or the functional integral [213, 214] methods. In the equations above, $\Delta_{\alpha\beta}(\boldsymbol{k})$ is the matrix element of $\tilde{\Delta}(\boldsymbol{k})$, $\xi_{\boldsymbol{k}} = \varepsilon_{\boldsymbol{k}} - \mu$ is the dispersion $\varepsilon_{\boldsymbol{k}}$ shifted by the chemical potential μ, $E_{\boldsymbol{k}} = \sqrt{\xi_{\boldsymbol{k}}^2 + \Delta_{\boldsymbol{k}}^2}$ is the quasiparticle excitation energy, and

$$\Delta_{\boldsymbol{k}}^2 \equiv \mathrm{Tr}\left[\tilde{\Delta}^\dagger(\boldsymbol{k})\tilde{\Delta}(\boldsymbol{k})\right]/2. \tag{5.111}$$

A Hartree shift is easily incorporated into the chemical potential, thus it is not explicitly written above.

The quasiparticle density of states (QDOS) can be obtained from the single particle Green's function Eq.(5.110) as

$$\mathcal{N}(\omega) = -\frac{1}{\pi}\mathrm{Tr}\sum_{\boldsymbol{k}} \mathrm{Im}\, G_{\alpha\beta}(\boldsymbol{k}, i\omega_n = \omega + i\delta). \tag{5.112}$$

The QDOS for the states ${}^3B_{3u}(a)$ (weak SOC) and A_{1u} (strong SOC) for various values of constants A, B, and C are compared in Fig. 5.14. The symmetry dependent features of the QDOS are manifested through the magnitude of the order parameter vector $\boldsymbol{d}(\boldsymbol{k})$, which takes the form

$$|\boldsymbol{d}(\boldsymbol{k})| \propto |\sin(k_x a)|, \tag{5.113}$$

for the ${}^3\mathrm{B}_{3u}(a)$ state (p_x-wave) and

$$|\boldsymbol{d}(\boldsymbol{k})| \propto \sqrt{A^2|\sin(k_x a)|^2 + B^2|\sin(k_y b)|^2 + C^2|\sin(k_z c)|^2}, \tag{5.114}$$

for the A_{1u} state. Only the case corresponding to $C = 0$, where $A \gg B$, $A = B$, and $A \ll B$ are studied in Figs. 5.14(b), (c) and (d), respectively. Notice that Fig. 5.14(d) is nearly identical to Fig. 5.14(a) given that $|\boldsymbol{d}(\boldsymbol{k})|$ is essentially the same in this case. Furthermore, notice that while the position of the peaks in Figs. 5.14(a), (b), (c) and (d) are essentially the same in scaled

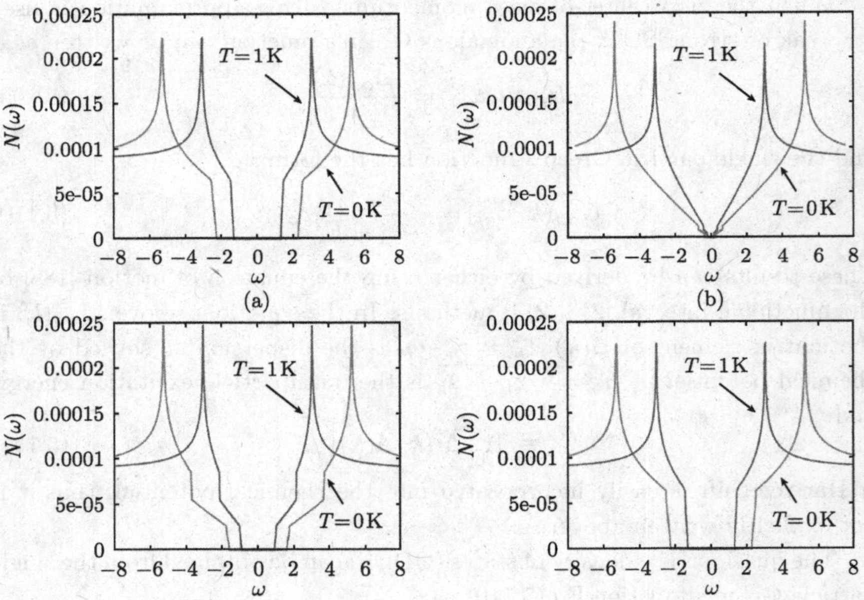

Figure 5.14 Quasiparticle density of states: weak SOC state ${}^3\mathrm{B}_{3u}$(a) is shown in (a); strong SOC state A_{1u} is shown in (b) for $A = 0.20$, $B = 1.40$, $C = 0$; in (c) for $A = B = 1.00$, $C = 0$; and in (d) for $A = 1.44$, $B = 0.10$, $C = 0$. The parameters used are $|t_x| = 5800\,\mathrm{K}$, $|t_y| = 1226\,\mathrm{K}$ and $|t_z| = 48\,\mathrm{K}$, and $\mu = -4101\,\mathrm{K}$, with $T_c = 1.5\,\mathrm{K}$. $N(\omega)$ is in inverse temperature units (K^{-1}) and ω is in temperature units (K). From Ref. [37].

units ($\omega_p/\Delta_0 \approx \pm 1.40$), the corresponding gap sizes in scaled units are respectively $\omega_g/\Delta_0 = 1.32, 0.01, 0.94, 1.32$. Gap sizes, peaks and the general shape of the QDOS are, in principle, identifiable in STM or photoemission experiments, pending of course on experimental resolution and clean material surface. Although these experiments have not yet been performed in $(TMTSF)_2ClO_4$ or $(TMSTF)_2PF_6$, this theoretical prediction can serve as qualitative guides for the extraction of gaps and symmetry dependent features of the experimental results when they become available. In particular, STM tunnelling or photoemission measured gaps can be compared with gaps measured thermodynamically (e.g., thermal conductivity [39] or specific heat). However, such experiments alone cannot uniquely determine the symmetry of the order parameter in triplet quasi-1D superconductors, since the QDOS depends only on the magnitude $|\boldsymbol{d}(\boldsymbol{k})|$.

The spin susceptibility tensor, on the other hand, explicitly depends both on the magnitude and direction of $\boldsymbol{d}$-vector. The general form of the spin susceptibility tensor is

$$\chi_{mn}(q_\mu) = -\mu_{\rm B}^2 (\sigma_m)_{\alpha\beta} (\sigma_n)_{\gamma\delta} \left[S_{\alpha\beta\gamma\delta}(q_\mu) + A_{\alpha\beta\gamma\delta}(q_\mu) \right], \tag{5.115}$$

where repeated greek indices indicate summation and the four-vector $q_\mu = (\boldsymbol{q}, i\nu)$ definition is used to shorten notations. The tensors

$$A_{\alpha\beta\gamma\delta}(\boldsymbol{q}, i\omega) = \beta^{-1} \sum_{\boldsymbol{k}, i\omega} F^\dagger_{\alpha\gamma}(\boldsymbol{k} - \boldsymbol{q}, i\nu - i\omega) F_{\beta\delta}(\boldsymbol{k}, i\omega), \tag{5.116}$$

$$S_{\alpha\beta\gamma\delta}(q_\mu) = \beta^{-1} \sum_{\boldsymbol{k}, i\omega} G_{\delta\alpha}(-\boldsymbol{k} + \boldsymbol{q}, -i\nu + i\omega) G_{\beta\gamma}(\boldsymbol{k}, i\omega) \delta_{\delta\alpha} \delta_{\beta\gamma} \tag{5.117}$$

contain the Green's functions described in Eqs.(5.109) and (5.110), respectively. In the limit of $\omega \to 0$ and $\boldsymbol{q} \to \boldsymbol{0}$, the spin susceptibility of a triplet superconductor (including lattice and particle-hole symmetry effects) is

$$\chi_{mn}(\boldsymbol{0}, 0) = \sum_{\boldsymbol{k}} \left[\chi_{mn,1}(\boldsymbol{k}) + \chi_{mn,2}(\boldsymbol{k}) \right], \tag{5.118}$$

where the $\boldsymbol{k}$-dependent tensors are

$$\chi_{mn,1}(\boldsymbol{k}) = \chi_{\parallel}(\boldsymbol{k}) \mathrm{Re} \left[\hat{\boldsymbol{d}}^*_m(\boldsymbol{k}) \hat{\boldsymbol{d}}_n(\boldsymbol{k}) \right], \tag{5.119}$$

$$\chi_{mn,2}(\boldsymbol{k})=\chi_{\perp}(\boldsymbol{k})\left[\delta_{mn}-\mathrm{Re}(\hat{\boldsymbol{d}}_m^*(\boldsymbol{k})\hat{\boldsymbol{d}}_n(\boldsymbol{k}))\right] \tag{5.120}$$

with $\hat{\boldsymbol{d}}_n(\boldsymbol{k}) = \boldsymbol{d}_n(\boldsymbol{k})/|\boldsymbol{d}_n(\boldsymbol{k})|$. Here, the parallel component is

$$\chi_{\parallel}(\boldsymbol{k}) = -2\mu_{\mathrm{B}}^2 \frac{\partial f(E_{\boldsymbol{k}})}{\partial E_{\boldsymbol{k}}}, \tag{5.121}$$

while the perpendicular component is

$$\chi_{\perp}(\boldsymbol{k}) = 2\mu_{\mathrm{B}}^2 \frac{\mathrm{d}}{d\xi_{\boldsymbol{k}}}\left[\frac{\xi_{\boldsymbol{k}}}{2E_{\boldsymbol{k}}}(1-2f(E_{\boldsymbol{k}}))\right]. \tag{5.122}$$

This result is slightly more general than the expression quoted by LMO [34] in a couple of ways. First, the expression derived in Eqs. (5.118), (5.121), and (5.122) includes particle-hole symmetry effects. Second, they are valid at finite temperatures. They do not include, however, standard Fermi liquid corrections [50], since they are expected to be small for $(TMTSF)_2PF_6$ at lower pressures (near the SDW phase). In this pressure regime this compound behaves like a good Fermi liquid. However, the same cannot be said at higher pressures, where large deviations from Fermi liquid behavior are reported [106].

In Fig. 5.15, the theoretical uniform χ_{mn} is shown only for the $^3\mathrm{B}_{3\mathrm{u}}(a)$ (i.e., p_x) and $\mathrm{A}_{1\mathrm{u}}$ states. The triangles correspond to χ_{xx}, circles to χ_{yy}, and squares to χ_{zz}. It is known experimentally (Knight shift) [33, 35] that the spin susceptibility of $(TMTSF)_2PF_6$ for $\boldsymbol{H} \parallel \boldsymbol{b}'$ and $\boldsymbol{H} \parallel \boldsymbol{a}$ are very close to the normal value χ_n, thus, reinforcing the idea of triplet superconductivity. For fields along the $\boldsymbol{c}^*$ direction the superconducting state can be easily destroyed, which makes the Knight shift measurement for ^{77}Se impractical, since it requires magnetic fields of the order of $1-2\,$T.

However, one has to be careful when comparing the theoretical calculation for the uniform spin susceptibility discussed above with experimental data, since the Knight shift measurements are performed for reasonably high magnetic fields ($H \approx 2.38\,$T along $\boldsymbol{b}'$ or $1.43\,$T along $\boldsymbol{a}$ directions [33, 35], as indicated in Sec. 3.2). In these field regimes the superconducting $(TMTSF)_2PF_6$ is in an inhomogeneous vortex state with rectangular Abrikosov lattice for ESTP or singlet pairing. Suppose that the magnetic field is applied along the μ direction (H_μ), and produces a rectangular vortex lattice with lattice constants a_{ν_1} and a_{ν_2}, with $\nu_1, \nu_2 \perp \mu$. Assuming that the vortices have a

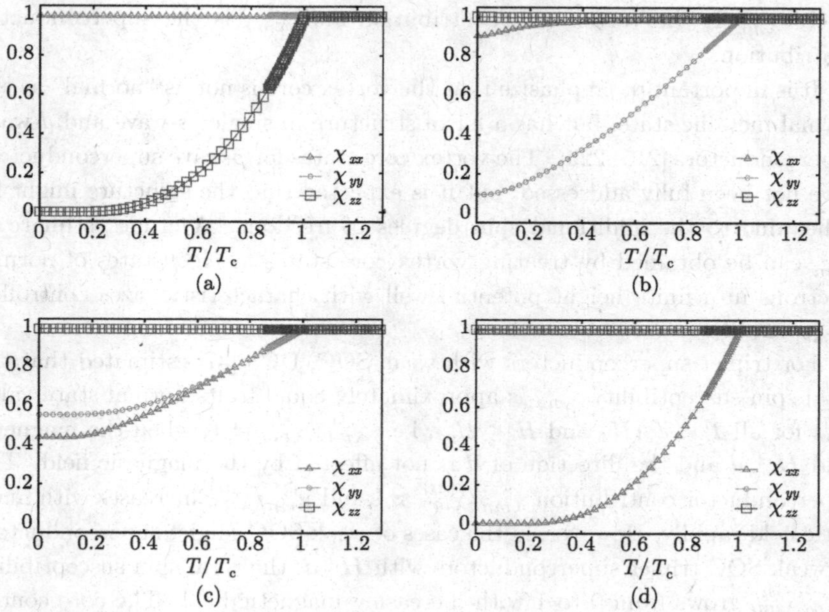

Figure 5.15 Plot of the theoretical uniform spin susceptibility tensor components χ_{xx} (triangles); χ_{yy} (circles); χ_{zz} (squares) at low temperatures. The weak SOC state ${}^3\mathrm{B}_{3u}(a)$ is shown in (a); strong SOC state A_{1u} is shown in (b) for $A = 0.20$, $B = 1.40$, $C = 0$; in (c) for $A = B = 1.00$, $C = 0$; and in (d) for $A = 1.41$, $B = 0.10$, $C = 0$. The parameters used are $|t_x| = 5800\,\mathrm{K}$, $|t_y| = 1216\,\mathrm{K}$ and $|t_z| = 48\,\mathrm{K}$, and $\mu = -4101\,\mathrm{K}$ with $T_c = 1.5\,\mathrm{K}$. From Ref. [37].

finite core size characterized by lengths ξ_{ν_1} and ξ_{ν_2}, and that the vortex core is normal, the normal fraction of the superconductor is

$$n_f = \frac{\xi_{\nu_1}\xi_{\nu_2}}{a_{\nu_1}a_{\nu_2}} = \frac{H_\mu}{H^\mu_{c2}}, \tag{5.123}$$

where H^μ_{c2} is the upper critical field along μ direction. If the superconducting and normal core states can be viewed as nearly independent, the total spin susceptibility tensor of the material in a magnetic field becomes the sum of the response of its individual component

$$\chi_{mn}(T, H_\mu) = n_f\chi^{\mathrm{C}}_{mn}(T, H_\mu) + (1 - n_f)\chi^{\mathrm{S}}_{mn}(T, H_\mu), \tag{5.124}$$

where χ^{C}_{mn} is the vortex core contribution and χ^{S}_{mn} is the superconductor contribution.

It is important to emphasize that the vortex core is not as "normal" as the normal metallic state, but has a lot of structure in singlet s-wave and d-wave superconductors [215–223]. The vortex core states for p-wave superconductors have not been fully addressed, but it is expected that the structure might be richer due to the additional spin degrees of freedom. A crude estimate of χ^{C}_{mn} can be obtained by treating vortex core states as eigenstates of normal electrons in a finite height potential well with characteristic sizes controlled by ξ_{ν_1} and ξ_{ν_2}.

For triplet superconductors with weak SOC, DCS [37] estimated that the total spin susceptibility χ_{mn} is approximately equal to its normal state value $\chi^{(n)}_{mn}$ for all $T < T_{\mathrm{c}}(H)$ and $H < H_{\mathrm{c}_2}$, i.e., $\chi_{mn}/\chi^{(n)}_{mn} \approx 1$, when the magnetic field $\boldsymbol{H} \perp \boldsymbol{d}$ and the direction of $\boldsymbol{d}$ is not affected by the magnetic field. The superconductor contribution $\chi^{\mathrm{S}}_{mn}/\chi^{(n)}_{mn} \approx 1$ and $\chi^{\mathrm{C}}_{mn}/\chi^{(n)}_{mn}$ increases with magnetic field rapidly. However, in the cases of weak SOC singlet superconductors or weak SOC triplet superconductors with $\boldsymbol{H} \parallel \boldsymbol{d}$, the total spin susceptibility $\chi_{mn}/\chi^{(n)}_{mn}$ grows from 0 to 1 with increasing magnetic field. The core contribution $\chi^{\mathrm{C}}_{mn}/\chi^{(n)}_{mn}$ increases rapidly as in the $\boldsymbol{H} \perp \boldsymbol{d}$ triplet case, but $\chi^{\mathrm{S}}_{mn}/\chi^{(n)}_{mn}$ decreases from 1 while reducing H and tends to zero as $H \to 0$, as discussed in Ref. [37].

This behavior of χ_{mn} is valid for triplet superconductors under the condition that the $\boldsymbol{d}$-vector direction is fixed. This assumption may be valid for strong SOC states (e.g., $\mathrm{A_{1u}}$), provided that the spin-orbit pinning field H_{so} is higher than H_{c_2} for any given direction. However, for weak SOC states [e.g., $\mathrm{B_{3u}}(a)$] with $H_{\mathrm{so}} < H_{\mathrm{c}_2}$ for some direction, the presence of magnetic field in the range $H_{\mathrm{so}} < H < H_{\mathrm{c}_2}$ can overcome the spin-orbit coupling and rotate the $\boldsymbol{d}$-vector to be perpendicular to $\boldsymbol{H}$ (barring any additional orbital effects due to the coupling of $\boldsymbol{H}$ with charge degrees of freedom). This flop transition may produce a discontinuous change in the spin susceptibility from smaller to larger (normal) values.

Consider for definiteness that the $\boldsymbol{d}$-vector is pinned along a given lattice direction $\boldsymbol{\mu}$, and that $\boldsymbol{H} \parallel \boldsymbol{\mu}$. The observability of this flop transition depends on how the magnitude H^{μ}_{so} compares to the magnitudes of the minimum field required to perform a Knight shift experiment (H^{μ}_{min}), and the lower ($H^{\mu}_{\mathrm{c}_1}$) and the upper ($H^{\mu}_{\mathrm{c}_2}$) critical fields along the μ direction. If $H^{\mu}_{\mathrm{so}} \ll H^{\mu}_{\mathrm{min}}$ then the transition cannot be observed in a Knight shift experiment. If $H^{\mu}_{\rm so} > H^{\mu}_{\rm min}$ but $H^{\mu}_{\rm so}$ is very close to $H^{\mu}_{\rm c2}$, the transition may not be resolvable due to a small susceptibility jump and the data would seem characteristic of a singlet response. The flop transition could be easily measured in the intermediate regime $H^{\mu}_{\rm min} \ll H^{\mu}_{\rm so} \ll H^{\mu}_{\rm c2}$.

A rough estimate of the flop transition field $H_{\rm so}$ can be obtained via a calculation similar to that performed for $\rm UPt_3$ [224]. In $\rm UPt_3$ an anisotropy inversion occurs between the upper critical fields $H^{a}_{\rm c2}$ and $H^{c}_{\rm c2}$ at $H_U \approx 1.9\,\rm T$ [225]. This anisotropy inversion suggests a strong spin-orbit coupling which locks the $\boldsymbol{d}$-vector to the $\boldsymbol{c}$ axis [224]. Noticing that the heaviest element in Bechgaard salts is selenium, thus the ratio between (atomic) spin-orbit couplings in $\rm (TMTSF)_2PF_6$ (for Se) and $\rm UPt_3$ (for U) is small: $r_{\rm so} = (Z_{\rm U}/Z_{\rm Se})^2 < 0.15$, where $Z_{\rm U}$ and $Z_{\rm Se}$ are the atomic numbers of uranium and selenium respectively. The $\boldsymbol{d}$-vector flop transition for $\rm (TMTSF)_2PF_6$ hence should occur at $H_{\rm so} = g_{\rm U}H_{\rm U}/g_{\rm Se}r_{\rm so}$, where $g_{\rm U}$ and $g_{\rm Se}$ are Lande g-factors. A theoretical estimate [37] gives $H_{\rm so} = 0.22\,\rm T$ for $\rm (TMTSF)_2PF_6$, which is much smaller than the anisotropy inversion field of $1.6\,\rm T$ [23]. Therefore, the physical origin of the H_{c_2} anisotropy inversion in $\rm (TMTSF)_2PF_6$ should not be the $\boldsymbol{d}$-vector flop transition, as suggested in the literature [34].

The analysis of DVS [36] and DCS [37] discussed above, lists all possibilities of orbital symmetry of the superconducting order parameter within an orthorhombic lattice structure, and distinguishes some candidates by quasiparticle density of states (QDOS) and uniform spin susceptibility calculations. Other thermodynamic quantities can be obtained theoretically for superconducting Bechgaard salts with various symmetries in an analogous way. For instance, Cherng and Sá de Melo [226] considered weak SOC triplet symmetries, p_x, p_y, p_z, and f_{xyz}, and calculated condensation energy, specific heat, and superfluid density for each candidate. This calculation can provide additional probe to distinguish superconducting symmetries when and if corresponding experiments become available.

Upon here, a careful reader must have raised the question regarding the lattice group $\rm G_c$ of the Bechgaard salt family. Indeed, as we stated in Chap. 2, the crystal structure of Bechgaard salts is truly triclinic with angles between crystallographic axes $\alpha = 83.39°$, $\beta = 86.27°$, and $\gamma = 71.01°$. The corresponding symmetry is represented by the $\rm C_i$ instead of the $\rm D_{2h}$ point group. Compared to $\rm D_{2h}$, $\rm C_i$ has much lower symmetry containing only two elements, the identity and inversion. Thus, the point group only differentiates between even and odd parity states, which can then be referred to as s-wave and p-wave states, respectively [210]. On the other hand, since the crystal has neither rotational nor reflecting symmetries, the superconducting states can no longer be categorized by their partial wave components. In other words, the terms of d-wave and s-wave for singlet pairing or f-wave and p-wave for triplet pairing are rather meaningless by assuming that the superconducting states do not break the symmetry of the underlying triclinic lattice.

With this assumption, Powell [210] conduct a group analysis for all possible symmetries of superconducting phases in Bechgaard salts, and comment on the compatibility with existing experiments for each candidate. In addition to superconducting states with even parity under frequency reversal, Powell [210] also include the discussion on the ones with odd frequency pairing, which are first proposed by Berezinskii in 1974 [227] but have not yet been confirmed experimentally in any systems. As the wave function of a fermionic system must be antisymmetric upon the exchange of all labels, the allowed states are then:

(i) even-frequency, s-wave, singlet;
(ii) even-frequency, p-wave, triplet;
(iii) odd-frequency, s-wave, triplet;
(iv) odd-frequency, p-wave, singlet.

For the singlet cases, the gap function of a translational invariant superconducting phase takes the form $\Delta(\boldsymbol{k}) = \Delta_s \psi_{\boldsymbol{k}}$, where the function $\psi_{\boldsymbol{k}}$ is the angular term with appropriate parity and Δ_s is the complex Ginzburg-Landau (GL) order parameter. Notice that there are only two symmetry distinct singlet paring states, including the conventional even-frequency, s-wave singlet and an exotic odd-frequency, p-wave singlet. For the triplet states, the gap function can be described by the $\boldsymbol{d}(\boldsymbol{k})$-vector as introduced in Eq.(5.108).

If the spin-orbit coupling is sufficiently weak, the spin and spatial degrees of freedom can be rotated independently, thus the $\boldsymbol{d}$-vector can be written as

$$\boldsymbol{d}(\boldsymbol{k}) = \Delta_t \phi_{\boldsymbol{k}}, \tag{5.125}$$

where Δ_t is the complex vector GL vector parameter and $\phi_{\boldsymbol{k}}$ is the parity determined angular-dependence. The GL free energy thus takes the following form,

$$\mathscr{F} = \mathscr{F}_n + \alpha|\Delta_t|^2 + \beta_1|\Delta_t|^4 + \beta_2|\Delta_t \cdot \Delta_t|^2. \tag{5.126}$$

The ground state for $\beta_2 > 0$ is $\Delta_t \propto (1, i, 0)$ and $\Delta_t \propto (1, 0, 0)$ up to arbitrary rotations in spin space. Powell [210] refers to these two possibilities as β-phase and polar phase, respectively, in analogy to the corresponding phases proposed in ^{3}He. The β-phase corresponds to pairing in a single spin channel, while the polar phase corresponds to pairing in both ESTP channels. Notice that both the β- and polar phases can either be even-frequency p-wave triplet or odd-frequency s-wave triplet states. Thus, there are four possible triplet states for weak SOC, as listed in Tab. 5.6.

Table 5.6 Summary of all eight symmetry distinct superconducting states allowed for a triclinic lattice. For triplet pairing states, the forms of $\boldsymbol{d}$-vector are also displayed where $\boldsymbol{\phi_{gk}}$ and $\boldsymbol{\phi_{uk}}$ are arbitrary functions with even and odd parities, respectively. [210].

State	Freq. Parity	SOC	$\boldsymbol{d}(\boldsymbol{k})$
s-singlet	Even	Any	N/A
β p-triplet	Even	Weak	$(1, i, 0)$
Polar p-triplet	Even	Weak	$(1, 0, 0)$
BW p-triplet	Even	Strong	$(A\phi^x_{uk}, B\phi^y_{uk}, C\phi^z_{uk})$
p-singlet	Odd	Any	N/A
β s-triplet	Odd	Weak	$(1, i, 0)$
Polar s-triplet	Odd	Weak	$(1, 0, 0)$
BW s-triplet	Odd	Strong	$(A\phi^x_{gk}, B\phi^y_{gk}, C\phi^z_{gk})$

If the spin-orbit coupling is strong, the rotations of the spin and spatial degrees of freedom are no longer independent. The $\boldsymbol{d}$-vector thus can be written as

$$\boldsymbol{d}(\boldsymbol{k}) = \Delta_t \left(A\phi^x_{\boldsymbol{k}} + B\phi^y_{\boldsymbol{k}} + C\phi^z_{\boldsymbol{k}}\right), \tag{5.127}$$

where $\phi^{i=x,y,z}_{\boldsymbol{k}}$ can be any functions that have the required parity for a translational invariant superconducting state. In this case, the prefactor Δ_t is the complex GL order parameter such that the GL free energy is

$$\mathscr{F} = \mathscr{F}_n + \alpha|\Delta_t|^2 + \beta|\Delta_t|^4. \tag{5.128}$$

Thus, there is only one symmetry distinct spin part of the wave function for triplet superconductivity with strong SOC. This symmetry is analogous to the BW phase in ^{3}He. Considering the frequency parity, two possibilities are present as an even-frequency p-wave triplet or an odd-frequency s-wave triplet state.

In Tab. 5.6, all eight symmetry distinct superconducting states for a triclinic lattice structure representing the Bechgaard salts family are listed [210]. Next, we analyze the properties of the eight superconducting states and compare with the known experimental facts. As there are apparent controversies for some aspects of superconducting states in $(TMTSF)_2PF_6$ and $(TMTSF)_2$ ClO_4, we do not assume *a priori* that the superconducting symmetries in these two sister compounds are the same.

Nodal structure- Since the extremely low symmetry of the C_i point group, none of the four even-frequency pairing states are required to have nodes, although they are perfectly allowed to do so. In fact, since the C_i group has only two elements of identity and inversion, the s-wave states can have finite gap everywhere while the p-wave states only need to vanish at the origin (inversion center). On the contrary, the four odd-frequency pairing states are intrinsically gapless, for the corresponding gap functions must change sign along the Fermi surfaces.

Experiments regarding the nodal structure, however, are not conclusive. While the thermal conductivity measurement on $(TMTSF)_2ClO_4$ suggests a nodeless gap [39], the more recent field-angle-resolved calorimetry experiment favors the existence of line nodes [40] in the same material. Meanwhile, the absence of Hebel-Slichter peak and the power-law temperature dependence of NMR relaxation rate T_1^{-1} could be suggestive of nodes in the gap at a first glance [13, 33, 35], but it can also be explained by effects of collective spin-wave [228] excitations in triplet superconductors with no nodes in presence.

Upper critical field- The Pauli paramagnetic limit occurs when the Zeeman energy penalty for forming $S_z = 0$ Cooper pairs cannot be compensated by the condensation energy gained by entering the superconducting state. Thus, this limit applies to singlet pairing state and to Cooper pairs in the $S_z = 0$ projection of a triplet state [229]. When SOC is weak, the spin part of the order parameter is free to rotate, such that a ground state with minimal magnetic energy is always favored with $\boldsymbol{d}(\boldsymbol{k})\perp\boldsymbol{H}$. Thus, the triplet phases for weak SOC will always be ESTP states in the lab frame in the presence of a magnetic field, hence are not Pauli limited. When SOC is strong, the spin axes are no longer free to rotate but are pinned to the sample axes. When a magnetic field exceeding the Pauli limit is applied to the system, it will completely suppress the pairing in the $S_z = 0$ channel, i.e., the projection of $\boldsymbol{d}(\boldsymbol{k})$ parallel to the field direction vanishes. However, the $\boldsymbol{d}(\boldsymbol{k})$-vector still have finite perpendicular components which are not affected by the field. Both the triplet states for strong SOC (see Tab. 5.6) can have finite perpendicular components to the conducting plane and hence are not Pauli limited.

Time-reversal symmetry. The even-frequency and odd-frequency pairing β phases obviously break the time reversal symmetry (TRS), hence are inconsistent with the zero-field-μSR measurements by Luke *et al.* [140], who find no signature for the broken of time reversal symmetry.

Spin susceptibility. The spin susceptibility χ_s strongly distinguishes between different pairing states. When a Cooper pair of $S_z = 0$ forms, it no longer contributes to the spin susceptibility. Hence, we expect that $\chi_\mathrm{s} \to 0$ as $T \to 0$ for a singlet superconductor or a triplet superconductor with $\boldsymbol{d}(\boldsymbol{k}) \parallel \boldsymbol{H}$. In contrast, the ESTP state does not affect the spin susceptibility. For a general triplet state which contains both ESTP and $S_z = 0$ components, the decrease in χ_s is proportional to the fraction of the $S_z = 0$ part. In the case of weak SOC, since $\boldsymbol{d}$-vectors are free to rotate such that an ESTP state is always favorable within magnetic fields, we expect $\chi_\mathrm{s}(T)/\chi_\mathrm{n} = 1$ for all $T < T_\mathrm{c}$, where χ_n is the normal-state spin susceptibility. When SOC is strong, the spin susceptibility is suppressed as

$$\frac{\chi_\mathrm{s}^i(T=0)}{\chi_\mathrm{n}} = 1 - \frac{\langle|\phi_{\boldsymbol{k}}^i|^2\rangle_\mathrm{FS}}{\langle|\phi_{\boldsymbol{k}}^x|^2 + |\phi_{\boldsymbol{k}}^y|^2 + |\phi_{\boldsymbol{k}}^z|^2\rangle_\mathrm{FS}}, \tag{5.129}$$

where $\langle\cdots\rangle_\mathrm{FS}$ denotes average over the Fermi surface, and $i = x, y, z$ indicates the direction of the applied field. If the averages over the three basis functions are the same, as they are in the BW phase of ^{3}He, then we would expect $\chi_\mathrm{s}(0)/\chi_\mathrm{n} = 2/3$ for all field directions.

Experiments on ^{77}Se Knight shift find no detectable decrease in spin susceptibility in superconducting $(\mathrm{TMTSF})_2\mathrm{PF}_6$ with a rather high field applied along either x [35] or y [33] directions. These results suggest $\chi_\mathrm{s}^x(T)/\chi_\mathrm{n} = \chi_\mathrm{s}^y(T)/\chi_\mathrm{n} = 1$, hence is compatible with all triplet pairing states in the case of weak SOC or BW phases with $\phi_{\boldsymbol{k}}^x = \phi_{\boldsymbol{k}}^y = 0$ along the entire Fermi surface in the case of strong SOC. Meanwhile, the same experiments on $(\mathrm{TMTSF})_2\mathrm{ClO}_4$ find a decrease of χ_s for $\boldsymbol{H} \parallel \boldsymbol{a}$ and $\boldsymbol{H} \parallel \boldsymbol{b}'$ at a much smaller magnetic field [38], hence suggest singlet pairing states or BW triplet phases with finite $\phi_{\boldsymbol{k}}^x = \phi_{\boldsymbol{k}}^y = 0$ in the case of strong SOC.

Table 5.7 Properties of all eight symmetry distinct superconducting states allowed for a triclinic lattice, as compared with existing experimental indications [210].

State	Requiring nodes	Breaking TRS	Pauli limited $H_{c_2}^a$	Pauli limited $H_{c_2}^{b'}$	$\frac{\chi_x^a(0)}{\chi_n}$	$\frac{\chi_x^{b'}(0)}{\chi_n}$	Non-mag. disorder suppresses T_c	Hebel-Slichter peak
s-singlet	No	No	Yes	Yes	0	0	No	Yes
β p-triplet	No	Yes	No	No	1	1	Yes	No
Polar p-triplet	No	No	No	No	1	1	Yes	No
BW p-triplet	No	No	No	No	$\leqslant 1$	$\leqslant 1$	Yes	No
p-singlet	Yes	No	Yes	Yes	0	0	Yes	No
β s-triplet	Yes	Yes	No	No	1	1	Yes	No
Polar s-triplet	Yes	No	No	No	1	1	Yes	No
BW s-triplet	Yes	No	No	No	$\leqslant 1$	$\leqslant 1$	Yes	No
$(TMTSF)_2PF_6$	No [230]	N/A	No [23]	No [23]	1* [35]	1* [33]	Yes [2, 4]	No [35, 33]
$(TMTSF)_2ClO_4$	Yes [40] or No [39]	No [140]	No [139]	No [138, 139]	$< 1^{\dagger}$ [38]	$< 1^{\dagger}$ [38]	Yes [136, 137]	No [13, 38]

* At high magnetic fields.

† At low magnetic fields.

In Tab. 5.7, we summarize the properties of all eight symmetry distinct superconducting states as listed in Tab. 5.6, while corresponding experimental suggestions are also shown for comparison. Notice that the assumption for the group analysis discussed in the section is that the superconducting phase does not break the underlying symmetry of the crystal lattice. Hence, a spatially modulated LOFF state is not considered as a possibility.

6
Inhomogeneous Pairing States

In the previous chapter, we have devoted much effort along the line to understand the experimental observation regarding the superconducting phase in Bechgaard salts and to unveil the symmetry of the order parameter. However, after going through the existing choices listed before, we have to admit that this long-lived question is not yet at the stage of reaching a satisfactory answer. Among the symmetry distinct states exhausted by the group analysis discussed in Sec. 5.7 for either an orthorhombic [36, 37] or a triclinic [210] lattice, one cannot single out a specific symmetry which can satisfy all experiments without any uncertainty. This is not even the whole list of candidates. In fact, the assumption of the group analysis is that the superconducting phase does not spontaneously break translational symmetry. Thus, all superconducting states we have investigated in the previous chapter are homogeneous in real space.

The study of possible inhomogeneous superconducting states has a long history since its original proposal by Larkin and Ovchinnikov [30] and Fulde and Ferrell [31]. Proposed as a compromise of superconductivity with finite magnetization, the so-called LOFF (or also widely used FFLO) state breaks both the translational and rotational symmetries by acquiring a spatial oscillation with finite wave vectors. In the most general case, a spatial-dependent superconducting order parameter can be expanded in momentum space as

$$\Delta(\boldsymbol{r}) = \sum_{\boldsymbol{k}} \Delta_{\boldsymbol{k}} e^{i\boldsymbol{k}\cdot\boldsymbol{r}}. \tag{6.1}$$

For order parameter in a BCS superconductor, only the zero-momentum component is nonzero, i.e.,

$$\Delta_{\boldsymbol{k}} = \Delta_0 \delta(\boldsymbol{k}), \tag{6.2}$$

where $\delta(\boldsymbol{k})$ is the three-dimensional delta function. This corresponds to the case where the two composite fermions within a Cooper pair have opposite momenta, hence add to a zero center-of-mass momentum. On the contrary, the LOFF state acquires an order parameter with finite momentum. A more careful definition should distinguish the LO and FF states, where

$$\Delta_{\boldsymbol{k}} = \Delta_{\boldsymbol{k}_0}\delta(\boldsymbol{k}-\boldsymbol{k}_0) \quad \text{(FF state)}, \tag{6.3a}$$

$$\Delta_{\boldsymbol{k}} = \Delta_{\boldsymbol{k}_0}\delta(\boldsymbol{k}-\boldsymbol{k}_0) \pm \Delta_{\boldsymbol{k}_0}\delta(\boldsymbol{k}+\boldsymbol{k}_0) \quad \text{(LO state)}. \tag{6.3b}$$

That is, the FF state has only spatial variation within the phase, while keeping the amplitude of the order parameter unchanged. Meanwhile, the order parameter has amplitude modulation in the LO state, and most importantly, presents gapless nodes in real space.

In this chapter, we investigate the possibility of a LOFF superconducting state in Bechgaard salts, especially in the presence of a high magnetic field. Since the LOFF state can accommodate finite magnetization, it is expected that such a state can sustain higher magnetic field than the conventional s-wave pairing state. In Sec. 6.1, we discuss the upper critical field of a LOFF state emerging from singlet pairing states, for magnetic field applied within the $\boldsymbol{a}$-$\boldsymbol{b}'$ plane. It is shown that both the excess of Pauli paramagnetic limit [23, 138, 139] and the in-plane anisotropy [139] can be understood within this framework at least in a qualitative level. Another possibility for a LOFF state is a mixing of singlet and triplet pairing symmetries. Discussion along this line has caught equal amount of attention, and will be introduced in Sec. 6.2. Within this picture, triplet pairing component will set in the singlet pairing background in hight magnetic fields, forming a singlet-triplet mixing LOFF state which can survive high magnetic field exceeding the Pauli limit. As before, we present both scenarios with an equal footing and leave the readers to reach their own judgement.

In addition to the LOFF state, where the inhomogeneity is induced by the presence of a magnetic field, the superconducting phase also becomes inhomogeneous when approaching the SDW phase, as observed in various experiments in $(\mathrm{TMTSF})_2\mathrm{PF}_6$ (see detailed discussion in Sec. 3.3) [76, 77, 78, 79]. In Sec. 6.3, we discuss theoretical proposals trying to understand the microscopic coexistence of SDW and superconductivity. In particular, Zhang and Sá de Melo [121, 122] find that the two orders indeed can coexist in a phase diagram containing a tetracritical point, and the superconducting phase becomes inhomogeneous induced by the intrinsic modulation of the SDW order. On the other hand, Gorkov and Grigoriev [231, 232] propose a soliton phase within the SDW regime where triplet superconductivity can emerge in the soliton wall. Both of the theories are qualitatively compatible with the observation of SDW-SC coexistence in a microscopic scale.

6.1 Upper critical field of a singlet LOFF state

After the NMR experiment in $(TMTSF)_2ClO_4$ which suggests a singlet pairing state at least at low magnetic fields [38] (see Sec. 4.4), a considerable amount of effort has been devoted to reconcile the singlet pairing scenario with the unusually high upper critical field observed in the same material [139]. Along this line, the possibility of an inhomogeneous superconducting, i.e., a LOFF state, draws much attention. In particular, Lebed [155, 233], and Lebed and Wu [154] studied the possibility of an inhomogeneous singlet d-wave-like nodal scenario and concluded that this pairing sate is consistent with the experimental observation of $H_{c_2}^{b'}, H_{c_2}^{a} \gtrsim 5\,\mathrm{T}$ [139]. Meanwhile, Croitoru, Houzet, and Buzdin (CHB) [156] considered a spatially modulated superconducting phase in layered s-wave superconductors to explain the angular dependence of H_{c_2} for magnetic field applied within the x-y plane [139]. In this section, we will discuss these proposals and investigate the upper critical field of a singlet LOFF state.

To study the effect of a magnetic field applied along the $\boldsymbol{b}'$ direction, Lebed [155, 233] employs the standard quasi-1D spectrum as in Eq.(2.1), but linearizes the spectrum near the two corrugated sheets of Fermi surfaces in a slightly different form compared to Eq.(2.2), leading to

$$\varepsilon_\alpha(\boldsymbol{k}) = v_{\mathrm{F}}(k_y)\left[\alpha k_x - k_{\mathrm{F}}(k_y)\right] - t_z \cos(k_z c). \tag{6.4}$$

Here, $\alpha = \pm$ labels the two sheets of the open Fermi surface as before, and the k_y-dependence is included in the momentum-dependent Fermi velocity $v_{\mathrm{F}}(k_y)$ and Fermi momentum $k_{\mathrm{F}}(k_y)$.

The real-space electronic wavefunction with eigenenergy E can be expressed in the following way

$$\Psi_\alpha(E, x, y, z, \sigma) = e^{ik_{\mathrm{F}}(k_y)x} e^{ik_y y} e^{ik_z z} \psi_\alpha(E, x, k_y, k_z, \sigma), \tag{6.5}$$

where $\sigma = \pm 1$ stands for spin up and spin down, respectively. By using the Peierls substitution with Landau gauge $\boldsymbol{A} = (0, 0, -Hx)$

$$\alpha k_{\mathrm{F}}(k_y) - k_x \to i\frac{d}{dx}, \quad k_z \to k_z - eA_z, \tag{6.6}$$

one can write down the Schrödinger's equation for the wavefunction ansatz, and obtain the corresponding solution as

$$\psi_\alpha(E, x, k_y, k_z, \sigma) = \frac{1}{\sqrt{2\pi v_{\mathrm{F}}(k_y)}} \exp\left[\frac{i\alpha(E - E_{\mathrm{F}})x}{v_{\mathrm{F}}(k_y)}\right] \exp\left[\frac{i\alpha\sigma\mu_{\mathrm{B}} Hx}{v_{\mathrm{F}}(k_y)}\right] \times \exp\left[\frac{i\alpha t_z}{v_{\mathrm{F}}(k_y)} \int_0^x \cos\left(k_z c + \frac{\omega_{\mathrm{c}}}{v_{\mathrm{F}}(0)} u\right) du\right]. \tag{6.7}$$

Here, μ_{B} is the Bohr magneton, and $\omega_{\mathrm{c}} = |e| H v_{\mathrm{F}}(0) c$ is the cyclotron frequency. The corresponding finite temperature Green's function can be obtained from the following equation

$$G_{i\omega_n}(x, x', k_y, k_z, \sigma) = \sum_E \frac{\psi^*(E, x, k_y, k_z, \sigma)\psi(E, x', k_y, k_z, \sigma)}{i\omega_n - E}, \tag{6.8}$$

leading to the result

$$G_{i\omega_n}(x, x', k_y, k_z, \sigma) = \frac{-i\mathrm{sgn}(\omega_n)}{v_{\mathrm{F}}(k_y)} \exp\left[-\frac{\alpha\omega_n(x - x')}{v_{\mathrm{F}}(k_y)}\right] \exp\left[\frac{i\alpha\sigma\mu_{\mathrm{B}} H(x - x')}{v_{\mathrm{F}}(k_y)}\right] \times \exp\left[\frac{i\alpha t_z}{v_{\mathrm{F}}(k_y)} \int_{x'}^x \cos\left(k_z c + \frac{\omega_{\mathrm{c}}}{v_{\mathrm{F}}(0)} u\right) du\right], \tag{6.9}$$

where ω_n is the Matsubara frequency. Notice that the Green's function in the form above acquires an explicit dependence on the momentum component k_y, via the Fermi velocity $v_{\mathrm{F}}(k_y)$ along conducting chains.

The singlet pairing order parameter is assumed to be in the following form

$$\Delta(x, k_y) = f(k_y b)\Delta(x), \tag{6.10}$$

where different orbital symmetries can be represented by different forms of function $f(s)$ with the normalization relation.

$$\int_0^{2\pi} f^2(s) ds = 2\pi. \tag{6.11}$$

In Lebed's discussion [155, 233], three possible order parameters are considered:

(i) s-wave, $f(s) = 1$;

(ii) d-wave-like nodal, $f(s) = \sqrt{2}\cos(s)$;

(iii) d-wave-like nodeless, $f(s) = \Theta(|s-\pi| - \pi/2)$ with $\Theta(s)$ the Heaviside function.

The linearized gap equation for all three possible singlet scenarios can be obtained using the Gorkov equation for non-uniform superconductivity, leading to

$$\begin{aligned}\Delta(x) = \tilde{g} \int \frac{dk_y}{v_{\mathrm{F}}(k_y)} \int_{|x-x'|>v_{\mathrm{F}}(k_y)/\Omega} \frac{2\pi T dx'}{v_{\mathrm{F}}(k_y)\sinh[2\pi T|x-x'|/v_{\mathrm{F}}(k_y)]} \\ \times J_0 \left\{ \frac{4t_z v_{\mathrm{F}}(0)}{\omega_{\mathrm{c}} v_{\mathrm{F}}(k_y)} \sin\left[\frac{\omega_{\mathrm{c}}(x-x')}{2v_{\mathrm{F}}(0)}\right] \sin\left[\frac{\omega_{\mathrm{c}}(x+x')}{2v_{\mathrm{F}}(0)}\right]\right\} \\ \times \cos\left[\frac{2\tilde{\beta}\mu_{\mathrm{B}} H(x-x')}{v_{\mathrm{F}}(k_y)}\right] f^2(k_y b)\Delta(x'),\end{aligned} \tag{6.12}$$

where $\tilde{g}$ is the effective electron-electron coupling constant, Ω is the cutoff energy, T is the temperature, J_0 is the zeroth order Bessel function, and $\tilde{\beta}$ is a phenomenological parameter assigned to represent possible deviations of the superconducting state from weak coupling limit. Notice that the term associated with J_0 describes the orbital effects against superconductivity in the presence of a magnetic field, and the term associated with $\tilde{\beta}$ resembles the paramagnetic effect.

As discussed in Chap. 2, the quasi-1D nature of Bechgaard salts provides an intriguing effect of field-induced dimensional crossover (FIDC) in the presence of a high magnetic field. Using the semiclassical approach outlined in Chap. 2, the characteristic length associated with electronic trajectory along the z direction at $H = 6\,\mathrm{T}$ reads

$$z_A = t_z c/\omega_{\mathrm{c}} \approx 0.48c \tag{6.13}$$

for d-wave-like nodal scenario of superconductivity, and

$$z_A = t_z c/\omega_{\mathrm{c}} \approx 0.68c \tag{6.14}$$

for d-wave-like nodeless or s-wave order parameters. These values suggest that both cases correspond to an emergence of FIDC with electronic motion is essentially confined within the x-y conducting layers. In this condition, superconductivity becomes almost 2D in nature and the Bessel function in Eq.(6.12) can be approximated as $J_0(z) \approx 1 - z^2/4$.

At zero temperature and within the quasi-2D condition $z_A < 1$, Lebed [155, 233] considers an order parameter ansatz

$$\Delta(x) = e^{ikx}\left[1 + \alpha_1 \cos\left(\frac{2\omega_c x}{v_F(0)}\right) + \alpha_2 \sin\left(\frac{2\omega_c x}{v_F(0)}\right)\right], \tag{6.15}$$

which corresponds to a LOFF state with both phase modulation (with wave vector k) and amplitude variation (up to first order expansion with coefficients $|\alpha_1|, |\alpha_2| \ll 1$). For such an order parameter, the gap equation (6.12) can be simplified under the condition $t_y/t_x \ll 1$, leading to

$$\frac{1}{g} = \int_{v_F(0)/\Omega}^{\infty} \frac{dz}{z} \cos\left(\frac{2\tilde{\beta}\mu_B H z}{v_F(0)}\right) \cos(kz)\left[J_0(\sqrt{2}t_y kz/t_z) - mJ_2(\sqrt{2}t_y kz/t_z)\right]$$
$$\times\left[1 - 2\frac{t_z^2}{\omega_c^2}\sin^2\left(\frac{\omega_c z}{2v_F(0)}\right)\right], \tag{6.16}$$

where g is the renormalized coupling constant, $m = 1$ for d-wave-like nodal superconducting order parameter, $m = 0$ for d-wave-like nodeless and s-wave scenarios, and $J_2(x)$ is the second order Bessel function. Notice that in the absence of paramagnetic effects (i.e., at $\tilde{\beta} = 0$), Eq.(6.16) can also describe the reentrant superconductivity with transition temperature being increased with a magnetic field, as discussed in Secs. 5.1 and 5.2.

The gap equation (6.16) can be further simplified by utilizing the following relation

$$\frac{1}{g} = \int_{v_F(0)/\Omega}^{\infty} \frac{2\pi T_c(0)dz}{v_F(0)\sinh\left[\frac{2\pi T_c(0)z}{v_F(0)}\right]}, \tag{6.17}$$

where $T_c(0)$ is the superconducting transition temperature in the absence of a magnetic field. As a result, we obtain

$$\ln\left(\frac{H_{c2}^{b'}}{H_*}\right) = \int_0^{\infty} \frac{dz}{z} \cos\left(\frac{2\tilde{\beta}\mu_B H z}{v_F(0)}\right) \times \left\{\cos(kz)\left[J_0(\sqrt{2}t_y kz/t_z)\right.\right.$$
$$\left.\left. - mJ_2(\sqrt{2}t_y kz/t_z)\right]\left[1 - 2\frac{t_z^2}{\omega_c^2}\sin^2\left(\frac{\omega_c z}{2v_F(0)}\right)\right] - 1\right\}, \tag{6.18}$$

where $H_* = \pi T_c(0)/2\mu_B\gamma$ with γ the Euler constant. By evaluating $H_{c_2}^{b'}$ as a function of k numerically, Lebed [155, 233] finds the optimal wave vector of the order parameter equation (6.15) for both $m = 1$ and $m = 0$. The results show that an upper critical field as high as $H_{c_2}^{b'} = 6$T can be obtained for d-wave-like nodal order parameter at $\tilde{\beta} \approx 0.85$, while for d-wave-like nodeless and s-wave order parameters at $\tilde{\beta} \approx 0.5$. Considering the fact that the g-factor in Bechgaard salts is very close to the free electron value $g = 2$, which corresponds to $\tilde{\beta} = 1$, the discussion by Lebed [155, 233] suggests that the d-wave-like nodal order parameter is much more consistent with the experimental data with $\tilde{\beta} \approx 0.85$ closer to the expected value $\tilde{\beta} = 1$.

Within the same scheme, Lebed and Wu [154] also study the upper critical field for magnetic field applied along the $\boldsymbol{a}$ axis, where only the d-wave-like nodal order parameter is considered

$$\Delta(x, y, k_y) = \sqrt{2}\cos(k_y b) e^{iqx} \Delta_q(y). \tag{6.19}$$

Here, the Cooper pairs are characterized by nonzero center-of-mass momentum $q \neq 0$ along the $\boldsymbol{a}$ axis, and $\Delta_q(y)$ is the spatially dependent order parameter. The gap equation thus reads the following form

$$\begin{aligned}\Delta_q(y) = \tilde{g} \int \frac{dk_y}{v_F(k_y)} \int_{|y-y'|>v_y(k_y)/\Omega} & \frac{2\pi T dy'}{v_y(k_y)\sinh\left[\frac{2\pi T|y-y'|}{v_y(k_y)}\right]} 2\cos^2(k_y b) \\ \times \cos & \left[\frac{2\tilde{\beta}\mu_B H(y-y')}{v_y(k_y)}\right] \cos\left[\frac{q v_y(k_y)(y-y')}{v_y(k_y)}\right] \\ \times J_0 & \left\{\frac{4t_z v_F(0)}{\omega_c v_y(k_y)} \sin\left[\frac{\omega_c(y-y')}{2v_F(0)}\right] \sin\left[\frac{\omega_c(y+y')}{2v_F(0)}\right]\right\} \Delta_q(y'), \end{aligned} \tag{6.20}$$

where $v_y(k_y) = t_y b \sin(k_y b)$ is the y-component of the Fermi velocity. For magnetic field in the range of ~ 1 T, the arguments of the sine functions in J_0 is small, such that the following approximation can be employed

$$\sin\left[\frac{\omega_c(y-y')}{2v_F(0)}\right] \sin\left[\frac{\omega_c(y+y')}{2v_F(0)}\right] \approx \frac{\omega_c^2[y^2 - (y')^2]}{4v_F^2(0)}. \tag{6.21}$$

Thus, the gap equation can be further simplified as

$$\begin{aligned}\Delta_q(y)=&\tilde{g}\int\frac{dk_y}{v_{\mathrm{F}}(k_y)}\int_{|z|>v_{\mathrm{F}}(0)/\Omega}\frac{2\pi T dz}{v_{\mathrm{F}}(0)\sinh(2\pi Tz/v_{\mathrm{F}}(0))}2\cos^2(k_y b)\\&\times\cos\left[\frac{2\tilde{\beta}\mu_{\mathrm{B}}Hz}{v_{\mathrm{F}}(0)}\right]\cos\left[\frac{qv_{\mathrm{F}}(k_y)z}{v_{\mathrm{F}}(0)}\right]\\&\times J_0\left\{\frac{t_z\omega_{\mathrm{c}}z}{v_{\mathrm{F}}^2(0)}\left[2y+\frac{v_{\mathrm{F}}(k_y)z}{v_{\mathrm{F}}(0)}\right]\right\}\Delta_q\left(y+\frac{v_{\mathrm{F}}(k_y)z}{v_{\mathrm{F}}(0)}\right).\end{aligned}\tag{6.22}$$

In Fig. 6.1, the numerical solution of the gap equation (6.22) for $\tilde{\beta}=0.9$ is shown, together with the experimental data (dots) from Ref. [139]. Here, the band parameters of $t_x = 2680\,\mathrm{K}$, $t_y = 268\,\mathrm{K}$, and $t_z = 5.2\,\mathrm{K}$ are used [154]. Notice that a fairly good agreement with experiments can be obtained for $\tilde{\beta}$ close to 1, indicating small deviations from a weak-coupling scenario of superconductivity. Thus, the picture of a LOFF phase arising from d-wave pairing states can be used to explain the excess of Pauli paramagnetic limit for field applied along both the $\boldsymbol{a}$ and $\boldsymbol{b}'$ directions.

In addition to the excess of Pauli limit, the upper critical field measurement also reveals an anomalous angular dependence for field applied within the $\boldsymbol{a}$-$\boldsymbol{b}'$ plane [139]. In the original experimental paper, Yonezawa *et al.* interpret

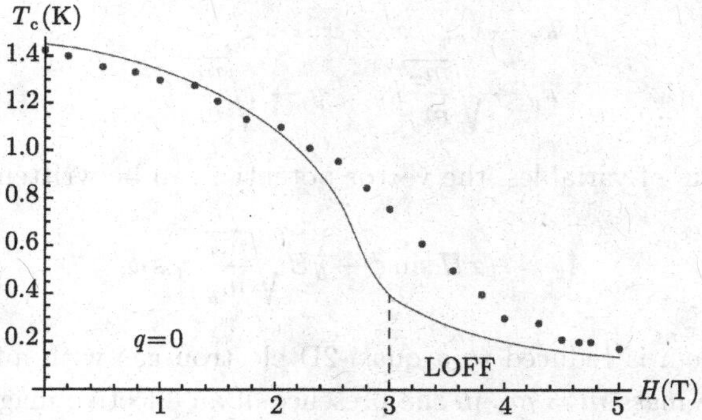

Figure 6.1 Superconducting transition temperature $T_{\mathrm{c}}(H)$ obtained by numerically solving the gap equation (6.22) (solid line). Notice that a phase transition from uniform superconductivity with center-of-mass momentum $q=0$ to the LOFF state with $q\neq 0$ is predicted in a magnetic field $H\approx 3\,\mathrm{T}$. Experimental results from Ref. [139] are included to show a fairly good agreement. From Ref. [154].

the in-plane angular dependence in the framework of a LOFF state sitting in a quasi-2D conducting layer. Motivated by these experimental findings, Croitoru, Houzet and Buzdin (CHB) analyze the in-plane anisotropy of the upper critical field of a spatially modulated superconducting phase in layered superconductors with s-wave singlet pairing [156].

In the theoretical work by CHB, the single electron spectrum of Bechgaard salts is approximated by that of a stack of two-dimensional electron layers

$$\varepsilon(\boldsymbol{k}) = \frac{k_x^2}{2m_x} + \frac{k_y^2}{2m_y} - t_z \cos(k_z c). \tag{6.23}$$

The in-plane dispersion is described within the effective mass approximation while the motion along the z direction is characterized by the tight-binding model. Under the symmetric gauge, the vector potential takes the form $\boldsymbol{A} = \boldsymbol{H} \times (x, y, 0) = (0, 0, A_z)$ with

$$A_z = -xH \sin\phi + yH \cos\phi, \tag{6.24}$$

where ϕ is the angle between the applied field and the $\boldsymbol{a}$ axis. The anisotropy of the dispersion relation Eq.(6.23) can be removed by employing a scale transformation [234]

$$\begin{aligned} &k_x \to k_x, \quad x \to x; \\ &k_y \to \sqrt{\frac{m_x}{m_y}} k_y, \quad y \to \sqrt{\frac{m_y}{m_x}} y. \end{aligned} \tag{6.25}$$

In this new sets of variables, the vector potential can be written as

$$A_z = -xH \sin\phi + yH \sqrt{\frac{m_x}{m_y}} \cos\phi, \tag{6.26}$$

and the problem is reduced to a quasi-2D electron gas with an isotropic in-plane effective mass $m = m_x$ in the presence of an effective magnetic field

$$\tilde{\boldsymbol{H}} = H \left(\sqrt{\frac{m_x}{m_y}} \cos\phi, \sin\phi, 0 \right). \tag{6.27}$$

In this case, the orientation of the LOFF wave vector $\boldsymbol{q}$ is free to rotate within the x-y plane. In realistic situation where the Fermi surface is not fully elliptical as approximated in the effective mass model, or in the presence of a spin-orbit coupling where spin and spatial axes are not free to rotate, the direction of the wave vector $\boldsymbol{q}$ will be fixed along an easy axis with modified amplitude. However, in the discussion of CHB [156], these effects are assumed to be small and their sole effect is to pin the direction of the vector $\boldsymbol{q}$ to be along an arbitrary, say the x-, direction.

Having in mind of the upper critical field $H_{c2}(T)$, it is proper to assume the superconducting order parameter to be small near the second-order phase transition line. Thus, the linearized Eilenberger equation for the anomalous Green's function $f(\omega_n, \hat{\boldsymbol{n}}, \boldsymbol{r}, k_z) = \Delta(\boldsymbol{r})$ can be written as

$$\left[\omega_n + i\mu_B H + \frac{\boldsymbol{v}_F \cdot \nabla}{2} + it_z \sin(k_z c)\sin(\boldsymbol{G}\cdot\boldsymbol{r})\right] f(\omega_n, \hat{\boldsymbol{n}}, \boldsymbol{r}, k_z) = \Delta(\boldsymbol{r}), \quad (6.28)$$

where ω_n is the positive Matsubara frequency, $\hat{\boldsymbol{n}} \parallel \boldsymbol{v}_F$ is the unit vector along the in-plane Fermi velocity, $\boldsymbol{r} = (x, y, 0)$ denotes the in-plane coordinate, and

$$\boldsymbol{G} = |e|Hc\left(-\sin\phi, \sqrt{m_x/m_y}\cos\phi, 0\right). \quad (6.29)$$

The order parameter is defined self-consistently as

$$\Delta(\boldsymbol{r}) = 2\pi T\tilde{g}\mathcal{N}_0 \,\mathrm{Re} \sum_{\omega_n > 0} \langle f(\omega_n, \hat{\boldsymbol{n}}, \boldsymbol{r}, k_z)\rangle, \quad (6.30)$$

where $\tilde{g}$ is the coupling constant, $\mathcal{N}_0$ is the density of states at the Fermi surface, and the bracket denotes averaging over $\hat{\boldsymbol{n}}$ and k_z. Here, the in-plane superconducting layer is assumed to be in the clean limit, with a mean free path much larger than the corresponding coherence length.

The solution of the Eilenberger equation (6.28) can be expanded in a basis of Bloch functions

$$f(\omega_n, \hat{\boldsymbol{n}}, \boldsymbol{r}, k_z) = e^{i\boldsymbol{q}\cdot\boldsymbol{r}} \sum_{\ell} e^{i\ell\boldsymbol{G}\cdot\boldsymbol{r}} f_\ell(\omega_n, \hat{\boldsymbol{n}}, k_z), \quad (6.31)$$

where ℓ are integers and $\boldsymbol{q}$ denotes the center-of-mass momentum of Cooper pairs. The order parameter $\Delta(\boldsymbol{r})$ can also be represented as

$$\Delta(\boldsymbol{r}) = e^{i\boldsymbol{q}\cdot\boldsymbol{r}} \sum_{\ell} e^{i2\ell\boldsymbol{G}\cdot\boldsymbol{r}} \Delta_{2\ell}. \quad (6.32)$$

Considering the purpose of investigating the upper critical field, the magnetic field $\boldsymbol{H}$ is taken to be in the high field regime, with

$$v_{\mathrm{F}}G = v_{\mathrm{F}}|e|Hc \gtrsim T_{\mathrm{c}}(H=0). \tag{6.33}$$

Besides, CHB [156] also assume that the inter-layer coupling t_z is small compared to $T_{\mathrm{c}}(H=0)$, i.e., $t_z \ll T_{\mathrm{c}}(0)$. Thus, one can reach the condition $\sqrt{t_z T_{\mathrm{c}}(0)} \ll v_{\mathrm{F}}G$. This condition allows us to consider only the terms up to the first harmonics in expansions of Eqs. (6.31) and (6.32), since all higher harmonics correspond to higher order terms of $(t_z/T_{\mathrm{c}}(0))$ after substituting into the Eilenberger equation (6.28). The resulting equations to this order read

$$\mathscr{L}(\boldsymbol{q})f_0 + \frac{t_z \sin(k_z c)}{2}(f_1 - f_{-1}) = \Delta_0, \tag{6.34a}$$

$$\mathscr{L}(\boldsymbol{q}\pm\boldsymbol{G})f_{\pm 1} \mp \frac{t_z \sin(k_z c)}{2} f_0 = 0, \tag{6.34b}$$

where $\mathscr{L}(\boldsymbol{q}) \equiv \omega_n + i\mu_{\mathrm{B}}H + iv_{\mathrm{F}}\cdot\boldsymbol{q}/2$. Notice that the equations above can describe the reentrant superconducting phase by neglecting the Zeeman term $\mu_{\mathrm{B}}H$ [16, 18, 235]. While keeping the terms up to second harmonics in Eqs. (6.31) and (6.32), the same procedure would yield the Lawrence-Doniach equation.

Inserting the solution of Eqs. (6.34) into the self-consistency equation (6.30), keeping only the terms up to the second order of $(t_z/T_{\mathrm{c}}(0))$, and subtracting with the gap equation for the case of zero magnetic field, one can obtain the following equation

$$\ln\frac{T_{\mathrm{c}}(H=0)}{T} = 2\pi T\ \mathrm{Re}\sum_{\omega_n>0}\Bigg\{\omega_n^{-1} - \langle\mathscr{L}^{-1}(\boldsymbol{q})\rangle$$

$$+\left\langle\frac{t_z^2\sin^2(k_z c)}{4\mathscr{L}^2(\boldsymbol{q})}\left(\mathscr{L}^{-1}(\boldsymbol{q}+\boldsymbol{G}) + \mathscr{L}^{-1}(\boldsymbol{q}-\boldsymbol{G})\right)\right\rangle\Bigg\}. \tag{6.35}$$

In the limit of $t_z \ll T_{\mathrm{c}}(0)$, the orbital effect can be neglected from the equation above, leading to the following reduced form after averaging over the Fermi surface

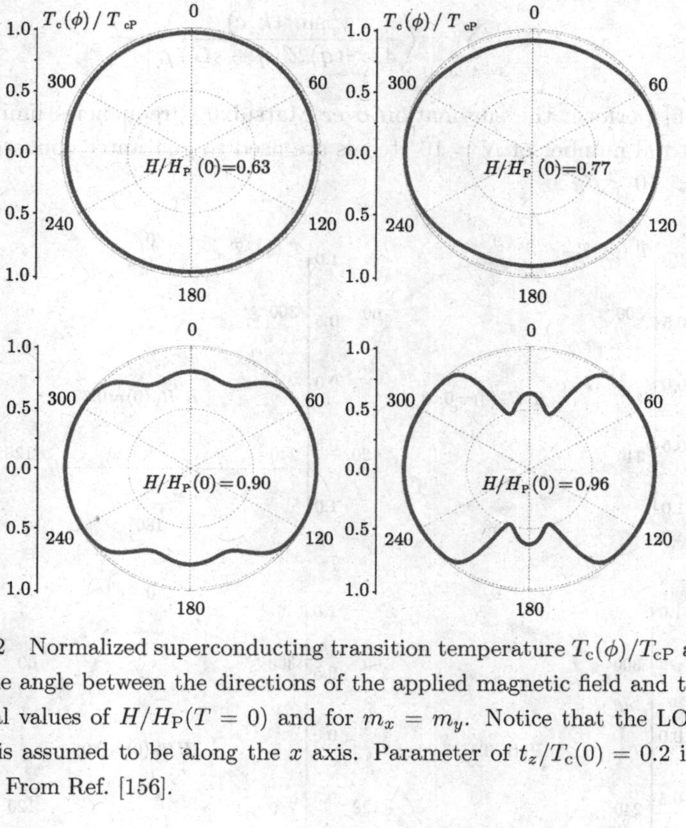

Figure 6.2 Normalized superconducting transition temperature $T_c(\phi)/T_{cP}$ as a function of the angle between the directions of the applied magnetic field and the x axis for several values of $H/H_P(T=0)$ and for $m_x = m_y$. Notice that the LOFF wave vector $\boldsymbol{q}$ is assumed to be along the x axis. Parameter of $t_z/T_c(0) = 0.2$ is used in this plot. From Ref. [156].

$$\ln \frac{T_c(0)}{T} = F(\tilde{h}, \tilde{q})$$
$$\equiv \sum_{n=0}^{\infty} \left\{ \left(n + \frac{1}{2}\right)^{-1} - \left[\left(n + \frac{1}{2} + i\tilde{h}\right)^2 + \frac{v_F^2 \tilde{q}^2}{4} \right]^{-1/2} \right\}, \quad (6.36)$$

where the reduced variables are $\tilde{h} = \mu_B H/2\pi T$ and $\tilde{q} = q/2\pi T$. By optimizing $\tilde{q}$ to reach a maximal value of transition temperature, one can define $T_{cP}(H)$ and $q_P(H)$ in the Pauli limit. Using these definitions, Eq.(6.35) can be written as

$$\frac{T_{cP} - T_c}{T_c} = \frac{2\pi T}{1 - \tilde{h}\partial F(\tilde{h}, \tilde{q})/\partial \tilde{h}}$$

$$\times \operatorname{Re} \sum_{s=\pm,\omega>0} \left\langle \frac{t_z^2 \sin^2(k_z c)}{4\mathscr{L}^2(\boldsymbol{q})\mathscr{L}(\boldsymbol{q}+s\boldsymbol{G})} \right\rangle \Bigg|_{T=T_{\mathrm{cP}}} . \tag{6.37}$$

CHB [156] perform the summation over Matsubara frequencies numerically, where a total number of $N = 10^4$ terms are used to guarantee convergency at $T/T_\mathrm{c}(0) > 10^{-2}$.

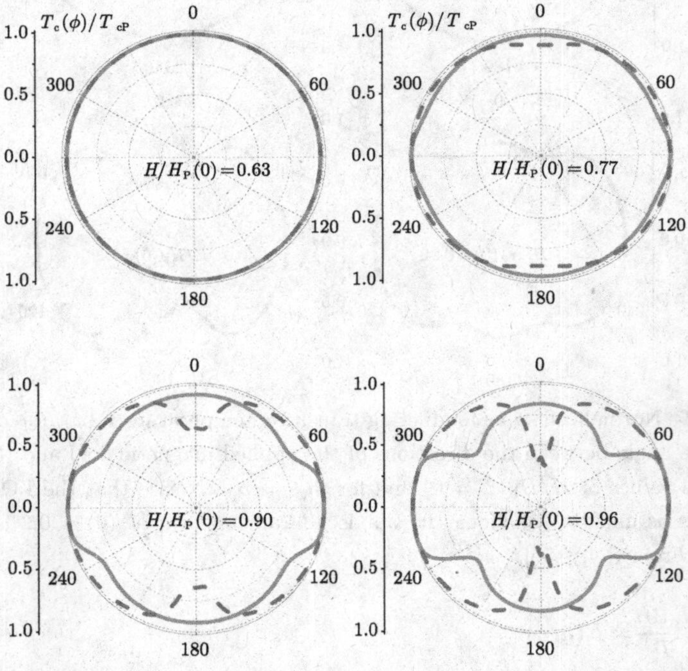

Figure 6.3 The same as in Fig. 6.2, but for $m_x = 10m_y$ (solid line) and $m_x = 0.1m_y$ (dashed line). From Ref. [156].

The superconducting transition temperature for magnetic field applied along different directions within the x-y plane is shown in Figs. 6.2 and 6.3, where the in-plane effective masses are assumed to isotropic ($m_x = m_y$) and anisotropic ($m_x = 10m_y$ and $m_x = 0.1m_y$), respectively. As stated above, the LOFF wave vector is taken to be along the x axis for all cases. In these polar plots, the direction of each point with respect to the vertical axis represents the magnetic-field direction with respect to the x axis, and the distance from the origin depicts the transition temperature T_c. Here, T_c is scaled with T_{cP} obtained from Eq.(6.36), and magnetic field is normalized with respect to zero-temperature Pauli paramagnetic limit $H_P(T = 0)$ of two-dimensional superconductors $H_P(0) = \Delta(T = 0)/\mu_B$ [236].

In the isotropic case of Fig. 6.2, the transition temperature is isotropic as expected for low fields. When increasing H above a critical value, a strong in-plane anisotropy of T_c emerges, and becomes pronounced at high fields. In particular, relatively strong dips at $\phi \approx \pm 18°$ and $\pm 162°$ and small peaks at $\phi = 0°$ and $180°$ develop with increasing magnetic field. Since the Fermi surface in this case is perfectly circular, this anisotropy of T_c originates solely from the angular effect induced by the LOFF wave vector $\boldsymbol{q} \parallel \hat{\boldsymbol{x}}$.

When the Fermi surface deviates from the isotropic case, the in-plane anisotropy of transition temperature acquires qualitatively different behavior, as shown in Fig. 6.3. Comparing the curves in Fig. 6.3, one can easily notice that angular dependence is completely different when the largest electron mass is along the x (solid line) or y directions (dashed line). For the former case with $m_x = 10m_y$, dips of T_c are present at $\phi \approx \pm 50°$ and $\pm 130°$. The positions of the small peaks are still the same as in the isotropic case, however, they are significantly broadened. For the case of opposite mass ratio $m_x = 0.1m_y$, the dips are strengthened and are found at $\phi \approx \pm 5°$ and $\pm 175°$. In spite of these peculiar angular dependence, one should stress that the highest T_c of the LOFF state is always located at $\phi = \pm 90°$, which corresponds to the case of $\boldsymbol{H} \parallel \boldsymbol{q}$. This is in clear contrast to the case of BCS superconductors, where the critical temperature is maximal for magnetic field pointing along the lightest mass direction.

6.2 Singlet-triplet mixing in a LOFF state

From a historical point of view, the possibility of a single-triplet mixing superconducting state is usually brought into consideration when singlet and triplet symmetries are both supported in different measurements or at different circumstances. In the context of Bechgaard salts, this is indeed the case where the Knight shift measurement finds no change of spin susceptibility in $(TMTSF)_2PF_6$ [33, 35] but observes noticeable decrease in $(TMTSF)_2ClO_4$ [38] Since the field used in $(TMTSF)_2ClO_4$ ($H \approx 1.38\,T$ for $\boldsymbol{H} \parallel \boldsymbol{a}$, and $H \approx 0.96\,T$

for $\boldsymbol{H} \parallel \boldsymbol{b}'$) is smaller than that used in $(TMTSF)_2PF_6$ ($H = 1.43\,\mathrm{T}$ for $\boldsymbol{H} \parallel \boldsymbol{a}$, and $H = 2.38\,\mathrm{T}$ for $\boldsymbol{H} \| \boldsymbol{b}'$), a singlet-triplet mixing state with dominant singlet portion at low fields and large triplet portion at high fields can explain the discrepancy in experiments, assuming the pairing states for the two sister compounds are of the same kind.

In this section, we discuss a proposal by Shimahara [237] regarding the possibility of a mixed single-triplet pairing state in the presence of a magnetic field. This work concentrates on quasi-2D layered superconductors without any in-plane anisotropy hence cannot be directly applied to Bechgaard salts. However, if we adopt an effective mass approximation as in Eq.(6.23) within Sec. 6.1, the in-plane effective mass anisotropy can be removed by a scaling transformation like Eq.(6.25). As a result, the original anisotropic system is mapped into an isotropic model with modified magnetic field in the form of Eq.(6.27). Thus, the discussion in this section, although focusing on isotropic 2D superconductors, is relevant in the context of quasi-1D Bechgaard salts. Since only isotropic 2D superconductors will be considered here, we will not specify the in-plane field direction within this section.

Following the study by Shimahara [237], we consider a system with both singlet and triplet pairing interactions. At zero and low magnetic fields, the singlet pairing interaction is assumed to be dominant. The pairing interactions can be written in a separated form

$$V(\boldsymbol{k}, \boldsymbol{k}') = -\sum_{\eta} g_\eta \Gamma_\eta(\boldsymbol{k}) \Gamma_\eta(\boldsymbol{k}'), \tag{6.38}$$

where η labels different pairing channels, g_η is the corresponding interaction strength, and $\boldsymbol{k} = (k_x, k_y)$ denotes in-plane momentum within this section. The symmetry functions $\Gamma_\eta(\boldsymbol{k})$ are defined as

$$s\text{-}wave: \;\; \Gamma_s(\boldsymbol{k}) = 1, \tag{6.39a}$$

$$p_i\text{-}wave: \;\; \Gamma_{p_i}(\boldsymbol{k}) = \hat{k}_i, \tag{6.39b}$$

$$d_{x^2\text{-}y^2}\text{-}wave: \;\; \Gamma_{d_{x^2-y^2}}(\boldsymbol{k}) = \hat{k}_x^2 - \hat{k}_y^2, \tag{6.39c}$$

$$d_{xy}\text{-}wave: \;\; \Gamma_{d_{xy}}(\boldsymbol{k}) = \hat{k}_x \hat{k}_y. \tag{6.39d}$$

To demonstrate the singlet-triplet mixing effect, the following two cases are considered as typical examples:

case 1: s-p mixing

$$g_s > g_{p_x} = g_{p_y} \equiv g_p > 0, \quad g_{d_{x^2-y^2}} = g_{d_{xy}} = 0; \tag{6.40}$$

case 2: d-p mixing

$$g_{d_{x^2-y^2}} = g_{d_{xy}} \equiv g_d > g_p > 0, \quad g_s = 0; \tag{6.41}$$

The superconducting gap function can be expanded using the same symmetry functions as

$$\Delta(\boldsymbol{k}) = \sum_\eta \Delta_\eta \Gamma_\eta(\boldsymbol{k}). \tag{6.42}$$

The gap equations in the vicinity of the second-order phase transition line can be written in the standard form

$$\Delta_\eta = g_\eta \sum_{\eta', \boldsymbol{k}', \sigma=\pm} \frac{\Gamma_\eta(\boldsymbol{k}')\Gamma_{\eta'}(\boldsymbol{k}')}{4\varepsilon_{\boldsymbol{k}'}} \tanh\left(\frac{\varepsilon_{\boldsymbol{k}'} + \sigma\zeta}{2T}\right) \Delta_{\eta'}, \tag{6.43}$$

where $\varepsilon_{\boldsymbol{k}'} = (k')^2/2m$ is the 2D dispersion. The parameter ζ is defined as $\zeta \equiv h(\bar{q}\cos\theta + 1)$, where $h = |\mu_e H|$ with μ_e the electron magnetic moment, $\bar{q} = qv_\mathrm{F}/2h$, and θ is the angle between $\boldsymbol{k}'$ and $\boldsymbol{q}$. Due to the symmetry in momentum and spin spaces, the superconducting order parameters with even and odd parities can mix with each other only in the case of $h \neq 0$ and $q \neq 0$. This requirement thus corresponds to a LOFF state in the presence of a magnetic field.

In the weak coupling limit, the gap equation (6.43) can be further simplified, leading to

$$\Delta_\eta \ln\left(\frac{T}{T_{\mathrm{c},\eta}(0)}\right) = -\sum_{\eta'} M_{\eta\eta'}\Delta_{\eta'}, \tag{6.44}$$

where $T_{c,\eta}(0)$ is the transition temperature at zero magnetic field in the η pairing channel, and

$$M_{\eta\eta'} = \int_0^{2\pi} \frac{d\phi}{2\pi} \frac{\mathscr{N}_{\eta\eta'}(0,\phi)}{\mathscr{N}_\eta(0)} \sinh^2\left(\frac{\beta\zeta}{2}\right) \Phi(\phi). \tag{6.45}$$

The functions in the equation above are defined as

$$\mathscr{N}_{\eta\eta'}(0,\phi) = \Gamma_\eta(\phi)\Gamma_{\eta'}(\phi)\mathscr{N}(0,\phi), \tag{6.46}$$

$$\mathcal{N}_\eta(0) = \int_0^{2\pi} \frac{d\phi}{2\pi} \mathcal{N}_{\eta\eta}(0, \phi), \tag{6.47}$$

$$\Phi(\phi) = \int_0^\infty dy \ln(y) \left\{ \frac{2\sinh^2(y)}{\left[\cosh^2(y) + \sinh^2(\beta\zeta/2)\right]^2} - \frac{1}{\cosh^2(y)\left[\cosh^2(y) + \sinh^2(\beta\zeta/2)\right]} \right\}. \tag{6.48}$$

Here, ϕ denotes the angle between $\boldsymbol{k}'$ and the k_x axis, and $\mathcal{N}(0, \phi)$ is the angle-dependent density of states at the Fermi level. The upper critical field $h = h(T)$ can then be obtained by solving Eq.(6.44) and optimizing with respect to the LOFF wave vector $\boldsymbol{q}$.

For case 1 with *s*-*p* mixing, the orientation of wave vector $\boldsymbol{q}$ can be set in an arbitrary direction. Without loss of generality, we can choose $\boldsymbol{q} \parallel \boldsymbol{k}_x$. Thus, the p_y-component of the order parameter Δ_{p_y} is not mixed with the *s*-wave component, while Δ_{p_x} can be mixed. In this case, the problem is reduced to only two partial waves, and one needs to calculate the smallest eigenvalue λ of the following 2×2 matrix

$$\begin{pmatrix} M_{ss} & M_{sp} \\ M_{ps} & M_{pp} + G_p \end{pmatrix}, \tag{6.49}$$

where the parameter

$$G_p \equiv \ln\left(\frac{T_{\mathrm{c},s}(0)}{T_{\mathrm{c},p}(0)}\right) = \frac{1}{g_p \mathcal{N}_p(0)} - \frac{1}{g_s \mathcal{N}_s(0)}. \tag{6.50}$$

The transition temperature is given by $T_\mathrm{c} = T_{\mathrm{c},s}(0) e^{-\lambda}$.

For case 2 with *d*-*p* mixing, a free rotation of the *d*-wave order parameter allows one to assume Δ_d is in the $d_{x^2\text{-}y^2}$ channel only. Thus, the problem would involve three pairing channels and one needs to find the smallest eigenvalue λ of a 3×3 matrix

$$\begin{pmatrix} M_{dd} & M_{dp_x} & M_{dp_y} \\ M_{p_x d} & M_{p_x p_x} + G_p & M_{p_x p_y} \\ M_{p_y d} & M_{p_y p_x} & M_{p_y p_y} + G_p \end{pmatrix}, \tag{6.51}$$

where the parameter G_p is defined by

$$G_p \equiv \ln\left(\frac{T_{c,d}(0)}{T_{c,p}(0)}\right) = \frac{1}{g_p\mathcal{N}_p(0)} - \frac{1}{g_d\mathcal{N}_d(0)}. \tag{6.52}$$

The transition temperature is thus given by $T_c = T_{c,d}(0)e^{-\lambda}$.

The numerical results for the maximal upper critical field $h(T, \boldsymbol{q}_0)$ with the optimal $\boldsymbol{q}_0$ are shown in Figs. 6.4 and 6.5 for case 1 of *s*-*p* mixing and case 2 of *d*-*p* mixing, respectively. In both cases, the singlet-triplet mixing LOFF state acquires a remarkable enhancement of the critical field for temperatures far above the transition temperature of the pure *p*-wave superconductivity in the ESTP state (vertical lines). In Fig. 6.4, we notice that the maximal $h(T, \boldsymbol{q}_0)$ of a singlet-triplet mixing LOFF state well exceeds both the Pauli paramagnetic limit (dashed line) and the critical field for a pure singlet LOFF state (dotted line) for the *s*-*p* pairing case with $T_{c,p}(0)/T_{c,s}(0) = 0.1$. This

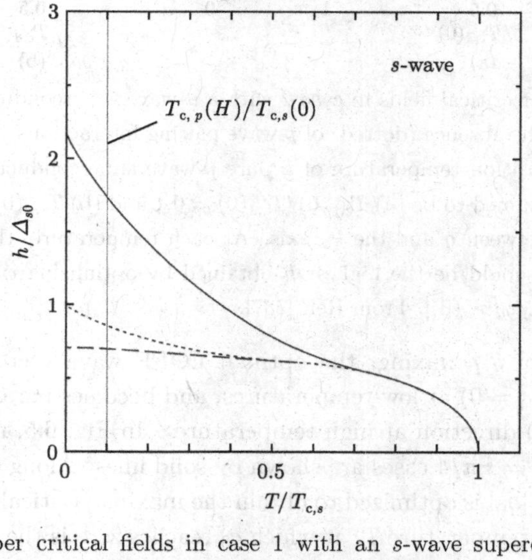

Figure 6.4 Upper critical fields in case 1 with an *s*-wave superconductor in the presence (solid) and the absence (dotted) of *p*-wave pairing interactions. The vertical line indicates the transition temperature of a pure *p*-wave superconductor in the ESTP state, which is assumed to be $T_{c,p}(0)/T_{c,s}(0) = 0.1$ in this plot. The dashed line shows the Pauli paramagnetic limit when the FFLO state is ignored. Here, $\Delta_{s0} \equiv 2\omega_c \exp[-1/g_s\mathcal{N}_s(0)]$ with ω_c the cyclotron frequency. From Ref. [237].

observation suggests that the triplet component plays an essential role in the high-field regime, although has little effect at zero or low magnetic fields. In the *d*-*p* mixing case, the effect of the triplet component is also prominent for an even smaller $T_{c,p}(0)/T_{c,d}(0) = 0.01$, as can be seen from Fig. 6.5(b).

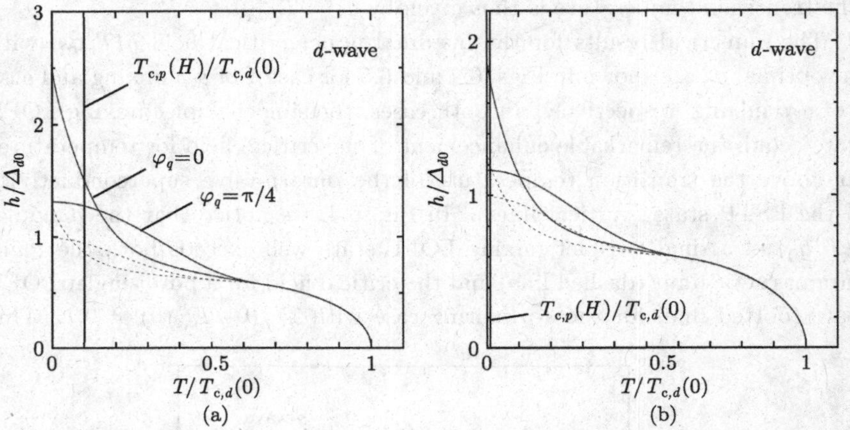

Figure 6.5 Upper critical fields in case 2 with a *d*-wave superconductor in the presence (solid) and the absence (dotted) of *p*-wave pairing interactions. The vertical line indicates the transition temperature of a pure *p*-wave superconductor in the ESTP state, which is assumed to be (a) $T_{c,p}(0)/T_{c,s}(0) = 0.1$ and (b) $T_{c,p}(0)/T_{c,s}(0) = 0.01$. ϕ_q is the angle between $\boldsymbol{q}$ and the k_x axis. At each temperature, the final result of the critical field should be the highest h obtained by optimizing ϕ_q and $|q|$. Here, $\Delta_{d0} \equiv 2\omega_c \exp[-1/g_d\mathcal{N}_d(0)]$. From Ref. [237].

For case 2 of *d*-*p* mixing, the optimal LOFF wave vector $\boldsymbol{q}_0$ is in the direction of k_x ($\phi = 0$) at low temperatures, and becomes parallel to the $k_x = \pm k_y$ ($\phi = \pm\pi/4$) direction at high temperatures. In Fig. 6.5, results for both the $\phi = 0$ and $\phi = \pm\pi/4$ cases are shown by solid lines. Along each curve, the magnitude $q_0 = |\boldsymbol{q}_0|$ is optimized to obtain the maximal critical field, as shown in Fig. 6.6. The temperature T_* at which q_0 reaches zero is the temperature of the tricritical point where the normal state, the BCS superconductivity with $q_0 = 0$, and the LOFF state with $q_0 \neq 0$ merge. Notice that T_* increases from $\approx 0.561T_{c,d}(0)$ to a higher value due to the triplet mixing. For example, $T_* \approx 0.668T_{c,d}(0)$ for $T_{c,p}(0)/T_{c,d}(0) = 0.1$. Shimahara [237] also confirms that a LOFF wave vector along any other direction is not preferable.

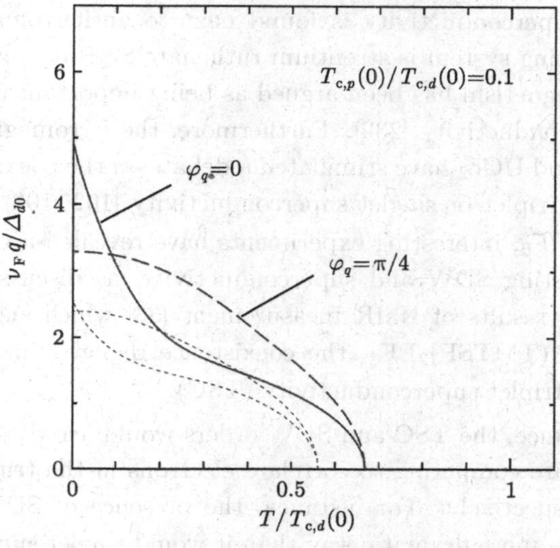

Figure 6.6 The magnitude of the optimal wave vector $\boldsymbol{q}$ at the upper critical fields for the *d*-*p* mixing case with $T_{c,p}(0)/T_{c,s}(0) = 0.1$ and $\phi_q = 0$ (solid) and $\phi_q = \pi/4$ (dashed). Results for pure *d*-wave pairing are also shown as dotted lines. From Ref. [237].

The discussion above suggests that in a 2D superconductor with coexisting singlet and triplet pairing interactions, a singlet-triplet mixing LOFF state is always favorable in the presence of an in-plane magnetic field. This LOFF state is characterized by a nontrivial wave vector $\boldsymbol{q}$, and acquires an upper critical field well above the Pauli paramagnetic limit. Although only a weak triplet pairing interactions is required to observe a sizable enhancement of critical field, the strength of this pairing component is assumed *a priori* in this proposal.

6.3 Microscopic coexistence of spin density wave and superconductivity

The competition or coexistence of magnetic order and superconductivity is a very important problem in condensed matter physics. There is a broad class of systems that presents a phase diagram with magnetic order and superconducting in close vicinity. One of the most important systems is copper oxides, where singlet superconductivity is found next to antiferromagnetism [238]. Another interesting system is strontium ruthenate Sr_2RuO_4, where the proximity to ferromagnetism has been argued as being important to the existence of triplet superconductivity [239]. Furthermore, the ferromagnetic superconductors $ZrZn_2$ and UGe_2 have stimulated a debate on the coexistence of ferromagnetism and triplet or singlet superconductivity [102, 103]. In Bechgaard salt $(TMTSF)_2PF_6$, interesting experiments have revealed a region of microscopically coexisting SDW and superconductivity, as discussed in Sec. 3.3. Considering the results of NMR measurement [33] which suggests a triplet pairing state in $(TMTSF)_2PF_6$, the coexistence region is likely to be built with SDW and triplet superconductivity (TSC).

At a first glance, the TSC and SDW orders would avoid coexistence since the two orders are competing to correlate electrons in the triplet and singlet spin sectors, respectively. For instance, the presence of SDW order would disrupt TSC in a more dramatic way than it would singlet superconductivity, while the ferromagnetic order would disrupt singlet superconductivity more than TSC. Therefore, it is easier to find in nature examples of coexistence of singlet superconductivity and SDW or TSC and ferromagnetism, while the conditions to find the coexistence of TSC and SDW are much more stringent. In this section, we introduce two theoretical proposals devoted to explain the TSC-SDW coexistence observed in experiments. In particular, Zhang and Sá de Melo [121, 122] propose a microscopic model where the two orders can coexist in a ϕ_1-ϕ_2 phase diagram [240]. On the other hand, Gorkov and Grigoriev [231, 232] account the coexistence region to a soliton phase within SDW, where triplet superconductivity can emerge within soliton walls as the most primary pairing instability.

The microscopic coexistence of SDW and TSC in $(TMTSF)_2PF_6$ has been discussed in a framework of strictly one-dimensional theory invoking SO(4) symmetry [119]. However, it is demonstrated that the degrees of freedom along the y and z directions in Bechgaard salts are truly important and the system can not be simply treated as a one-dimensional electron gas. Indeed, the curvature of the Fermi surface induced by the k_y- and k_z-dependent dispersion spectrum qualitatively modify the density of states, results in nontrivial effect on the SDW and superconductivity orders. To fill this gap, Zhang and Sá de Melo [122] derive a microscopic model starting from the tight-binding dispersion relation as in Eq.(2.1), with interacting Hamiltonian

$$\mathscr{H}_{\text{int}} = \sum_{\boldsymbol{k},\boldsymbol{k}',\boldsymbol{p}} \sum_{\alpha\beta\gamma\delta} V(\boldsymbol{k},\boldsymbol{k}') \boldsymbol{d}^{\dagger}_{\alpha\beta}(\boldsymbol{k},\boldsymbol{p}) \cdot \boldsymbol{d}_{\gamma\delta}(\boldsymbol{k}',\boldsymbol{p})$$

$$+ \sum_{\boldsymbol{k},\boldsymbol{k}',\boldsymbol{q}} \sum_{\alpha\beta\gamma\delta} J(\boldsymbol{q}) \boldsymbol{s}^{\dagger}_{\alpha\beta}(\boldsymbol{k},\boldsymbol{q}) \cdot \boldsymbol{s}_{\gamma\delta}(\boldsymbol{k}',\boldsymbol{q}). \tag{6.53}$$

The first and second terms in the Hamiltonian above describe interactions in TSC and SDW channels, respectively. These interactions allow for the possibility of competition or coexistence of TSC and SDW instabilities at low temperatures. Here, Greek letters indicate spin indices and $\boldsymbol{k}$, $\boldsymbol{k}'$, $\boldsymbol{p}$, and $\boldsymbol{q}$ represent three-dimensional linear momenta. The vector order parameters are defined as

$$\boldsymbol{d}^{\dagger}_{\alpha\beta}(\boldsymbol{k},\boldsymbol{p}) = c^{\dagger}_{\boldsymbol{k}+\boldsymbol{p}/2,\alpha} \hat{\boldsymbol{v}}_{\alpha\beta} c^{\dagger}_{-\boldsymbol{k}+\boldsymbol{p}/2,\beta}, \tag{6.54a}$$

$$\boldsymbol{s}^{\dagger}_{\alpha\beta}(\boldsymbol{k},\boldsymbol{q}) = c^{\dagger}_{\boldsymbol{k}-\boldsymbol{q}/2,\alpha} \hat{\sigma}_{\alpha\beta} c_{\boldsymbol{k}+\boldsymbol{q}/2,\beta}, \tag{6.54b}$$

where $\hat{\sigma} = (\sigma_x, \sigma_y, \sigma_z)$ are Pauli matrices, and $\hat{\boldsymbol{v}} = i\hat{\sigma}\sigma_y$. In the case of weak spin-orbit coupling, the TSC interaction $V(\boldsymbol{k},\boldsymbol{k}')$ can be written in a separate form

$$V(\boldsymbol{k},\boldsymbol{k}') = V h_{\Gamma}(\boldsymbol{k},\boldsymbol{k}') \phi_{\Gamma}(\boldsymbol{k}) \phi_{\Gamma}(\boldsymbol{k}'), \tag{6.55}$$

where V is a prefactor with a dimension of energy, $h_{\Gamma}(\boldsymbol{k},\boldsymbol{k}')$ characterizes the momentum dependence, and $\phi_{\Gamma}(\boldsymbol{k})$ is the symmetry basis function for an irreducible representation Γ of the lattice orthorhombic D_{2h} group [36, 37]. Having a nodeless triplet pairing states in mind, Zhang and Sá de Melo [122] consider only the p_x symmetry with $\phi_{\Gamma}(\boldsymbol{k}) = \sin(k_x a)$, and take $h_{\Gamma}(\boldsymbol{k},\boldsymbol{k}') = 1$ without loss of generality.

The order parameters for the TSC and SDW are then defined as

$$\boldsymbol{D}(\boldsymbol{p}) = \left\langle \sum_{\boldsymbol{k},\alpha,\beta} V \phi_{\Gamma}(\boldsymbol{k}) \boldsymbol{d}_{\alpha\beta}(\boldsymbol{k},\boldsymbol{p}) \right\rangle, \tag{6.56a}$$

$$\boldsymbol{N}(\boldsymbol{q}) = J(\boldsymbol{q}) \left\langle \sum_{\boldsymbol{k},\alpha,\beta} \boldsymbol{s}_{\alpha\beta}(\boldsymbol{k},\boldsymbol{q}) \right\rangle, \tag{6.56b}$$

where $\langle \cdots \rangle$ denotes ensemble average. With these definitions, the effective action can be obtained by writing the Hamiltonian in a quadratic form such that the fermions can be integrated out. The quadratic terms of the effective action takes the form

$$S_2^{\mathrm{TSC}} = \sum_{\boldsymbol{p}} A(\boldsymbol{p})\boldsymbol{D}^\dagger(\boldsymbol{p}) \cdot \boldsymbol{D}(\boldsymbol{p}), \tag{6.57a}$$

$$S_2^{\mathrm{SDW}} = \sum_{\boldsymbol{q}} B(\boldsymbol{q})\boldsymbol{N}^\dagger(-\boldsymbol{q}) \cdot \boldsymbol{N}(\boldsymbol{q}), \tag{6.57b}$$

where coefficients $A(\boldsymbol{p})$ and $B(\boldsymbol{q})$ can be obtained from their corresponding diagrams

$$A(\boldsymbol{p}) = -2V^{-1} - 2T\sum_{\boldsymbol{k}} G(\boldsymbol{k}+\boldsymbol{p}/2,\omega_n)\phi_\Gamma(\boldsymbol{k}+\boldsymbol{p}/2)$$

$$\times G(-\boldsymbol{k}+\boldsymbol{p}/2,-\omega_n)\phi_\Gamma(-\boldsymbol{k}+\boldsymbol{p}/2), \tag{6.58a}$$

$$B(\boldsymbol{q}) = -2J^{-1}(\boldsymbol{q}) - 2T\sum_{\boldsymbol{k}} G(\boldsymbol{k},\omega_n)G(\boldsymbol{k}+\boldsymbol{q},\omega_n). \tag{6.58b}$$

Here, T is the temperature, $G(\boldsymbol{k},\omega_n) = 1/(i\omega_n + \varepsilon_{\boldsymbol{k}} - \mu)$ is the fermion propagator, ω_n is Matsubara frequency, and $\sum_k = \sum_{\boldsymbol{k},\omega_n}$ is used to simplify notation. Notice that the two order parameters $\boldsymbol{D}(\boldsymbol{p})$ and $\boldsymbol{N}(\boldsymbol{q})$ do not couple to quadratic order, because TSC and SDW are instabilities in the particle-particle and particle-hole channels, respectively. In addition, the small anisotropy in the SDW order parameter [241] is not included in this level, but is leaving for discussion in a later stage.

The quadratic effective action can be further simplified within the following standard assumptions. First, the saddle point TSC order parameter is assumed to be dominated by the zero center-of-mass momentum component $\boldsymbol{D}_0 \equiv \boldsymbol{D}(\boldsymbol{p}=0)$. Second, the saddle point SDW order parameter $\boldsymbol{N}$ is taken to be real in $\boldsymbol{r}$-space, and have Fourier components determined by Fermi surface nesting vectors $\boldsymbol{q} = \boldsymbol{Q}_i = (\pm Q_a, \pm Q_b, \pm Q_c)$ [52]. In this case, the coefficients $B(\boldsymbol{Q}_i)$ are identical for all $\boldsymbol{Q}_i$'s, since the lattice dispersion in invariant under reflections and inversions compatible with the $D_{2\mathrm{h}}$ group. In addition, the coefficients of all higher order terms involving $\boldsymbol{N}(\boldsymbol{Q}_i)'s$ share the same properties. Given that $\boldsymbol{N}(\boldsymbol{r})$ is real, and that there is periodic boundary conditions, it is possible to choose a specific reference phase where $\boldsymbol{N}(\boldsymbol{Q}_i)$ are real and identical. Thus, by defining $\boldsymbol{N}_0 \equiv \boldsymbol{N}(\boldsymbol{Q}_i)$ for all i, the quadratic terms are dominated in the long wavelength limit by

$$S_2^{\mathrm{TSC}} \approx A(0)|\boldsymbol{D}_0|^2, \tag{6.59a}$$

$$S_2^{\mathrm{SDW}} \approx (m/2)B(\boldsymbol{Q}_1)|\boldsymbol{N}_0|^2, \tag{6.59b}$$

where m is the number of nesting vectors, and $\boldsymbol{Q}_1 = (Q_a, Q_b, Q_c)$ is chosen for definiteness.

Although the order parameters for TSC and SDW do not couple to quadratic order, they do couple in the fourth order effective action, which is given by

$$S_4^{\mathrm{C}} = (C_1 + C_2/2)|\boldsymbol{D}_0|^2|\boldsymbol{N}_0|^2 - C_2|\boldsymbol{D}_0 \cdot \boldsymbol{N}_0|^2, \tag{6.60}$$

where the coefficients C_1 and C_2 can be obtained from their corresponding diagrams, leading to

$$C_1 = mT\sum_{k} G(\boldsymbol{k}, \omega_n)G^2(-\boldsymbol{k}, -\omega_n)G(-\boldsymbol{k}+\boldsymbol{Q}_1, -\omega_n)\phi_\Gamma(-\boldsymbol{k})\phi_\Gamma(\boldsymbol{k}), \tag{6.61a}$$

$$C_2 = mT\sum_{k} G(\boldsymbol{k}, \omega_n)G(\boldsymbol{k}+\boldsymbol{Q}_1, \omega_n) \times G(-\boldsymbol{k}-\boldsymbol{Q}_1, \omega_n)G(-\boldsymbol{k}, -\omega_n)\phi_\Gamma(-\boldsymbol{k}-\boldsymbol{Q}_1)\phi_\Gamma(\boldsymbol{k}). \tag{6.61b}$$

Notice that the second term in Eq.(6.60) can be parametrized as $-C_2\cos^2(\theta)$ $|\boldsymbol{D}_0|^2|\boldsymbol{N}_0|^2$, where

$$\cos^2\theta \equiv |\boldsymbol{D}_0 \cdot \boldsymbol{N}_0|^2/|\boldsymbol{D}_0|^2|\boldsymbol{N}_0|^2 \leqslant 1 \tag{6.62}$$

is independent of $|\boldsymbol{D}_0|$ and $|\boldsymbol{N}_0|$. Since $\boldsymbol{D}_0$ is unitary, its global phase can be eliminated in S_4^{C}, and θ can be regarded as the angle between $\boldsymbol{D}_0$ and $\boldsymbol{N}_0$. The coefficient C_2 for $(TMTSF)_2PF_6$ is positive definite, indicating that $\boldsymbol{D}_0$ and $\boldsymbol{N}_0$ tend to be aligned ($\theta = 0$) or anti-aligned ($\theta = \pi$). Additional fourth order terms are

$$S_4^{\mathrm{TSC}} = D_1|\boldsymbol{D}_0|^4; \tag{6.63a}$$

$$S_4^{\mathrm{SDW}} = D_2|\boldsymbol{N}_0|^4, \tag{6.63b}$$

where the coefficients D_1 and D_2 are obtained diagrammatically as

$$D_1 = T\sum_k G^2(\boldsymbol{k},\omega_n)G^2(-\boldsymbol{k},-\omega_n)\phi_\Gamma^2(\boldsymbol{k})\phi_\Gamma^2(-\boldsymbol{k}); \tag{6.64a}$$

$$\begin{aligned} D_2 = {} & (m/2)T\sum_k G(\boldsymbol{k},\omega_n)G^2(\boldsymbol{k}+\boldsymbol{Q}_1,\omega_n)[G(\boldsymbol{k},\omega_n)+G(\boldsymbol{k}+2\boldsymbol{Q}_1,\omega_n)] \\ & +(m/4)T\sum_k G(\boldsymbol{k},\omega_n)G(\boldsymbol{k}+\boldsymbol{Q}_1,\omega_n) \\ & \times \sum_i{}' G(\boldsymbol{k}+\boldsymbol{Q}_1+\boldsymbol{Q}_i,\omega_n)[G(\boldsymbol{k}+\boldsymbol{Q}_i,\omega_n)+G(\boldsymbol{k}+\boldsymbol{Q}_1,\omega_n)]. \end{aligned} \tag{6.64b}$$

Here, $\sum_i'$ represents $\sum_{\boldsymbol{Q}_i \neq \pm \boldsymbol{Q}_1}$.

Combining the results above, the resulting effective action is

$$S_{\text{eff}} = S_0 + S_2^{\text{TSC}} + S_2^{\text{SDW}} + D_1|\boldsymbol{D}_0|^4 + D_2|\boldsymbol{N}_0|^4 + C(\theta)|\boldsymbol{D}_0|^2|\boldsymbol{N}_0|^2, \tag{6.65}$$

where S_0 is the normal state contribution, and $C(\theta) = C_1 + C_2/2 - C_2\cos^2\theta$. The phase diagram that emerges from this action leads to either bicritical or tetracritical points as illustrated in Fig. 6.7. If $R = C^2(0)/(4D_1D_2) > 1$, the critical point (P_c, T_c) is bicritical and there is a first order transition line at $(m/2)B(\boldsymbol{Q}_1) = A(0)$ when both $B(\boldsymbol{Q}_1) < 0$ and $A(0) < 0$, as seen in Fig. 6.7(a). However, if $R < 1$, (P_c, T_c) is tetracritical and a coexistence region for TSC and SDW orders occurs when both $B(\boldsymbol{Q}_1) < 0$ and $A(0) < 0$, as seen in Fig. 6.7(b). The action S_{eff} obtained in three dimensions is not SO(4) invariant, indicating that the SO(4)-symmetry-based theory [119] can only be applied to one-dimensional systems, but not to the highly anisotropic but three-dimensional Bechgaard salts.

The ratio $R \approx 0.12$ for the Bechgaard salt $(\text{TMTSF})_2\text{PF}_6$ around (P_c, T_c), when the interaction strengths V and J are chosen to give the same $T_c = 1.2\,\text{K}$ at quarter filling for parameters $|t_x| = 5800\,\text{K}$, $|t_y| = 1226\,\text{K}$, $|t_z| = 58\,\text{K}$, used in combination with $\phi_\Gamma(\boldsymbol{k}) = \sin(k_x a)$ (p_x-symmetry for TSC) and the nesting vectors $\boldsymbol{Q} = (\pm\pi/2a, \pm\pi/2b, 0)$ ($m = 4$). This shows that $(\text{TMTSF})_2\text{PF}_6$ has a TSC-SDW coexistence region as suggested by experiments discussed in Sec. 3.3.

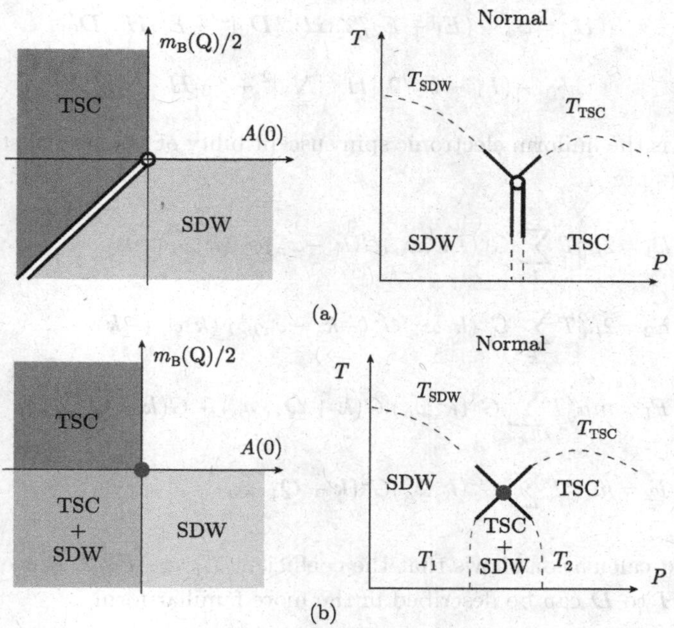

Figure 6.7 *P*-*T* phase diagrams indicating (a) first order transition line with no coexistence phase, and (b) two second order lines with coexistence region between TSC and SDW phases. From Ref. [122].

The magnetic field effect of this coexistence region can be studied by considering a uniform magnetic field $\boldsymbol{H}$ which couples with charge via the Peierls substitution $\boldsymbol{k} \to \boldsymbol{k} - e\boldsymbol{A}$ in the dispersion spectrum with $\boldsymbol{A}$ the vector potential, and couples with spin via the paramagnetic term

$$\mathscr{H}_{\mathrm{P}} = -\mu_0 \boldsymbol{H} \cdot \sum_{\boldsymbol{k},\alpha\beta} c^{\dagger}_{\boldsymbol{k}\alpha} \hat{\boldsymbol{\sigma}}_{\alpha\beta} c_{\boldsymbol{k}\beta}, \tag{6.66}$$

where μ_0 is the effective magnetic moment. Upon integration over fermions, the corresponding effective action is

$$S_{\text{eff}}(\boldsymbol{H}) = S_0(\boldsymbol{H}) + S_2^{\text{TSC}}(\boldsymbol{H}) + S_2^{\text{SDW}}(\boldsymbol{H}) + S_4(\boldsymbol{H}), \tag{6.67}$$

where $S_2^{\text{TSC}}(\boldsymbol{H})$ and $S_2^{\text{SDW}}(\boldsymbol{H})$ are obtained from S_2^{TSC} and S_2^{SDW} by the Peierls substitution, respectively. The other terms in $S_{\text{eff}}(\boldsymbol{H})$ are

$$S_0(\boldsymbol{H}) = S_0 + \frac{|\boldsymbol{H}|^2}{8\pi} - \frac{\chi_n |\boldsymbol{H}|^2}{2}, \tag{6.68a}$$

$$S_4(\boldsymbol{H}) = S_4 + (E_1 + E_2/2)|\boldsymbol{H}|^2|\boldsymbol{D}_0|^2 - E_2|\boldsymbol{H}\cdot\boldsymbol{D}_0|^2$$

$$+(F_1 - F_2/2)|\boldsymbol{H}|^2|\boldsymbol{N}_0|^2 + F_2|\boldsymbol{H}\cdot\boldsymbol{N}_0|^2, \qquad (6.68b)$$

where χ_n is the uniform electronic spin susceptibility of the normal state. The coefficients are

$$E_1 = 2\mu_0^2 T \sum_{\boldsymbol{k}} G^3(-\boldsymbol{k}, \omega_n) G(\boldsymbol{k}, -\omega_n)\phi_\Gamma(\boldsymbol{k})\phi_\Gamma(-\boldsymbol{k}), \qquad (6.69a)$$

$$E_2 = 2\mu_0^2 T \sum_{\boldsymbol{k}} G^2(\boldsymbol{k}, \omega_n) G^2(-\boldsymbol{k}, -\omega_n)\phi_\Gamma(\boldsymbol{k})\phi_\Gamma(-\boldsymbol{k}), \qquad (6.69b)$$

$$F_1 = m\mu_0^2 T \sum_{\boldsymbol{k}} G^3(\boldsymbol{k}, \omega_n)[G(\boldsymbol{k}+\boldsymbol{Q}_1, \omega_n) + G(\boldsymbol{k}-\boldsymbol{Q}_1, \omega_n)], \qquad (6.69c)$$

$$F_2 = m\mu_0^2 T \sum_{\boldsymbol{k}} G^2(\boldsymbol{k}, \omega_n) G^2(\boldsymbol{k}+\boldsymbol{Q}_1, \omega_n). \qquad (6.69d)$$

A detailed calculation shows that the coefficient $E_1 = -E_2/2$, hence the coupling of $\boldsymbol{H}$ to $\boldsymbol{D}$ can be described in the more familiar form

$$F_\mathrm{M} - \sum_{\mu\nu} H_\mu \chi_{\mu\nu} H_\nu/2, \qquad (6.70)$$

where $\chi_{\mu\nu} = \chi_n\delta_{\mu\nu} + E_2 D_\mu^* D_\nu$.

For Bechgaard salts, the coefficients $E_2 < 0$ and $F_2 > 0$ indicating that $\boldsymbol{D}$ and $\boldsymbol{N}$ prefer to be perpendicular to the magnetic field $\boldsymbol{H}$. These conditions, when combined with $C_2 > 0$ in Eq. (6.60), indicate that $\boldsymbol{D}$ and $\boldsymbol{N}$ prefer to be parallel to each other, but perpendicular to $\boldsymbol{H}$. However, the relative orientation of these vectors in small fields is affected by spin anisotropy effects which were already observed in $(\mathrm{TMTSF})_2\mathrm{PF}_6$, where the easy axis for $\boldsymbol{N}$ is the $\boldsymbol{b}'$ direction [241]. Such an anisotropy effect can be described by adding a quadratic term $-u_N N_{b'}^2$ with $u_N > 0$, which favors $\boldsymbol{N} \| \boldsymbol{b}'$. Similarly, the $\boldsymbol{D}$ vector also has anisotropic effect caused by spin-orbit coupling, and can be described by adding a quadratic term $-u_\mathrm{D} D_i^2$, where i is the easy axis for TSC. (Quartic TSC and SDW terms also become anisotropic.)

However, a sufficiently large $\boldsymbol{H} \| \boldsymbol{b}'$ can overcome spin anisotropy effects, and drive the $\boldsymbol{N}$ vector to flop onto the $\boldsymbol{a}$-$\boldsymbol{c}^*$ plane. This canting (flop) transition was reported [241] in $(\mathrm{TMTSF})_2\mathrm{PF}_6$ for $H \approx 1\,\mathrm{T}$ at zero pressure and

$T = 8\,\text{K}$. If such a spin-flop transition persists near the TSC-SDW critical point (P_c, T_c) as suggested in our discussion, then the flop transition of the $\boldsymbol{N}$ vector forces the $\boldsymbol{D}$ vector to flop as well, and has potentially serious consequences to the superconducting state. Schematic phase diagrams are shown in Figs. 6.8(a) and 6.8(b). For $P < P_c$, if a flop transition occurs for $H_F < H_1(0)$ [see Fig. 6.8(a)], then $\boldsymbol{N}$ flops both in the pure SDW and in the TSC-SDW coexistence phases, in which case it forces $\boldsymbol{D}$ vector to flop as well. If the flop transition occurs for $H_{SDW}(0) < H_F < H_1(0)$ (not shown) then only the pure SDW phase is affected. This situation is qualitatively different for $P > P_c$. In the zero (weak) spin-orbit coupling limit the $\boldsymbol{D}$ vector is free to rotate in a magnetic field and tends to be perpendicular to $\boldsymbol{H}$ in order to minimize its magnetic free energy F_M. Thus, for $\boldsymbol{H} \parallel \boldsymbol{b}'$ and $|\boldsymbol{H}| > H_2$, the $\boldsymbol{D}$ vector lies in the $\boldsymbol{a}$-$\boldsymbol{c}^*$ plane since there is no SDW order. However, at lower temperatures and small magnetic fields when TSC and SDW orders coexist, the spin anisotropy field forces $\boldsymbol{N}$ to be along $\boldsymbol{b}'$ and $\boldsymbol{N}$ forces $\boldsymbol{D}$ to flop from the $\boldsymbol{a}$-$\boldsymbol{c}^*$ plane to $\boldsymbol{b}'$ direction. This canting transition occurs at $H_F < H_2(0)$ [see Fig. 6.8(b)], when $\boldsymbol{N}$ flops in the TSC-SDW coexistence phase, and forces the $\boldsymbol{D}$ vector to flop as well.

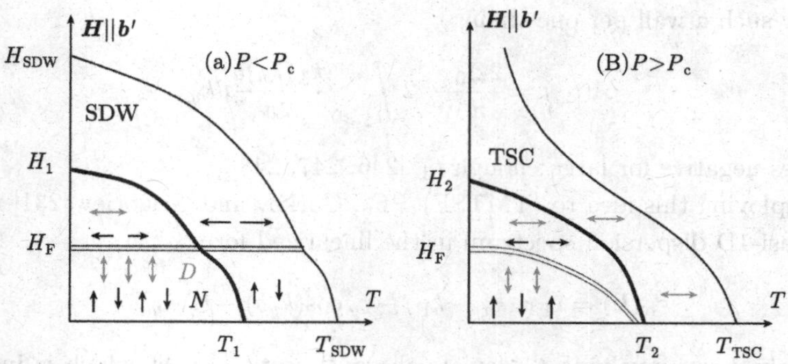

Figure 6.8 H-T phase diagrams showing the TSC-SDW coexistence region (thick solid line) and canting transitions (double line) for (a) $P < P_c$ and (b) $P > P_c$. From Ref. [122].

From the discussion above, we have shown that a unified microscopic theory can be used to understand the coexistence of TSC and SDW order parameters in the P-T phase diagram for quasi-1D organic conductors. The coexistence phase is non-uniform and is in a qualitative agreement with the existing experiments discussed in Sec. 3.3. At roughly the same time, another theory has been proposed by Gorkov and Grigoriev [231, 232], which tries to account the coexistence of SDW and superconductivity to a soliton phase appeared in the SDW background with superconductivity established in the soliton walls. Since the soliton walls are perpendicular to the most conducting $\boldsymbol{a}$ axis, the superconductivity in this coexistence region resembles layered superconductors in the $\boldsymbol{b}'$-$\boldsymbol{c}^*$ plane.

The soliton excitations are known to be of fundamental importance in strictly one-dimensional models of CDW or SDW. Comparing to an electron-hole pair which has excitation energy $\sim 2\Delta_0$, a soliton separating two degenerate ground states acquires lower energy [242, 243, 244, 245]. To be more specific, a simple estimation of the energy cost for one soliton on a single one-dimensional chain is $E_s = (2/\pi)\Delta_0$ [245], while numerical minimization gives a slightly smaller value $E_s \approx 0.6\Delta_0$ [243]. Here, Δ_0 is the gap of the underlying CDW or SDW phase. For chains packed along the y direction with inter-chain spacing b and tunneling $-t_\perp(\boldsymbol{k}_y)$, solitons would form extended states by creating a band in the transverse direction. In this case, solitons on different chains may be able to form a soliton wall, provided that the energy cost for such a wall per one chain

$$A(t_\perp) = \frac{2\Delta_0}{\pi} - 2\int_{t_\perp \geqslant 0} \frac{t_\perp(k_y)b}{2\pi} dk_y \tag{6.71}$$

becomes negative for large enough $t_\perp$ [246, 247, 248].

Employing this idea to $(\mathrm{TMTSF})_2\mathrm{PF}_6$, Gorkov and Grigoriev [231] model the quasi-1D dispersion spectrum in the linearized form

$$\varepsilon_\alpha(\boldsymbol{k}) = v_\mathrm{F}(\alpha k_x - k_\mathrm{F}) - t_y \cos(k_y b) - t'_y(k_y), \tag{6.72}$$

where the term involving t'_y depicts the anti-nesting effect which is induced by applying pressure, and the three dimensional k_z-dependence is not taken into account. This spectrum can be further simplified with the aid of a gauge transformation of the electron wavefunctions [247],

$$\psi^\mathrm{R}_{\boldsymbol{k},s}(x) = e^{ix(k_x - \tilde{\varepsilon}_-(k_y)/v_\mathrm{F})}\psi^\mathrm{R}_s(x), \tag{6.73a}$$

$$\psi^\mathrm{L}_{\boldsymbol{k}-\boldsymbol{Q},s}(x) = e^{-ix(k_x - \tilde{\varepsilon}_-(k_y)/v_\mathrm{F})}\psi^\mathrm{L}_s(x), \tag{6.73b}$$

where $s = \uparrow, \downarrow$ stands for spin, and the superscripts R and L label electrons moving along the right and left sheets of the Fermi surface, respectively. Here, the function

$$\tilde{\varepsilon}_{\pm} = \frac{t_y \cos(k_y b) \pm t_y \cos[(k_y - Q_y)b]}{2}. \tag{6.74}$$

This procedure removes the large nesting term $t_y \cos(k_y b)$ from the original dispersion Eq.(6.72). As a result, the transverse tunneling in Eq.(6.71) is reduced to $t_\perp(k_y) = t'_y(k_y)$.

In the limit of large distance between the walls, the linear energy density of the soliton phase [247] at zero temperature takes the form

$$W_{\text{sw}} = \frac{\Delta_0^2}{2\pi v_{\text{F}}} + n_{\text{sw}} A(t'_y) + n_{\text{sw}} E_-^2 B(t'_y), \tag{6.75}$$

where n_{sw} is the linear density of soliton walls, and the last term represents the exponentially decaying interaction between the walls. The function B is

$$B(t'_y) = \frac{1}{2\pi\Delta_0} - \frac{b}{2\pi} \frac{1}{|dt'_y(k_y)/dk_y|_0}, \tag{6.76}$$

where $|dt'_y(k_y)/dk_y|_0$ is the value of the transverse velocity taken at the positions of k_y where $t'_y(k_y) = 0$. From the energy density Eq. (6.75), one can easily read out that the case with $B(t'_y) > 0$ corresponds to a second-order transition between the uniform SDW and the soliton walls lattice state occurring at $A(t'_y) = 0$. On the other hand, when $B(t'_y) > 0$, the phase transition is of first order and happens as $A(t'_y)$ takes a finite negative value.

In the dilute limit of soliton walls, the density n_{sw} can be approximated by

$$n_{\text{sw}} = \frac{E_+}{v_{\text{F}} K \sqrt{1 - E_-^2/E_+^2}} \approx \frac{\Delta_0}{v_{\text{F}} \ln(4\Delta_0/E_-)}, \tag{6.77}$$

where $E_+ \approx \Delta_0$ at $n \to 0$ and $K(s)$ is the complete elliptic integral of the first kind. From Eq.(6.77), one can reach

$$E_- \approx 4\Delta_0 \exp\left(-\frac{\Delta_0}{n_{\text{sw}} v_{\text{F}}}\right). \tag{6.78}$$

For the detailed form of the anti-nesting term $t'_y(k_y)$, Gorkov and Grigoriev [231] invoke the following two models:

model 1 - periodic step-like function:

$$t'_y(k_y) = \begin{cases} t'_y, & 0 < k_y < \pi/b; \\ -t'_y, & \pi/b < k_y < 2\pi/b; \end{cases} \tag{6.79}$$

model 2 - tight-binding dispersion:

$$t'_y(k_y) = t'_y \cos(2k_y b). \tag{6.80}$$

We stress that the first model is an oversimplified model, while the second one corresponds to next-nearest neighbor hopping and is more realistic.

By numerically evaluating and minimizing the energy Eq.(6.75) with optimal value of E_- (or equivalently n_{sw}), Gorkov and Grigoriev [231] determine the position where $A(t'_y)$ changes sign at $t_{\mathrm{c1}} \approx 0.64\Delta_0$. This point corresponds to the second-order transition [with $B(t'_y) > 0$] above which the uniform SDW phase gives way to the soliton phase. By comparing the energy density of the soliton phase with that of the normal phase

$$W_n = -\frac{1}{\pi v_{\mathrm{F}}} \int \frac{b[t'_y(k_y)]^2 dk_y}{2\pi}, \tag{6.81}$$

the transition point between soliton and normal phases is also obtained at $t_{\mathrm{c}} \approx 1.08\Delta_0$. As a consequence, there exists a parameter window $t_{\mathrm{c1}} < t'_y < t_{\mathrm{c}}$ for a stable soliton phase in the framework of model 1 dispersion Eq.(6.79).

For model 2 with spectrum Eq.(6.80), a similar calculation gives equal values for t_{c1} and t_{c}, i.e., $t_{\mathrm{c1}} = t_{\mathrm{c}} = \Delta_0$, indicating that a stable region for soliton phase is absent. In fact, it can be proven that model 1 dispersion corresponds to the largest possible interval of a soliton phase. In realistic materials, the soliton phase will exist in parameter regions between model 1 and model 2. Assuming the anti-nesting intensity t'_y is monotonically increased with pressure, Gorkov and Grigoriev [231] attribute the coexistence region of SDW and metal phases to the emergence of a soliton phase. Within this soliton phase, a stack of soliton walls may be treated as metallic sheets. These layered conductors will overlap with increasing number of soliton walls, and eventually form a three-dimensional metallic band that lies inside the SDW gap. In a subsequent publication, Gorkov and Grigoriev [232] analyze the pairing instability in this layered conducting states, and conclude that a triplet superconducting state can be stabilized as the primary pairing instability.

To describe the spatial modulation of SDW in the presence of soliton walls, Gorkov and Grigoriev [232] consider a general case of SDW order parameter

$$\Delta_{\mathrm{SDW}}(\boldsymbol{r}) = \Delta_{\mathrm{SDW}}(x) \cos(\boldsymbol{Q} \cdot \boldsymbol{r}) (\hat{\boldsymbol{\sigma}} \cdot \hat{\boldsymbol{m}}), \tag{6.82}$$

where $\boldsymbol{Q}$ is the SDW nesting vector with $\boldsymbol{Q} = (\pm 2k_{\mathrm{F}}, \pi/b, 0)$, and $\hat{\boldsymbol{m}}$ stands for the direction of the staggered magnetization. The SDW variation in the presence of soliton walls is described by the periodic function $\Delta_{\mathrm{SDW}}(x)$, which can be solved exactly for the commensurate case [249] and in the limit of a single soliton wall [244]. Here, we also drop the third dimensional k_z-dependence out of the problem for brevity. Adding this extra degrees of freedom causes no changes in neither the derivation nor the qualitative conclusion. Thus, the Schrödinger's equation for free electrons reads $\mathscr{H}_{\boldsymbol{k}}^0 \Psi_{\boldsymbol{k}} = E(\boldsymbol{k}) \Psi_{\boldsymbol{k}}$, where the noninteracting Hamiltonian is

$$\mathscr{H}_{\boldsymbol{k}}^0 = \begin{pmatrix} \varepsilon_\alpha(\boldsymbol{k}_\perp) - iv_{\mathrm{F}}\partial_x & \Delta_{\mathrm{SDW}}(x)\,(\hat{\boldsymbol{\sigma}}\cdot\hat{\boldsymbol{m}}) \\ \Delta^*_{\mathrm{SDW}}(x)\,(\hat{\boldsymbol{\sigma}}\cdot\hat{\boldsymbol{m}}) & \varepsilon_\alpha(\boldsymbol{k}_\perp - \boldsymbol{Q}_\perp) + iv_{\mathrm{F}}\partial_x \end{pmatrix} \tag{6.83}$$

with $\boldsymbol{k}_\perp = (0, k_y, 0)$. Notice that the dispersion $\varepsilon_{\alpha=+}(\boldsymbol{k}_\perp) = \varepsilon_{\alpha=-}(\boldsymbol{k}_\perp)$ for $k_x = 0$. The four-component wavefunction are defined as

$$\Psi_{\boldsymbol{k}} = \left(\psi^{\mathrm{R}}_{\boldsymbol{k},\uparrow}(x), \psi^{\mathrm{R}}_{\boldsymbol{k},\downarrow}(x), \psi^{\mathrm{L}}_{\boldsymbol{k}-\boldsymbol{Q},\uparrow}(x), \psi^{\mathrm{L}}_{\boldsymbol{k}-\boldsymbol{Q},\downarrow}(x)\right)^{\mathrm{T}}, \tag{6.84}$$

where T stands for matrix transpose. In comparison to the normal state, the presence of the SDW background introduces extra terms to the pairing instability from the off-diagonal Green's functions G_{RL} and G_{LR}, which correspond to the mixing of the two sheets of the Fermi surface. The 4×4 matrix Green's functions can be written as

$$\hat{G}\left(\boldsymbol{r}, \boldsymbol{r}', i\omega_n\right) = -\sum_{\boldsymbol{k}_\perp, \lambda} \frac{\Psi^\dagger_{\boldsymbol{k}}(\boldsymbol{r}') \otimes \Psi_{\boldsymbol{k}}(\boldsymbol{r})}{i\omega_n - E_\lambda(\boldsymbol{k}_\perp)}, \tag{6.85}$$

where the summation over λ runs over all eigenstates of the Schrödinger's equation.

The superconducting state is characterized by the anomalous Green's function

$$f^{\mathrm{LR}}_{ss'}(\boldsymbol{r}) = \langle \psi^{\mathrm{L}}_s(\boldsymbol{r}) \psi^{\mathrm{R}}_{s'}(\boldsymbol{r}) \rangle, \tag{6.86a}$$

$$f^{\mathrm{RL}}_{ss'}(\boldsymbol{r}) = \langle \psi^{\mathrm{R}}_s(\boldsymbol{r}) \psi^{\mathrm{L}}_{s'}(\boldsymbol{r}) \rangle. \tag{6.86b}$$

Upon spatial inversion symmetry, the pairing states can be classified into two symmetry distinct states:

singlet state:

$$f^{\mathrm{LR}}_{ss'} = f^{\mathrm{RL}}_{ss'}, \tag{6.87}$$

triplet state:

$$f^{\mathrm{LR}}_{ss'} = -f^{\mathrm{RL}}_{ss'}, \tag{6.88}$$

After summing over the Cooper's ladder diagrams, the equation for the anomalous Green's function $f^{\mathrm{LR}}_{ss'}$ reads

$$\begin{aligned} f^{\mathrm{LR}}_{ss'} = -Tg \sum_{\boldsymbol{k},i\omega_n} \Big[& G^{\mathrm{RR}}(\boldsymbol{k}, i\omega_n) f^{\mathrm{RL}}_{ss'} G^{\mathrm{LL}}(-\boldsymbol{k}, -i\omega_n) \\ & + (\hat{\boldsymbol{\sigma}} \cdot \hat{\boldsymbol{m}})\, G^{\mathrm{LR}}(\boldsymbol{k}, i\omega_n) f^{\mathrm{LR}}_{ss'} (\hat{\boldsymbol{\sigma}} \cdot \hat{\boldsymbol{m}})^{\mathrm{T}}\, G^{\mathrm{RL}}(-\boldsymbol{k}, -i\omega_n) \Big], \end{aligned} \tag{6.89}$$

where g is the inter-electron interaction, and $G^{\alpha\alpha'}(\boldsymbol{k}, i\omega_n)$ is the corresponding component of the matrix Green's function Eq.(6.85) after Fourier transform to the momentum space by assuming translational symmetry. The equation for $f^{\mathrm{RL}}_{ss'}$ can be inferred similarly by assigning the corresponding parity for singlet or triplet states.

For the singlet pairing state, the spin structure of the anomalous Green's function is

$$f^{\mathrm{LR}}_{ss'} = f^{\mathrm{RL}}_{ss'} = i\,(\sigma_y)_{ss'}\, f^{\mathrm{LR}}_0. \tag{6.90}$$

Considering the identity $\sigma_y\,(\hat{\boldsymbol{\sigma}} \cdot \hat{\boldsymbol{m}}) = -\,(\hat{\boldsymbol{\sigma}} \cdot \hat{\boldsymbol{m}})^{\mathrm{T}}\,\sigma_y$, Eq.(6.89) can be rewritten as

$$\begin{aligned} 1 = -Tg \sum_{\boldsymbol{k},i\omega_n} \Big[& G^{\mathrm{RR}}(\boldsymbol{k}, i\omega_n) G^{\mathrm{LL}}(-\boldsymbol{k}, -i\omega_n) \\ & - G^{\mathrm{LR}}(\boldsymbol{k}, i\omega_n) G^{\mathrm{RL}}(-\boldsymbol{k}, -i\omega_n) \Big]. \end{aligned} \tag{6.91}$$

The minus sign before the second term on right-hand-side of the equation above originates from the spin structure of the background SDW phase. As a consequence, the logarithmic divergence from the two terms cancel with each other, leading to a decrease of the intensity of the pairing instability in the singlet channel.

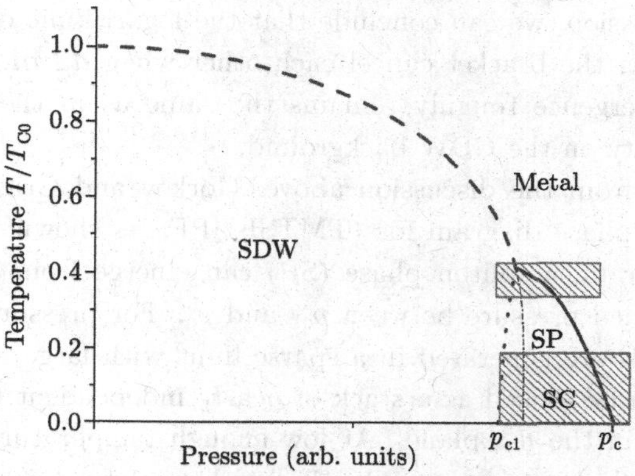

Figure 6.9 A schematic phase diagram for $(TMTSF)_2PF_6$. The solid line indicates the first-order transition between the metal and the soliton phase (SP), while the dotted line stands for the second-order transition between uniformly gapped SDW phase and the SP. The dashed line represents the line of absolute instability of metallic phase towards the formation of SDW. At high temperatures this line represents the second-order transition line between the SDW and metallic phases. The lower filled bar shows the region where superconductivity appears. The upper filled bar shows the high-temperature region of soliton phase where substantial effects are expected due to the thermal excitations in the SDW and the SP. The detailed shape of all lines is sensitive to the choice of the "anti-nesting" term $t'_y(k_y)$. From Ref. [231].

On the other hand, this cancellation may not occur in the triplet pairing case. For such a state, the spin structure of the anomalous Green's function becomes

$$f_{ss'}^{\mathrm{LR}} = -f_{ss'}^{\mathrm{RL}} = \left(\hat{\boldsymbol{\sigma}} \cdot \hat{\boldsymbol{d}} \sigma_y\right)_{ss'} f_0^{\mathrm{LR}}, \tag{6.92}$$

where $\boldsymbol{d}$ denotes the spin direction of the triplet Cooper pair. Using the identity

$$(\hat{\boldsymbol{\sigma}} \cdot \hat{\boldsymbol{m}}) \left(\hat{\boldsymbol{\sigma}} \cdot \hat{\boldsymbol{d}}\right) \sigma_y (\hat{\boldsymbol{\sigma}} \cdot \hat{\boldsymbol{m}})^{\mathrm{T}} = \left(\hat{\boldsymbol{\sigma}} \cdot \hat{\boldsymbol{d}}\right) \sigma_y - 2\left(\hat{\boldsymbol{d}} \cdot \hat{\boldsymbol{m}}\right) (\hat{\boldsymbol{\sigma}} \cdot \hat{\boldsymbol{m}}) \sigma_y, \tag{6.93}$$

equation (6.89) becomes

$$\begin{aligned} 1 = -Tg \sum_{\boldsymbol{k}, i\omega_n} \Big\{ & G^{\mathrm{RR}}(\boldsymbol{k}, i\omega_n) \left(\hat{\boldsymbol{\sigma}} \cdot \hat{\boldsymbol{d}}\right) G^{\mathrm{LL}}(-\boldsymbol{k}, -i\omega_n) \\ & - \left[\left(\hat{\boldsymbol{\sigma}} \cdot \hat{\boldsymbol{d}}\right) - 2\left(\hat{\boldsymbol{d}} \cdot \hat{\boldsymbol{m}}\right) (\hat{\boldsymbol{\sigma}} \cdot \hat{\boldsymbol{m}})\right] G^{\mathrm{LR}}(\boldsymbol{k}, i\omega_n) G^{\mathrm{RL}}(-\boldsymbol{k}, -i\omega_n) \Big\}. \end{aligned} \tag{6.94}$$

From this expression, we can conclude that the logarithmic divergences from the two terms in the bracket cancel each other when $\boldsymbol{d} \perp \boldsymbol{m}$. For $\boldsymbol{d} \| \boldsymbol{m}$, the intensity of divergence roughly remains the same as in the case of singlet superconductivity on the CDW background.

Concluding from the discussion above, Gorkov and Grigoriev [231, 232] propose a *P*-*T* phase diagram for $(TMTSF)_2PF_6$ as shown in Fig. 6.9. On this phase diagram, a soliton phase (SP) can emerge from the SDW background for applied pressure between p_{c1} and p_c. For pressure slightly above p_{c1}, soliton walls are generated in a sparse limit with large mutual distance. The system can be looked as a stack of nearly independent two-dimensional metallic sheets in the *y*-*z* plane. At low enough temperatures, these metallic soliton walls become superconducting with triplet pairing symmetry. As pressure gets further increased, more soliton walls appear and start to interact with each other. In this case, the superconducting states within each sheet are connected via Josephson effect, and the system enters the three-dimensional superconducting state.

7
Origins of Cooper Pairing

In addition to the pairing symmetry discussed in the previous chapters, another fundamental question in research on the superconducting states in Bechgaard salts is the pairing mechanism. Due to the quasi-one-dimensional nature of the crystallographic structure, early discussion focuses on a low energy effective theory called g-ology for pure one-dimensional interacting electron gas, i.e., the Luttinger liquid model [71]. In the g-ology phase diagram, four phases are present, including the SDW, CDW, singlet superconducting (SSC), and triplet superconducting (TSC) states, as illustrated in Fig. 7.1. The TSC and SDW phases are separated by a second order phase transition for positive background scattering. This observation intrigues people to relate this model with quasi-1D Bechgaard salts as discussed in some studies [41]. Then, the TSC–SDW phase boundary has been discussed as having SO(4) symmetry [119, 120]. Nevertheless, since exact numerical studies on the pure-1D extended Hubbard model, where the on-site repulsive U and the nearest neighbor V is considered [250], show that superconductivity is not present in a realistic parameter region, it is likely that some kind of attractive interaction should be necessary within the g-ology picture.

In conventional BCS theory, the attractive interaction between electrons that have opposite spins is mediated by electron-phonon interaction. A naive generalization of this mechanism is to pair electrons with arbitrary spins and assume that the effective electron-electron interaction is not spin dependent, as long as the spin of a single electron does not change before and after scattering. If the phonon-mediated interaction is isotropic, it is natural to expect that the isotropic s-wave singlet pairing state would win for it minimizes kinetic energy. However, an observation of anisotropic pairing does not necessarily exclude a phonon-mediated pairing mechanism. For example, Foulkes and Gyorffy showed that the short-range Coulomb interaction can suppress the s-wave pairing and favor the p-wave state in metals such as Rh, W and Pd [251]. In connection with high-T_c superconductors, Abrikosov proposed an anisotropic s-wave pairing state originated from phonon-mediated interaction with weak Coulomb screening [252, 253]. This anisotropic s-wave pairing will give its way to d-wave superconductivity if interactions mediated by antiferromagnetic (AF) spin fluctuation is also included into consideration [254, 255]. The phonon mediated pairing mechanism has also been discussed in the context of quasi-2D organic superconductor κ-BEDT-TTF [256], and heavy fermion compound UGe_2 [257]. Moreover, it is also proposed that spin-triplet superconductivity can be stabilized by the phonon-mediated interaction in strontium ruthenate [258].

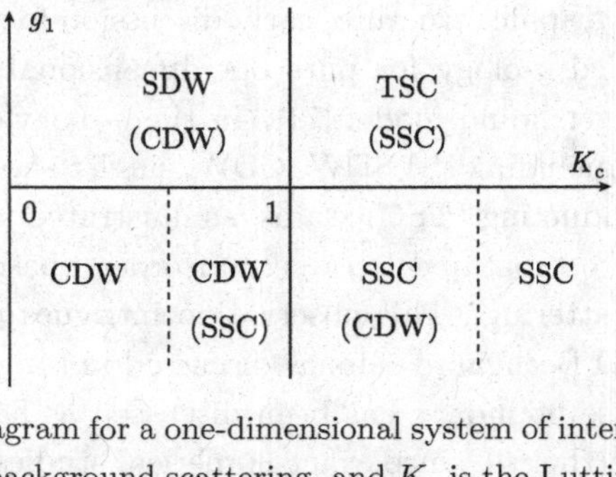

Figure 7.1 Phase diagram for a one-dimensional system of interacting spin-1/2 fermions. Here, g_1 is the background scattering, and K_c is the Luttinger parameter in the charge sector.

For quasi-1D Bechgaard salts, the pairing symmetry within the phonon-mediated interactions is investigated by several authors, where specific attention has been paid on the possibility of triplet pairing states. Kohmoto and Sato (KS) [260] have shown that a nodeless spin-triplet p_x-wave superconducting state would become stable with a phonon mediated pairing interaction and AF spin fluctuations. Meanwhile, by considering screened electron-phonon interactions, Suginishi and Shimahara (SS) [261] also concluded that the p_x-wave state is more favorable than the p_y-wave state and the d-wave state.

Another possible pairing mechanism is the AF spin fluctuation, which is originally discussed by Emery in 1986 in the context of high-T_c cuprates where an AF state is found next to the superconducting state on the phase diagram [262]. Then, it is natural to consider the same pairing scenario for Bechgaard salts for the system also presents a phase diagram of neighboring superconducting and SDW phases. Such a possibility has been discussed by several theoretical proposals using random phase approximation (RPA) [263], fluctuation exchange approach (FLEX) [264], third order perturbation theory [265], and quantum Monte Carlo method (QMC) [266] based on a quasi-1D single-band Hubbard model, and all suggest a singlet d-wave pairing state mediated by $2k_F$ spin fluctuation.

This resulting singlet d-wave pairing is apparently inconsistent with the NMR measurement in $(TMTSF)_2PF_6$ which suggests a triplet superconducting state, as discussed in Sec. 3.2. In order to solve this puzzle, Kuroki, Arita, and Aoki (KAA) [43] proposed a pairing mechanism which is mediated by coexisting spin and charge fluctuations, and suggested a nodeless p_x-wave symmetry for $(TMTSF)_2ClO_4$ and a nodal f-wave symmetry for $(TMTSF)_2PF_6$.

In addition, other proposals have also been raised as possible mechanisms for triplet pairing state. For example, Rozhkov and Millis studied isolated electron chains with Ising anisotropy, and found that a spin gap will open at high energies [47]. This high energy gap will dramatically change the intra-chain interactions and particle hopping at low temperatures, such that the single-particle hopping will be suppressed and the physics of coupled chains is dominated by a competition between pair hopping and exchange interaction. As a result, they found a phase diagram where SDW and triplet superconductivity are separated by a first-order transition. Besides, Ohta proposed a pairing mechanism based on the fact that in a triangle lattice consisting of three sites with two electrons, a ferromagnetic interaction arises by considering a consecutive exchange of the positions of electrons [48]. If ferromagnetic spin fluctuation arises due to this "ring-exchange" scenario, triplet superconductivity may be stabilized. However, both these proposals can not be directly applied to Bechgaard salts. In the theory of Rozhkov and Millis, a prerequisite is a strong Ising anisotropy of spin-spin interaction, which, however, is quite weak in Bechgaard salts [267]. Meanwhile, the calculation of Ohta's proposal is only restricted to a pure one-dimensional electron chain, although extension to quasi-1D cases may be feasible.

In this chapter, we first introduce the phonon-mediated pairing mechanisms in Sec. 7.1. In Sec. 7.2, the effect of antiferromagnetic spin fluctuation is discussed. The coexistence of spin and charge fluctuations and its resulting superconducting states are studied in Sec. 7.3. Finally, we investigate the Ising anisotropy and "ring-exchange" mechanisms in Sec. 7.4.

7.1 Phonon mediated attraction

In the discussion by Kohmoto and Sato (KS) [260], an effective interaction Hamiltonian with phonon-mediated interaction (ph) of the following form is considered

$$\mathscr{H}_{\rm int}^{\rm ph} = -\sum_{\boldsymbol{k}\boldsymbol{k}'\boldsymbol{q}}\sum_{\alpha\beta} V(\boldsymbol{q}) c^\dagger_{\boldsymbol{k}+\boldsymbol{q},\alpha} c_{\boldsymbol{k},\alpha} c^\dagger_{\boldsymbol{k}'-\boldsymbol{q},\beta} c_{\boldsymbol{k}',\beta}, \tag{7.1}$$

where α and β are spin indices. Considering Cooper pairs with zero center-of-mass momentum, this expression can be rewritten as

$$\mathscr{H}_{\rm int}^{\rm ph} = -\frac{1}{2}\sum_{\boldsymbol{k}\boldsymbol{q}} V(\boldsymbol{q}) \left[\Delta^\dagger(\boldsymbol{k}+\boldsymbol{q})\Delta(\boldsymbol{k}) + \boldsymbol{D}^\dagger(\boldsymbol{k}+\boldsymbol{q})\cdot\boldsymbol{D}(\boldsymbol{k})\right], \tag{7.2}$$

where

$$\Delta(\boldsymbol{k}) = \sum_{\alpha\beta} c_{\boldsymbol{k},\alpha}\left(\sigma_y\right)_{\alpha\beta} c_{-\boldsymbol{k},\beta}, \tag{7.3}$$

$$\boldsymbol{D}(\boldsymbol{k}) = \sum_{\alpha\beta} c_{\boldsymbol{k},\alpha}\left(\sigma_y\hat{\sigma}\right)_{\alpha\beta} c_{-\boldsymbol{k},\beta} \tag{7.4}$$

are singlet and triplet order parameters, respectively. KS assume that the interaction $V(\boldsymbol{q})$ is peaked at $\boldsymbol{q} = \boldsymbol{0}$, as in conventional BCS theory, and approximate Eq. (7.2) by

$$\mathscr{H}_{\rm int}^{\rm ph} = -\frac{1}{2}\sum_{\boldsymbol{k}} V_0 \left[\Delta^\dagger(\boldsymbol{k})\Delta(\boldsymbol{k}) + \boldsymbol{D}^\dagger(\boldsymbol{k})\cdot\boldsymbol{D}(\boldsymbol{k})\right]. \tag{7.5}$$

This result shows that the phonon-mediated interaction Eq. (7.1) has the same strength in the singlet and triplet pairing channels. Therefore, a superconducting state with no nodes will be favored since it minimizes the kinetic energy. In a system of connected Fermi surfaces, a spin-triplet order parameter must have nodes due to requirements of continuity, parity, and Pauli principle, thus s-wave singlet superconductivity is more favorable. However, quasi-1D systems have disconnected Fermi surfaces as shown in Fig. 2.4. Thus the order parameter can be nearly a constant along each Fermi sheet, but change sign from one sheet to another, as in the case of p_x-state. This nodeless p_x-wave (triplet) state hence is as favorable as s-wave (singlet) superconductivity, and in this case, the competition between singlet and triplet channels becomes subtle.

The discussion above is oversimplified since only a spin-independent effective coupling mediated by phonons was considered. It is ignored that the superconducting state of quasi-1D systems neighbors an antiferromagnetic (AF) state characterized by an SDW with $\boldsymbol{Q} \approx 2\boldsymbol{k}_{\mathrm{F}}$, as seen on the generic phase diagram Fig. 2.4. Adding to the interaction Hamiltonian a term due to AF fluctuations [260]

$$\mathscr{H}_{\mathrm{int}}^{\mathrm{AF}} = -\sum_{\boldsymbol{k}\boldsymbol{k}'\boldsymbol{q}} \sum_{\alpha\beta\gamma\delta} J(\boldsymbol{q}) \left[c^{\dagger}_{\boldsymbol{k}+\boldsymbol{q},\alpha} \hat{\sigma}_{\alpha\beta} c_{\boldsymbol{k},\beta}\right] \cdot \left[c^{\dagger}_{\boldsymbol{k}'-\boldsymbol{q},\gamma} \hat{\sigma}_{\gamma\delta} c_{\boldsymbol{k}',\delta}\right] \tag{7.6}$$

can help distinguish between the singlet and triplet channels. If the coupling constant $J(\boldsymbol{q}) > 0$ is peaked at a finite wave vector $\boldsymbol{Q} \approx 2\boldsymbol{k}_F$ (nesting vector), the AF term can be approximated by [260]

$$\mathscr{H}_{\mathrm{int}}^{\mathrm{AF}} = \sum_{\boldsymbol{k}\approx\boldsymbol{k}_{\mathrm{F}}} J(\boldsymbol{Q}) \left[3\Delta^{\dagger}(\boldsymbol{Q}-\boldsymbol{k})\Delta(\boldsymbol{k}) + \boldsymbol{D}^{\dagger}(\boldsymbol{Q}-\boldsymbol{k}) \cdot \boldsymbol{D}(\boldsymbol{k})\right]. \tag{7.7}$$

Therefore, in the case of nodeless order parameters (s- or p_x-wave), AF fluctuations tend to suppress more the spin-singlet than the spin-triplet state, as can be seen in Eq. (7.7). For strictly one-dimensional systems where $t_y = t_z = 0$, the nesting vector $\boldsymbol{Q} = (Q_x, 0, 0)$, and hence the triplet (p_x-wave) order parameter is non-zero along each independent Fermi surface, and has opposite signs on the left and right sheets. Given the topology of the Fermi surface, the superconducting p_x-state is nodeless just like the s-wave state. Based on this, the kinetic energy of the two cases is similar, however as discussed above, the potential energy for AF fluctuations is more detrimental to s-wave. Thus, the combined analysis of kinetic energy and potential energy terms leads to the conclusion that the p_x-state is more favorable for strictly one-dimensional systems.

The mean field calculation discussed above provides a phonon-mediated pairing mechanism that favors triplet pairing, but it also requires assistance from AF spin fluctuations to suppress singlet pairing. A relevant question here is whether the phonon-mediated interaction alone can stabilize triplet pairing state. To address this issue, Suginishi and Shimahara (SS) consider phonon-mediated interactions with three different models, and investigate the stability and priority of different possible pairing symmetries [261]. By considering moderately screened phonons with corrections from charge fluctuations, they obtain an attractive pairing interaction that has a large magnitude around $\boldsymbol{q} \sim 0$ and a smaller peak around $\boldsymbol{q} \sim \boldsymbol{Q}$, where $\boldsymbol{Q}$ is the nesting vector. By further considering the finite range Coulomb pseudo-potential, which suppresses only the s-wave pairing, they conclude that the p_x-wave pairing state dominates in a parameter region which covers the realistic material case.

The three models discussed by SS are:

model a - conventional phonon-mediated pairing interaction

$$V_{\text{eff}}(\boldsymbol{q}) = -g \max_{\boldsymbol{K}} \frac{q_s^2}{|\boldsymbol{q}-\boldsymbol{K}|^2 + q_s^2}; \tag{7.8}$$

model b - effective interaction with corrections due to charge fluctuations

$$V_{\text{eff}}(\boldsymbol{q}) = -g \max_{\boldsymbol{K}} \frac{q_s^2}{|\boldsymbol{q}-\boldsymbol{K}|^2 + q_s^2 \chi_0(\boldsymbol{q}) \mathcal{N}^{-1}(0)}; \tag{7.9}$$

model c - effective interaction with corrections due to charge fluctuations and short-range Coulomb interaction U

$$V_{\text{eff}}(\boldsymbol{q}) = -\frac{g}{\kappa_U^2(\boldsymbol{q})} \max_{\boldsymbol{K}} \frac{q_s^2}{|\boldsymbol{q}-\boldsymbol{K}|^2 + q_s^2 \chi_U(\boldsymbol{q}) \mathcal{N}^{-1}(0)}. \tag{7.10}$$

Here, the factor g denotes the interaction strength, q_s^{-1} represents the range of the pairing interaction, $\boldsymbol{K}$ is the reciprocal lattice vector, $\mathcal{N}(0)$ is the density of states at Fermi energy, and the dielectric function $\kappa_U(\boldsymbol{q}) = 1 + U\chi_0(\boldsymbol{q})$. The static susceptibility with (χ_U) and without (χ_0) correction from Coulomb interaction U are

$$\chi_U(\boldsymbol{q}) = \frac{\chi_0(\boldsymbol{q})}{1+U\chi_0(\boldsymbol{q})}, \tag{7.11a}$$

$$\chi_0(\boldsymbol{q}) = \frac{1}{N} \sum_{\boldsymbol{k}} \frac{n_f(\xi_{\boldsymbol{k}}) - n_f(\xi_{\boldsymbol{k}+\boldsymbol{q}})}{\xi_{\boldsymbol{k}+\boldsymbol{q}} - \xi_{\boldsymbol{k}}} \tag{7.11b}$$

with $n_f(x)$ the Fermi function, N the number of lattice sites, and $\xi_{\boldsymbol{k}} = \varepsilon_{\boldsymbol{k}} - \mu$.

The range of the effective interaction q_s^{-1} relies on the detail of the mechanism of the electron-ion interaction. But it should be on the order of the screening length since it must reflect the range of the Coulomb interaction between electrons and ions. In real materials, the screening charges could not exist within the distance shorter than the intermolecular distance. Therefore, it is natural to assume $q_s^{-1} \gtrsim a$ in the application to $(TMTSF)_2X$ systems with lattice constant a along the easy conducting axis.

The gap equation reads

$$\Delta(\boldsymbol{k}) = -\frac{1}{N}\sum_{\boldsymbol{k}'} V_{\text{eff}}(\boldsymbol{k}-\boldsymbol{k}')\frac{\tanh(E_{\boldsymbol{k}'}/2T)}{2E_{\boldsymbol{k}'}}\Delta(\boldsymbol{k}'), \tag{7.12}$$

where $E_{\boldsymbol{k}'} = \sqrt{\xi_{\boldsymbol{k}'}^2 + |\Delta(\boldsymbol{k}')|^2}$ is the quasiparticle dispersion, and the summation over $\boldsymbol{k}'$ is restricted to $|\xi_{\boldsymbol{k}'}| < \omega_{\text{D}}$ near the Fermi surface with ω_{D} the Debye frequency. By neglecting the z-dependence in both $\Delta(\boldsymbol{k})$ and $V_{\text{eff}}(\boldsymbol{q})$, and adopting the linearized dispersion near the Fermi surface, the gap equation can be transferred into the matrix form near the superconducting transition temperature T_{c}

$$\begin{pmatrix} \Delta^{(+)}(k_y) \\ \Delta^{(-)}(k_y) \end{pmatrix} = -\frac{1}{\lambda}\int_{-\pi/b}^{\pi/b}\frac{b\,dk_y'}{2\pi}\mathscr{N}(k_y') \times \begin{pmatrix} V_{\text{eff}}^{(++)}(k_y,k_y') & V_{\text{eff}}^{(+-)}(k_y,k_y') \\ V_{\text{eff}}^{(-+)}(k_y,k_y') & V_{\text{eff}}^{(--)}(k_y,k_y') \end{pmatrix}\begin{pmatrix} \Delta^{(+)}(k_y') \\ \Delta^{(-)}(k_y') \end{pmatrix}, \tag{7.13}$$

where the dimensionless coupling constant λ is defined as

$$\frac{1}{\lambda} = \ln\left(\frac{2e^{\gamma}\omega_{\text{D}}}{\pi T_{\text{c}}}\right), \tag{7.14}$$

$\mathscr{N}(k_y') = (1/2\pi t_x)\sin[k_{Fx}(k_y')a]$ is the density of states at position k_y' on the Fermi surface, $k_{\text{F}x}$ is the x-component of the Fermi wave vector, $\Delta^{(\pm)}(k_y)$ are order parameters along the two open Fermi sheets, and γ is the Euler constant. The matrix element $V_{\text{eff}}^{(\pm\pm)}$ are defined as

$$V_{\text{eff}}^{(++)}(k_y,k_y') = V_{\text{eff}}^{(--)}(k_y,k_y') \equiv V(k_{\text{F}x}(k_y) - k_{\text{F}x}(k_y'), k_y - k_y'), \tag{7.15a}$$

$$V_{\text{eff}}^{(+-)}(k_y,k_y') = V_{\text{eff}}^{(-+)}(k_y,k_y') \equiv V(k_{\text{F}x}(k_y) + k_{\text{F}x}(k_y'), k_y - k_y'). \tag{7.15b}$$

By solving the eigenvalue λ from the matrix-form gap equation (7.13), one can obtain the superconducting transition temperature T_{c} via

$$T_{\text{c}} = \frac{2e^{\gamma}}{\pi}\omega_{\text{D}}e^{-1/\lambda}. \tag{7.16}$$

Thus, a higher value of λ corresponds to a higher transition temperature of the pairing state, which is hence more favorable compared with other possible pairing states.

In the context of Bechgaard salts, Suginishi and Shimahara [261] solve the gap equation (7.13) numerically for several different pairing symmetries including singlet s-, d-, and d_{xy}-wave, and triplet p_x- and p_y-wave states. They set $\omega_D = 200\text{K}$ and $t_y/t_x = 0.1$, and consider all three models in the description of phonon-mediated pairing. In Fig. 7.2, the resulting dimensionless coupling constants λ are plotted as functions of the lattice constants ratio a/b, where other parameters are taken as $q_s a = 1$, $U/t_x = 0.75$, $n = 1/4$ and $g/t_x = 2.5$. Notice that the s-wave pairing interaction is reduced by incorporating the charge fluctuation and the short-range Coulomb interaction as in model (c). As a consequence, the p_x-wave state becomes favorable for $a/(2b) \lesssim 1.4$, which covers the case of $(\text{TMTSF})_2\text{X}$ with $a/(2b) \sim 0.468$. For even larger values of $a/(2b) \gtrsim 1.4$, the p_y-wave state is favored with highest transition temperature.

In Fig. 7.3, zero temperature phase diagrams are shown by varying the on-site Coulomb interaction U, the electron-phonon coupling g, the filling factor n, and the hopping integral anisotropy t_y/t_x. It is found that a triplet pairing p_x-wave state is favored for strong short-range interaction, weak electron-phonon coupling, large number density and enhanced quasi-one-dimensional anisotropy. In all these phase diagrams, the range of the pairing interaction is assumed to be $q_s a = 1.0$ as a typical example.

The theoretical proposals by Kohmoto and Sato [260], and Suginishi and Shimahara [261] discuss the possibility of spin-triplet pairing state based on

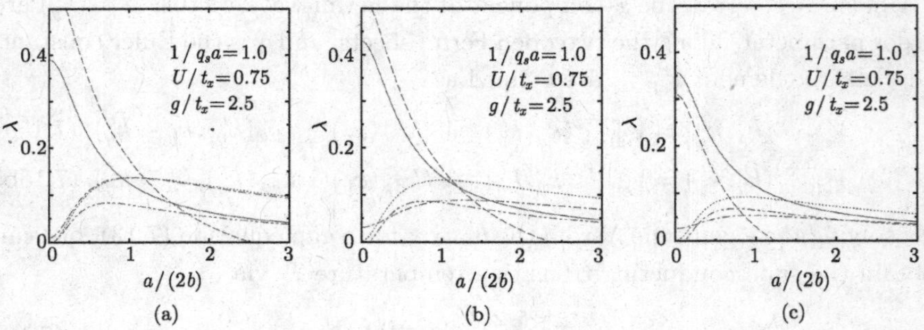

Figure 7.2 Dimensionless coupling constants λ as functions of the ratio a/b. The panels (a), (b), and (c) show results for the corresponding models. The solid, dotted, dashed, dotted-dashed, and double-dotted-dashed curves represent outcomes for p_x-, p_y-, s-, d- and d_{xy}-wave states, respectively. From Ref. [261].

interaction mediated by phonons. These results should be contrasted with scaling arguments and exact solutions in one dimension, when the two infrared instabilities (SDW nesting and Cooper pairing) are treated on the same footing in perturbation theory [71]. In the latter case, at variance with the mean-field described, the p_x triplet state is completely suppressed in the presence of singular antiferromagnetic correlations (and vice-versa) in the physically relevant sector of interactions with weak spin anisotropy.

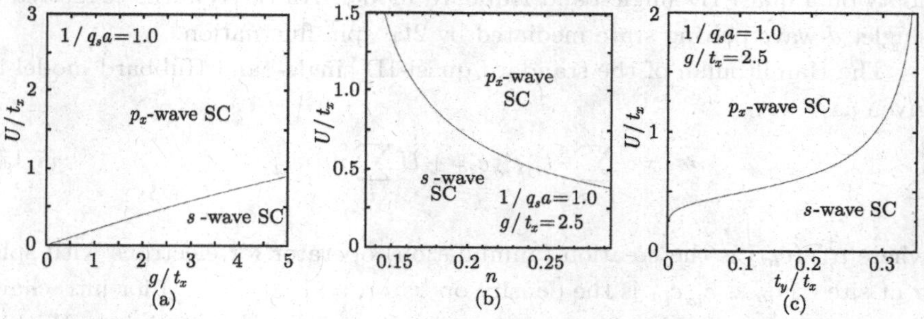

Figure 7.3 Phase diagram in the (a) g–U plane, (b) n–U plane, and (c) t_y–U plane at zero temperature. From Ref. [261].

Furthermore, additional care is required when applying mean field results to quasi-1D systems. For instance, repulsive intrachain interactions, which favor antiferromagnetic SDW order, are well known to promote unconventional singlet superconducting states [118, 262, 268]. For quasi-1D crystals with orthorhombic lattices, singlet d-wave pairing mediated by AF spin fluctuations is naturally expected [262, 263, 264, 266, 269, 270, 271]. While the singlet d-wave pairing scenario contradicts with the experimental facts pointing towards spin-triplet pairing, especially for $(TMTSF)_2PF_6$, a considerable amount of effort has been paid in attempting to accommodate triplet pairing state within the framework of AF fluctuation theories, as will be discussed in the next section.

7.2 Spin fluctuation mediated interaction

The mechanism of spin fluctuation mediated Cooper pairing has been originally discussed by Emery in 1986, motivated by the existence of antiferromagnetism in the close vicinity of superconducting state in high-T_c cuprates [262]. It is then natural for researchers to consider the same pairing mechanism for Bechgaard salts for the system presents the same phase diagram of neighboring superconducting and SDW phases. Along this line, several theoretical studies using random phase approximation (RPA) [263], fluctuation exchange approach (FLEX) [264], third order perturbation [265], or quantum Monte Carlo method (QMC) [266] have been performed to investigate superconductivity on a quasi-1D single-band Hubbard model. All these works suggested a singlet d-wave pairing state mediated by $2k_F$ spin fluctuation.

The Hamiltonian of the standard quasi-1D single-band Hubbard model is given as

$$\mathscr{H} = \sum_{<i,j>,\sigma} t_{ij} c_{i\sigma}^{\dagger} c_{j\sigma} + U \sum_{i} n_{i\uparrow} n_{i,\downarrow}, \tag{7.17}$$

where $c_{i\sigma}^{\dagger}$ ($c_{i\sigma}$) is the creation (annihilation) operator for electrons with spin σ at site i, $n_{i\sigma} \equiv c_{i\sigma}^{\dagger} c_{i\sigma}$ is the density operator, and $t_{ij} = t_x(t_y)$ for intrachain (interchain) hopping between nearest neighbors denoted by $< i, j >$. Within the framework of RPA, the pairing interactions in singlet and triplet channels can be written in the following form [272, 273],

$$\begin{aligned} V^{\mathrm{s}}(\boldsymbol{q}) &= U + \frac{3}{2} U^2 \chi_{\mathrm{s}}(\boldsymbol{q}) - \frac{1}{2} U^2 \chi_{\mathrm{c}}(\boldsymbol{q}), \\ V^{\mathrm{t}}(\boldsymbol{q}) &= -\frac{1}{2} U^2 \chi_{\mathrm{s}}(\boldsymbol{q}) - \frac{1}{2} U^2 \chi_{\mathrm{c}}(\boldsymbol{q}), \end{aligned} \tag{7.18}$$

where U is the on-site Hubbard repulsive interaction. The spin and charge susceptibilities χ_s and χ_c are given as

$$\begin{aligned} \chi_{\mathrm{s}}(\boldsymbol{q}) &= \frac{\chi_0(\boldsymbol{q})}{1 - U\chi_0(\boldsymbol{q})}, \\ \chi_{\mathrm{c}}(\boldsymbol{q}) &= \frac{\chi_0(\boldsymbol{q})}{1 + U\chi_0(\boldsymbol{q})}. \end{aligned} \tag{7.19}$$

Here, χ_0 is the bare susceptibility determined as

$$\chi_0(\boldsymbol{q}) = \frac{1}{N} \sum_{\boldsymbol{k}} \frac{n_{\mathrm{F}}(\varepsilon_{\boldsymbol{k}+\boldsymbol{q}}) - n_{\mathrm{F}}(\varepsilon_{\boldsymbol{k}})}{\varepsilon_{\boldsymbol{k}} - \varepsilon_{\boldsymbol{k}+\boldsymbol{q}}}, \tag{7.20}$$

where $n_F(x)$ is the Fermi distribution function. The linearized gap equation can be derived within the BCS theory, leading to

$$\lambda\Delta(\boldsymbol{k}) = -\sum_{\boldsymbol{k}'} V^{\mathrm{s,t}}(\boldsymbol{k}-\boldsymbol{k}')\frac{\tanh(\beta\varepsilon_{\boldsymbol{k}'}/2)}{2\varepsilon_{\boldsymbol{k}'}}\Delta(\boldsymbol{k}'). \tag{7.21}$$

By solving this equation, one can obtain the gap function $\Delta(\boldsymbol{k})$ as the eigenfunction, and the transition temperature T_c as the temperature where the eigenvalue λ reaches unity. Notice that the summation over $\boldsymbol{k}'$ on the right-hand-side of the gap equation is dominated by the contribution from $\boldsymbol{k}'$ on the Fermi surface. Thus, by multiplying $\Delta(\boldsymbol{k})$ on both sides of the gap equation and summing over $\boldsymbol{k}$ on the Fermi surface, one can easily draw the conclusion that the effective interaction

$$V_{\mathrm{eff}}^{\mathrm{s,t}} \equiv -\frac{\sum_{\boldsymbol{k},\boldsymbol{k}'} V^{\mathrm{s,t}}(\boldsymbol{k}-\boldsymbol{k}')\Delta(\boldsymbol{k})\Delta(\boldsymbol{k}')}{\sum_{\boldsymbol{k}}[\Delta(\boldsymbol{k})]^2} \tag{7.22}$$

must be large positive in order to have large λ, and hence to support superconductivity.

One intriguing feature of the quasi-1D band structure is the good nesting at quarter filling. As a consequence, the bare susceptibility $\chi_0(\boldsymbol{q})$ peaks at the nesting vector $\boldsymbol{q} = \boldsymbol{Q}_{2k_F}$. For repulsive Hubbard interaction $U > 0$, the spin susceptibility $\chi_s(\boldsymbol{q})$ becomes large at $\boldsymbol{q} = \boldsymbol{Q}_{2k_F}$, while the charge susceptibility $\chi_c(\boldsymbol{q})$ remains small at all $\boldsymbol{q}$. Within the RPA framework, the SDW transition takes place at temperature where $U\chi_0(\boldsymbol{Q}_{2k_F})$ reaches unity. Thus, in the vicinity of the SDW transition temperature, the pairing interactions in the singlet and triplet channels roughly satisfy the following relation [273]

$$V^{\mathrm{s}}(\boldsymbol{Q}_{2k_F}) = -3V^{\mathrm{t}}(\boldsymbol{Q}_{2k_F}) > 0. \tag{7.23}$$

Since both the singlet and triplet pairing interactions have large absolute values at nesting vectors, the condition for $V_{\mathrm{eff}}^{\mathrm{s,t}}$ in Eq. (7.22) to be large positive can be approximately written as

$$V^{\mathrm{s,t}}(\boldsymbol{Q}_{2k_F})\Delta(\boldsymbol{k})\Delta(\boldsymbol{k}+\boldsymbol{Q}_{2k_F}) < 0 \tag{7.24}$$

for both $\boldsymbol{k}$ and $\boldsymbol{k}+\boldsymbol{Q}_{2k_F}$ residing on the Fermi surface.

By combining the two conditions of Eqs. (7.23) and (7.24) together, one can see that the gap function needs to change sign between $\boldsymbol{k}$ and $\boldsymbol{k}+\boldsymbol{Q}_{2k_F}$ for singlet pairing, and keep the sign across the nesting vector for triplet pairing. Besides, since the pairing wavefunction also needs to be antisymmetric with respect to the exchange of electrons, the gap function must acquire even parity with $\Delta^s(\boldsymbol{k}) = \Delta^s(-\boldsymbol{k})$ for singlet pairing, and odd parity with $\Delta^t(\boldsymbol{k}) = -\Delta^t(-\boldsymbol{k})$ for triplet pairing. The gap functions fulfilling all these conditions are schematically shown in Fig. 7.4(a) and 7.4(b) for singlet and triplet cases, respectively. The singlet pairing state is usually referred as *d*-wave state in the sense that the gap changes sign as $+-+-$ along the Fermi surface. The triplet pairing state is called *f*-wave state since the gap changes sign as $+-+-+-$. Since the pairing interaction in the singlet channel is three times larger than that in the triplet channel, the *d*-wave state is more favorable over the *f*-wave state [263].

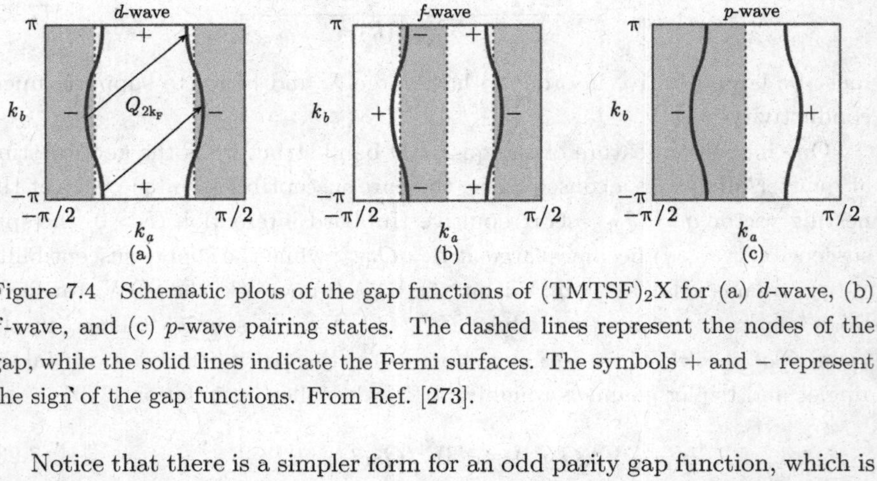

Figure 7.4 Schematic plots of the gap functions of $(TMTSF)_2X$ for (a) *d*-wave, (b) *f*-wave, and (c) *p*-wave pairing states. The dashed lines represent the nodes of the gap, while the solid lines indicate the Fermi surfaces. The symbols + and − represent the sign of the gap functions. From Ref. [273].

Notice that there is a simpler form for an odd parity gap function, which is referred as *p*-wave pairing state and changes sign as $+-$ along the Fermi surface. A schematic drawing of this state is shown in Fig. 7.4(c). However, this state does not satisfy the condition of Eq.(7.24) because the triplet interaction in this channel is negative. Thus, the *p*-wave pairing state is not favorable for the Hubbard model at least within RPA.

The conclusion of a stable *d*-wave pairing state can also be obtained via other approaches as summarized at the beginning of this section. For example, Kino and Kontani apply fluctuation exchange (FLEX) method to a single-band Hubbard model at half-filling, which corresponds to the strong dimerization limit, and obtain a finite T_c for the *d*-wave pairing [264]. Kuroki, Arita, and Aoki adopt the same approach to a two-band model with finite dimerization and with next nearest neighbor interchain hopping, and draw the same conclusion that the d-wave pairing state is most favorable [43]. The FLEX method will also be discussed with more details in the next section.

Another approach which has been applied to investigate the pairing states is the perturbation method. Nomura and Yamada consider a quasi-1D Hubbard model in the half-filled dimerization limit with next nearest neighbor interchain hopping [265]. By including perturbations to the self-energy up to third order of on-site repulsive interaction U, they also conclude that a singlet d-wave pairing state is more favorable than the triplet f-wave state. In Fig. 7.5, the eigenvalues λ of the gap equation (7.21) for singlet and triplet pairing states are plotted as functions of temperature. From these results, the transition temperature for singlet pairing state can be determined as the position where λ reaches unity.

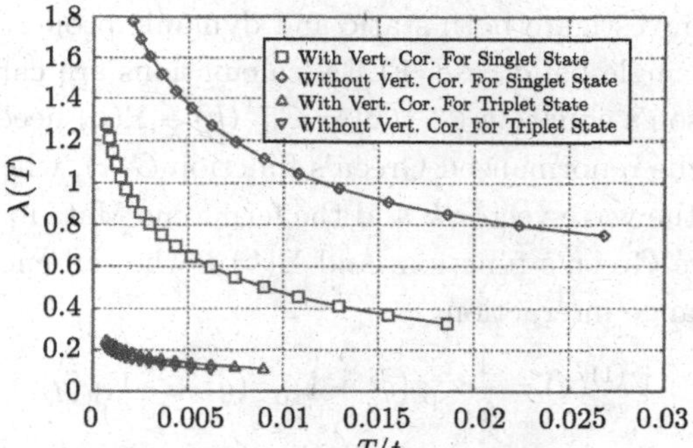

Figure 7.5 Optimal eigenvalues $\lambda(T)$ solved from the gap equation (7.21) via a third order perturbation method with (or without) vertex corrections for singlet and triplet pairing states. Note that the value of λ for triplet pairing state remains far below unity till lowest attainable temperature. From Ref. [265].

7.3 Coexistence of spin and charge fluctuations

As discussed in the previous section, AF fluctuations favor a singlet pairing state with a d-wave-like symmetry. This conclusion, however, is clearly inconsistent with the outcome of existing Knight shift and NMR experiments [33, 35], which suggests a triplet superconducting state in $(TMTSF)_2PF_6$ (see Sec.

3.2). In order to fulfill this gap, Kuroki, Arita and Aoki (KAA) [43] investigate the effects of the spin fluctuation anisotropy and the coexistence of spin and charge orders, and argue that a triplet state would be more favorable against singlet ones if both factors are taken into account. Based on their theory, KAA propose that the superconducting state of $(TMTSF)_2ClO_4$ has nodeless p_x-wave symmetry, while that of $(TMTSF)_2PF_6$ has f-wave symmetry with nodes.

In their proposal, KAA perform a fluctuation exchange (FLEX) calculation for a single-band Hubbard model

$$\mathscr{H} = \sum_{<i,j>,\sigma} t_{ij} c^{\dagger}_{i\sigma} c_{j\sigma} + U \sum_i n_{i\uparrow} n_{i\downarrow} \tag{7.25}$$

in an anisotropic two-dimensional rectangular lattice. Here, $t_{ij} = t_a$ $(= t_b)$ for i and j are nearest neighbors along $\boldsymbol{a}$ $(\boldsymbol{b}')$ direction. FLEX is a self-consistent method used to investigate both static and dynamic properties in interacting systems. In the single-band case, FLEX calculations are carried out in three steps. First, Dyson's equation $G^{-1}(k) = G_0^{-1}(k) - \Sigma(k)$ needs to be solved in order to obtain the renormalized Green's function $G(k)$, where $k = (\boldsymbol{k}, i\omega_n)$ is a shorthand for the wave vector $\boldsymbol{k}$ and the fermionic Matsubara frequency ω_n, $G_0(k)$ is the bare Green's function, and $\Sigma(k)$ is the self-energy. Second, the fluctuation-exchange interaction

$$V^{(1)}(q) = \frac{1}{2} V_{\mathrm{sp}}^{\boldsymbol{zz}}(q) + V_{\mathrm{sp}}^{+-}(q) + \frac{1}{2} V_{\mathrm{ch}}(q), \tag{7.26}$$

consisting of contributions from longitudinal ($\boldsymbol{zz}$) and transverse ($+-$) spin fluctuations (sp), and from charge fluctuations (ch) needs to be calculated. Notice that the boldfaced $\boldsymbol{z}$ is used here to denote the spin quantization axis, but not the crystallographic $\boldsymbol{c}^*$ (or z) axis. For the Hubbard model Eq. (7.25), KAA [43] find that the interaction terms correspond to $V_{\mathrm{sp}}^{\boldsymbol{zz}} = V_{\mathrm{sp}}^{+-} (\equiv V_{\mathrm{sp}}) = U^2 \chi_{\mathrm{sp}}$ and $V_{\mathrm{ch}} = U^2 \chi_{\mathrm{ch}}$, where the spin and charge susceptibilities are given by

$$\chi_{\mathrm{sp}}(q) = \chi_{\mathrm{ir}}(q)/[1 - U\chi_{\mathrm{ir}}(q)], \tag{7.27}$$

$$\chi_{\mathrm{ch}}(q) = \chi_{\mathrm{ir}}(q)/[1 + U\chi_{\mathrm{ir}}(q)], \tag{7.28}$$

respectively. The irreducible susceptibility is defined as

$$\chi_{\mathrm{ir}}(q) = -\frac{T}{N} \sum_k G(k+q) G(k), \tag{7.29}$$

where N is the number of points in the $\boldsymbol{k}$-space mesh, and $\sum_k = \sum_{\boldsymbol{k},i\omega_n}$. Here, q is a shorthand notation for $(\boldsymbol{q}, i\nu_m)$ where ν_m is a bosonic Matsubara frequency. Third, the self-energy

$$\Sigma(k) = \frac{T}{N}\sum_q G(k-q)V^{(1)}(q), \tag{7.30}$$

needs to be substituted back into Dyson's equation, which is solved self-consistently for $G(k)$ until the desired convergence is achieved.

Using the Green's function $G(k)$, the order parameter equation can be written as

$$\Delta_\eta(k) = -\lambda_\eta \frac{T}{N}\sum_{k'} |G(k')|^2 V_\eta^{(2)}(k-k')\Delta_\eta(k'), \tag{7.31}$$

where $V_\eta^{(2)}(q)$ is the pairing interaction for singlet ($\eta = s$) or triplet ($\eta = t$) states. In the singlet channel, $V_\eta^{(2)}(q)$ becomes

$$V_s^{(2)}(q) = \frac{1}{2}V_{\text{sp}}^{zz}(q) + V_{\text{sp}}^{+-}(q) - \frac{1}{2}V_{\text{ch}}(q), \tag{7.32}$$

while in the triplet channel it takes the form

$$V_{t\perp}^{(2)}(q) = -\frac{1}{2}V_{\text{sp}}^{zz}(q) - \frac{1}{2}V_{\text{ch}}(q) \tag{7.33}$$

for equal spin pairing with $\boldsymbol{d} \perp \boldsymbol{z}$ and total spin projection $S_z = \pm 1$, or

$$V_{t\|}^{(2)}(q) = \frac{1}{2}V_{\text{sp}}^{zz}(q) - V_{\text{sp}}^{+-}(q) - \frac{1}{2}V_{\text{ch}}(q) \tag{7.34}$$

for $\boldsymbol{d} \parallel \boldsymbol{z}$ and $S_z = 0$.

In the on-site Hubbard model with isotropic spin fluctuations, the conditions $V_{\text{sp}}^{zz} = V_{\text{sp}}^{+-}$ and $V_{\text{sp}} \gg V_{\text{ch}}$ are satisfied, hence $|V_t^{(2)}| \approx (1/3)|V_s^{(2)}|$ holds with $V_{t\|}^{(2)} = V_{t\perp}^{(2)} \equiv V_t^{(2)}$. Therefore, a triplet state with either $\boldsymbol{d} \perp \boldsymbol{z}$ or $\boldsymbol{d} \parallel \boldsymbol{z}$ is not favorable as far as the on-site Hubbard model is concerned, as discussed in the previous section. If $(TMTSF)_2PF_6$ is indeed in a triplet state, deviations from the Hubbard model need to be considered in order to favor the triplet state ($|V_t^{(2)}| > |V_s^{(2)}|$). KAA [43] consider this problem by comparing quasi-1D systems to another well known candidate for triplet superconductivity, Sr_2RuO_4 (strontium ruthenate), where SDW fluctuations may be important [274]. For strontium ruthenate, Takimoto [275] propose that charge fluctuations (or more precisely orbital fluctuations) could arise from repulsions between degenerate $4d$ orbitals, and that the coexistence of spin and charge fluctuations could lead to triplet pairing. Furthermore, another theory for strontium ruthenate was suggested by Sato and Kohmoto [276], and Kuwabara and Ogata [277], where the presence of the spin fluctuation anisotropy [278] may also lead to triplet pairing. Coincidentally, both the spin fluctuation anisotropy [241] and the coexistence of charge (CDW) and spin (SDW) density waves [279] have been observed experimentally in $(TMTSF)_2PF_6$. Thus, KAA [43] adapt the theories for ruthenates, and propose a mechanism for triplet pairing in Bechgaard salts based on spin fluctuation anisotropy and coexistence of CDW and SDW.

The coexistence of a $2k_F$–CDW and a $2k_F$–SDW suggests an enhancement of charge fluctuations $V_{ch}(\boldsymbol{q})$ with respect to the value found for the on-site Hubbard model (7.25). If the enhancement of $V_{ch}(\boldsymbol{Q})$ is dramatic such that the condition $V_{sp}(\boldsymbol{Q}) \approx V_{ch}(\boldsymbol{Q})$ is satisfied, the triplet state becomes as favorable as the singlet state since the effective interaction $|V_t^{(2)}(\boldsymbol{Q})| \approx |V_s^{(2)}(\boldsymbol{Q})|$, as shown by Eqs. (7.32) and (7.33).

Assume that the spin fluctuation anisotropy implies $V_{sp}^{zz}(\boldsymbol{Q}) > V_{sp}^{+-}(\boldsymbol{Q})$ for the quantization axis $\boldsymbol{z} \parallel \boldsymbol{b}'$. When the condition $V_{sp}^{zz} < (2V_{sp}^{+-} + V_{ch})$ is satisfied, triplet pairing is more favorable in the equal spin pairing channel given that $|V_{t\perp}^{(2)}| > |V_{t\parallel}^{(2)}|$. As a result, the triplet state dominates over the singlet when $|V_{t\perp}^{(2)}(\boldsymbol{Q})| > V_s^{(2)}(\boldsymbol{Q})$, that is

$$V_{ch}(\boldsymbol{Q}) > V_{sp}^{+-}(\boldsymbol{Q}). \tag{7.35}$$

Notice that the $V_{t\perp}^{(2)}(\boldsymbol{Q})$ is attractive (negative) in the equal spin pairing state, so that the order parameter should satisfy $\Delta_{t\perp}(\boldsymbol{k} + \boldsymbol{Q}) = \Delta_{t\perp}(\boldsymbol{k})$. Since the triplet order parameter satisfies $\Delta_t(\boldsymbol{k}) = -\Delta_t(-\boldsymbol{k})$, the nesting condition above can be satisfied by adding extra nodal lines along $k_x \sim 0, \pi/a$. This pairing is denoted as f-wave in the sense that the order parameter behaves like "$+ - + - +-$" when the Fermi surface is sampled counter-clockwise, as schematically shown in Fig. 7.4(b) for a non-dimerized rectangular lattice with only nearest neighbor electron transfer. [See Fig. 7.8(d) for the result of a dimerized lattice with next-to-nearest neighbor electron transfer]. In KAA's two-dimensional theory, this triplet f-wave order parameter has the same number of nodes as in the singlet d-wave case [see Fig. 7.4(a)]. Thus, the kinetic energy of these two states is comparable and the f-wave state is more favorable when condition (7.35) is satisfied.

If the spin fluctuation anisotropy is sufficiently large that $V_{\rm sp}^{zz} > (2V_{\rm sp}^{+-} + V_{\rm ch})$, another triplet state with p-wave symmetry could emerge. In this case, the effective interaction $V_{t\|}^{(2)}(\boldsymbol{Q})$ $(\boldsymbol{d} \| \boldsymbol{z})$ becomes repulsive (positive), as shown in Eq. (7.34), and can mediate triplet pairing only when $\Delta_{t\|}(\boldsymbol{k}+\boldsymbol{Q}) = -\Delta_{t\|}(\boldsymbol{k})$. This requirement, along with the triplet parity condition $\Delta_t(\boldsymbol{k}) = -\Delta_t(-\boldsymbol{k})$, can be satisfied by putting nodes only at $k_x \simeq 0$ and $k_x \simeq \pi/a$, hence making the order parameter nodeless on the open Fermi surface of quasi-1D systems [280, 281]. In particular, when the ratios $V_{\rm ch}/V_{\rm sp}^{zz}$ and $V_{\rm sp}^{+-}/V_{\rm sp}^{zz}$ are both sufficiently small, the pairing interactions $V_s^{(2)}$, $V_{t\perp}^{(2)}$, and $V_{t\|}^{(2)}$ favor d-wave, f-wave, and p-wave, respectively. The resulting phase diagram is shown in Fig. 7.6, where the regions of s-, p-, d- and f-wave pairing are indicated as a function of the ratios $V_{\rm sp}^{+-}/V_{\rm sp}^{zz}$ and $V_{\rm ch}/V_{\rm sp}^{zz}$. Notice that the region where p-wave (nodeless state) is favored requires small ratios $V_{\rm sp}^{+-}/V_{\rm sp}^{zz}$, which means large spin anisotropies. However, the experimentally observed spin anisotropy [241] in $(\rm TMTSF)_2ClO_4$ does not seem to be that large. Otherwise, for small $V_{\rm ch}/V_{\rm sp}^{zz}$ and small spin anisotopries $V_{\rm sp}^{+-}/V_{\rm sp}^{zz} \approx 1$, the preferred state is d-wave. Notice in addition, that for small spin anisotropies $V_{\rm sp}^{+-}/V_{\rm sp}^{zz} \approx 1$ and increasing ratio $V_{\rm ch}/V_{\rm sp}^{zz}$ the sequence of favored states is d-wave $\rightarrow$ f-wave $\rightarrow$ s-wave.

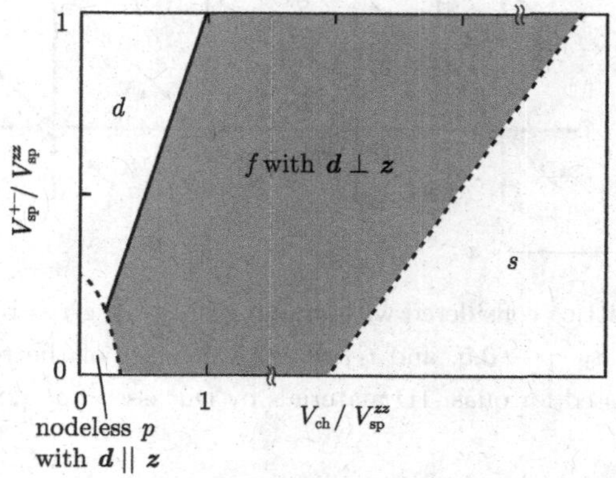

Figure 7.6 Phase diagram of $V_{\rm ch}/V_{\rm sp}^{zz}$ (charge/spin ratio) versus $V_{\rm sp}^{+-}/V_{\rm sp}^{zz}$ (spin anisotropy). The solid line is according to Eq. (7.35), while the dashed lines are schematic. From Ref. [43].

KAA [43] also consider the on-site Hubbard model with dimerization along the $\boldsymbol{a}$ direction, as shown in Fig. 7.7. The dimerization provides two inequivalent sites A and B along each chain, and generates two electron bands within the folded Brillouin zone along the k_x direction. The one-band FLEX procedure discussed above can be generalized into the two-band version by writing G, χ, and Σ as 2×2 matrices, with indices α and β to denote the A or B sites. This site representation may be converted to the band representation via a unitary transformation. Since the Fermi surface lies in the lower band for quarter filling, KAA [43] concentrate on the Green's function and the order parameter in that band, denoted as G, Δ_s and Δ_t, respectively. As for the spin susceptibility, they diagonalize the 2×2 matrix $\chi_{\rm sp}$ and denote the larger eigenvalue as χ. Using this notation, the discussion above about the competition of order parameter symmetries in systems without dimerization can be also used in the case with dimerization, where only the lowest dimerized band is considered.

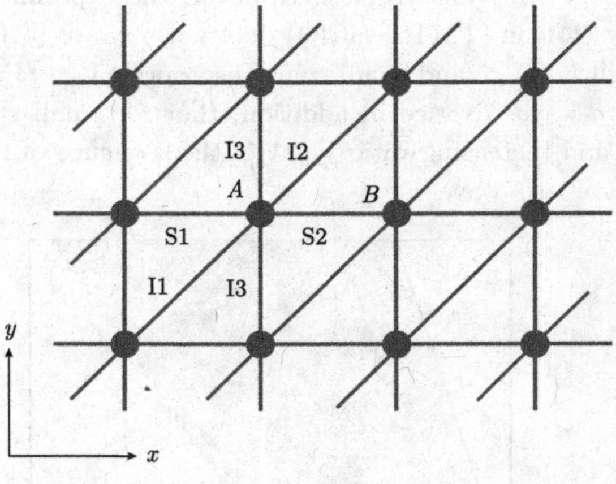

Figure 7.7 The lattice considered with hopping integrals given by $t_{\rm S1} = -2.8$, $t_{\rm S2} = -2.5$, $t_{\rm I1} = +0.2$, $t_{\rm I2} = +0.5$, and $t_{\rm I3} = -0.5$ in units of the typical energy scale 100 meV, as suggested for quasi-1D materials by Ducasse *et al.* [282]. From Ref. [43].

In Fig. 7.8, we show contour plots of (a) $|G(\boldsymbol{k}, i\pi k_{\rm B}T)|^2$, (b) $\chi(\boldsymbol{k}, 0)$, (c) $\Delta_s(\boldsymbol{k}, i\pi k_{\rm B}T)$, and (d) $\Delta_t(\boldsymbol{k}, i\pi k_{\rm B}T)$ for the dimerized Hubbard model at temperature $T = 0.015t_x$. The Fermi surface as identified from the ridge in $|G(\boldsymbol{k})|^2$

is a pair of warped quasi-1D pieces. The spin susceptibility $\chi(\boldsymbol{q}, 0)$ has peaks at $\boldsymbol{Q} \simeq (\pm\pi/2a, \pm\pi/2b)$, as expected from the nesting vector arguments and in good agreement with experiments [283, 284].

Based on these results, KAA [43] suggest that the f-wave and nodeless p-wave states are candidates for $(TMTSF)_2PF_6$ and $(TMTSF)_2ClO_4$, respectively. Thus, this proposal implies that PF_6 and ClO_4 are both triplet superconductors, but with different orbital symmetries.

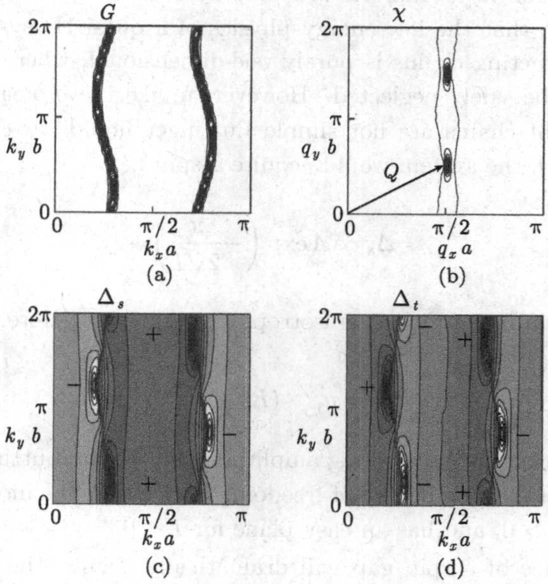

Figure 7.8 Contour plots for various quantities: (a) $|G(\boldsymbol{k}, i\pi k_{\mathrm{B}}T)|^2$; (b) $\chi(\boldsymbol{q}, 0)$; (c) $\Delta_s(\boldsymbol{k}, i\pi k_{\mathrm{B}}T)$; and (d) $\Delta_t(\boldsymbol{k}, i\pi k_{\mathrm{B}}T)$ for a quarter filled system using $U = 8$, $T = 0.015$, and parameters specified in Fig. 7.7. From Ref. [43].

7.4 Other pairing mechanisms

In addition to the pairing mechanisms outlined in the sections above, other proposals have also been raised as microscopic models for the presence of superconducting orders in quasi-1D organic superconductors. In particular, Rozhkov and Millis studied isolated electron chains with Ising anisotropy, and found that a spin gap Δ_s will open at high energies [47]. This high energy gap will drastically alter the intrachain interactions and particle hopping at the temperature lower than Δ_s, such that the single-particle hopping will be suppressed and the physics of coupled chains is dominated by a competition between pair hopping and exchange interaction. As a consequence, a phase diagram where spin-density-wave and triplet superconductivity are separated by a first-order phase transition is found.

The essence of the mechanism proposed by Rozhkov and Millis based upon the observation that the low energy physics of a quasi-1D system consisting of weakly interacting chains is purely one-dimensional, where the interchain couplings can be safely neglected. However, unlike the g-ology picture, it is also agreed that chains are not simple Luttinger liquids in the high-energy regime. Instead, the system would acquire a spin gap

$$\Delta_s \propto \Lambda \exp\left(-\frac{\pi}{2\sqrt{I}}\right) \tag{7.36}$$

as the intrachain interaction is anisotropic in spin space. Here, the parameter I is defined as

$$I = g_1^2 - (K_s - 1)^2 , \tag{7.37}$$

with g_1 the background scattering amplitude and K_s the Luttinger parameter associated with the spin degree-of-freedom. Notice that the interaction has an easy axis for $I > 0$, and has an easy plane for $I < 0$.

The presence of a spin gap will dramatically change the physics at low temperatures. In fact, an addition of a single electron or hole to a chain with a spin gap creates a state with a spin boson soliton. The energy associated with such a soliton is at least in the order of Δ_s. Thus, if we consider a quasi-1D system of weakly interacting chains, the single-particle interchain hopping will be significantly suppressed. Meanwhile, it is possible to add a pair of particles or a particle and a hole in such a way that the soliton is not created and the spin bosons are left undisturbed. This observation implies that the hopping of particle-particle (Cooper pair) or particle-hole (exchange) pairs could be well survived or even dominant in the presence of the spin gap [267].

By considering the effect of interchain coupling between 1D electron chains, Rozhkov and Millis derived an effective Hamiltonian density [47]

$$\mathscr{H}_{\rm eff} = \frac{v_{\rm c}}{2} \sum_j [K_{\rm c}^{-1}(\nabla \Phi_{{\rm c}j})^2 + K_{\rm c}(\nabla \Theta_{{\rm c}j})^2]$$
$$+ \frac{2}{a_{\rm c}} \sum_{ij} \Big[J_{ij}^{\rm SDW} \left(\eta^i_{L\uparrow}\eta^i_{R\uparrow}\eta^j_{R\uparrow}\eta^j_{L\uparrow} + \eta^i_{L\downarrow}\eta^i_{R\downarrow}\eta^j_{R\downarrow}\eta^j_{L\downarrow} \right) \cos\sqrt{2\pi}(\Phi_{{\rm c}i} - \Phi_{{\rm c}j})$$
$$- J_{ij}^{\rm TSC} \left(\eta^i_{L\uparrow}\eta^i_{R\downarrow}\eta^j_{R\downarrow}\eta^j_{L\uparrow} + \eta^i_{L\downarrow}\eta^i_{L\uparrow}\eta^j_{R\uparrow}\eta^j_{L\downarrow} \right) \cos\sqrt{2\pi}(\Theta_{{\rm c}i} - \Theta_{{\rm c}j}) \Big], \quad (7.38)$$

where $\eta_{p=L,R;\sigma=\uparrow,\downarrow}$ are Klein factors associated with left-going (L) and right-going (R) electrons, $v_{\rm c}$ is the charge velocity, and $\Theta_{\rm c}$ and $\Phi_{\rm c}$ are the charge boson field and dual charge field operators within the bosonization description, respectively. Here, $J_{ij}^{\rm SDW}$ is the effective exchange coupling and $J_{ij}^{\rm TSC}$ is the effective Josephson coupling, and the ultraviolet cutoff for this effective theory is equal to $a_{\rm c}^{-1} = \Delta_{\rm s}/v_{\rm c}$.

The first line in this Hamiltonian describes the intrachain dynamics of the charge bosons. The second term acts to order $\Phi_{\rm c}$, which corresponds to the phase of the SDW order. When $\exp(i\sqrt{2\pi}\Phi_{\rm c})$ acquires finite expectation value, the superconducting order parameter is zero and the ground state becomes SDW. The last term describes the Josephson coupling between pairs of the chains. The field $\Theta_{\rm c}$ is associated with the phase of the superconducting order parameter. If $\Theta_{\rm c}$ is ordered, the SDW order parameter has zero expectation value and the system is a triplet superconductor. Thus, the TSC and SDW orders are mutually excluded in this theory. The phase diagram of the effective Hamiltonian can be obtained by the mean-field approach. By assuming the exchange constants $J^{\rm SDW}$ equal to the Josephson coupling constants $J^{\rm TSC}$, and by considering only the coupling between nearest neighbors (J_1) and next-nearest neighbors (J_2) with $|J_1| > |J_2|$, Rozhkov and Millis mapped out the phase diagram as illustrated in Fig. 7.9. In this figure, the Luttinger parameter $K_{\rm c}$ is confined within the interval $1/2 < K_{\rm c} < 1$ to comply with the requirement of a repulsive Hubbard model.

As shown in Fig. 7.9, the emergence of superconductivity can be explained by an Ising-anisotropy-induced spin gap in the quasi-1D system with purely repulsive interactions. The superconducting phase has a lower energy because the energy of transverse hopping is reduced comparing with the normal phase. This is in clear contrast to the BCS theory where the energetic advantage is achieved by gaining the condensation energy of electron pairs.

A crucial prerequisite for the theory discussed above is a strong Ising anisotropy of spin-spin interaction, which, however, seems to be quite weak in Bechgaard salts [285]. This observation brings much complication to a direct application of this theory to quasi-1D organic superconductors, because the size of the spin gap Eq. (7.36) is exponentially small at weak anisotropy. An extension of this theory to the case of small spin gap is very challenging due to the presence of unquenched operator of the single-electron transversal hopping [286]. When expressed in terms of the bosonic fields $\Theta_{\mathrm{s,c}}$ and $\Phi_{\mathrm{s,c}}$, this operator has distinct nonlocal structure such that kinks in both $\Theta_{\mathrm{s,c}}$ and $\Phi_{\mathrm{s,c}}$ fields are created simultaneously. This makes it impossible to directly apply the theory discussed above to such systems.

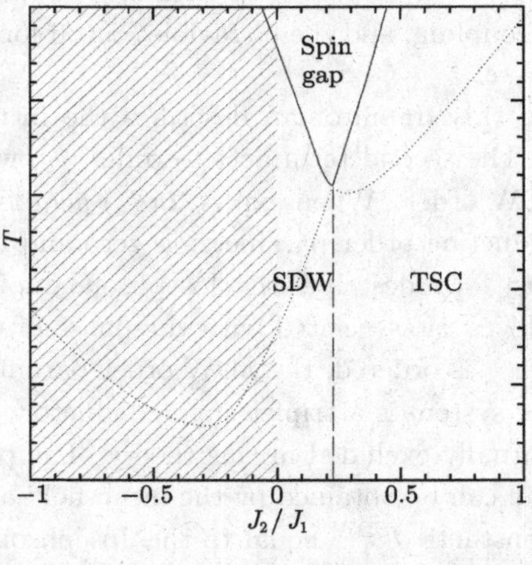

Figure 7.9 Mean-field phase diagram of the model proposed by Rozhkov and Millis [47]. Solid and dash lines correspond to second-order and first-order phase transitions, respectively. The region labeled with "Spin gap" denotes the phase in which the charge bosons are disordered. The gap in the charge sector are shown by dotted lines. The shaded region is the unphysical regime where $J_2 < 0$. From Ref. [47].

Another theoretical proposal about the pairing mechanism was introduced by Ohta *et al.*, under which a spin-triplet pairing state can be stabilized via a "ring exchange" process [48]. In this work, Ohta modeled the quasi-1D organic superconductors as a set of parallel conducting chains with zigzag interchain hopping, as depicted in Fig. 7.10(a). The interchain hopping changes sign alternatively, while the sign for intrachain hopping remains positive definite [282]. After a particle-hole transformation, the hopping integral is negative along the chains, and still alternating signs along the zigzag bonds. Within the hole notation, the system becomes a triangular-lattice where the three hopping integrals of each triangle satisfies the ferromagnetic sign rule $t_1t_2t_3 > 0$. Under this condition, it is known that two spins on a triangle tend to form a ferromagnetic state with full polarization due to a ferromagnetic interaction introduced by a ring-exchange mechanism if a sufficiently large Hubbard U is present [287, 288, 289, 290]. Ohta *et al.* then suggested that for an intermediate, and also realistic, interaction strength of U, a spin-triplet superconducting state may be stabilized via a short-range spin-triplet correlation present within this ferromagnetic background.

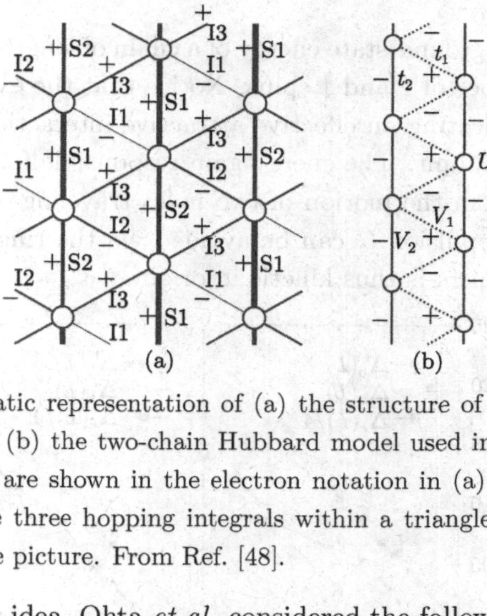

Figure 7.10 Schematic representation of (a) the structure of hopping integrals for Bechgaard salts and (b) the two-chain Hubbard model used in Ref. [48]. The signs of hopping integrals are shown in the electron notation in (a) and hole notation in (b). Notice that the three hopping integrals within a triangle satisfy the sign rule $t_1t_2t_3 > 0$ in the hole picture. From Ref. [48].

To validate this idea, Ohta *et al.* considered the following two-chain Hubbard model [see Fig. 7.10(b)]

$$\mathscr{H} = \sum_{<i,j>\sigma} t_{ij} \left(c_{i\sigma}^{\dagger} c_{j\sigma} + \text{H.C.} \right) + U \sum_{i} n_{i\uparrow} n_{i\downarrow}, \tag{7.39}$$

where $c_{i\sigma}^{\dagger}$ ($c_{i\sigma}$) are the creation (annihilation) operator of a hole with spin σ at site i, $n_{i\sigma}$ is the number operator, $< i, j >$ stands for nearest-neighbor pairs, and U is the on-site repulsive interaction. The hopping integral $t_{ij} = t_1$ for interchain zigzag hopping and $t_{ij} = t_2$ for intrachain hopping. By using the density matrix renormalization group (DMRG) method, Ohta *et al.* calculated several properties of the ground state for clusters of length $L \leqslant 128$. In their calculation, the energy unit is chosen as the hopping integral along the chain (i.e., $|t_2| = 1$), the discarded weights are typically of the order $10^{-7} \sim 10^{-6}$ to obtain the ground-state energy within an accuracy of ~ 0.001, and up to $m \approx 4500$ density-matrix eigenstates are kept.

In Fig. 7.11, the binding energy of holes Δ_b and the charge gap Δ_c are shown for $|t_1| = 0.5$ with varying system's length L. The quantities are defined as

$$\begin{aligned} \Delta_b &= \lim_{L\to\infty} \Delta_{\rm b}^{\pm}(L), \text{ with } \Delta_{\rm b}^{\pm}(L) = E_L(N \pm 2) + E_L(N) - 2E_L(N \pm 1), \\ \Delta_c &= \lim_{L\to\infty} \Delta_{\rm c}(L), \text{ with } \Delta_{\rm c}(L) = E_L(N+2) + E_L(N-2) - 2E_L(N), \end{aligned} \tag{7.40}$$

where $E_L(N)$ is the ground-state energy of a chain of length L with N electrons with an equal number of $\uparrow$ and $\downarrow$ spins. Notice that the extrapolated value of $\Delta_{\rm b}$ is negative, indicating an effective attractive interaction between holes in the thermodynamic limit. The energy gain responsible for the negative value of $\Delta_{\rm b}$ may come from the motion of two holes traveling around the triangle, where the on-site repulsion U can be avoided via the ring-exchange of holes. The pairing mechanism is thus kinetic in origin.

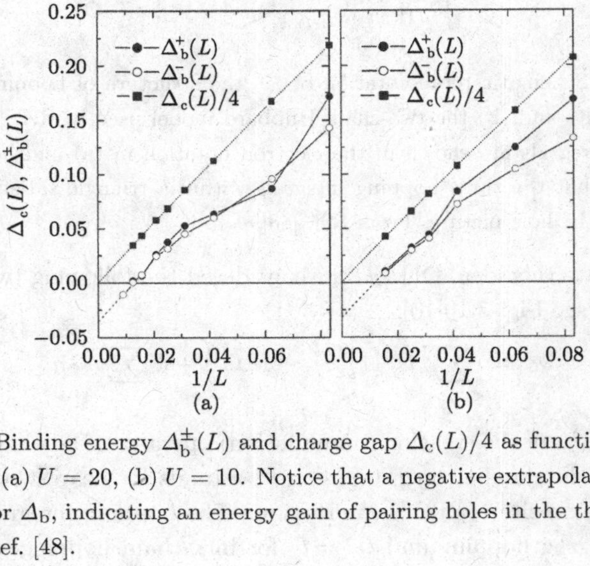

Figure 7.11 Binding energy $\Delta_{\rm b}^{\pm}(L)$ and charge gap $\Delta_{\rm c}(L)/4$ as functions of $1/L$ for $|t_1| = 0.5$ and (a) $U = 20$, (b) $U = 10$. Notice that a negative extrapolation value can be obtained for $\Delta_{\rm b}$, indicating an energy gain of pairing holes in the thermodynamic limit. From Ref. [48].

To characterize the property of this pairing state, Ohta *et al.* also calculated the pair correlation function $D(l) = \langle \Delta_{i+l}^{\dagger} \Delta_i \rangle$ with

$$\Delta_i \equiv c_{i\uparrow}c_{i+r\downarrow} - c_{i\downarrow}c_{i+r\uparrow}$$

for singlet pairs, and

$$\Delta_i \equiv c_{i\uparrow}c_{i+r\downarrow} + c_{i\downarrow}c_{i+r\uparrow}$$

for triplet pairs. Here, $i+r$ denotes the nearest-neighbor sites of i. The results for the $L = 128$ system are show in Fig. 7.12. Notice that the pair correlation function features a power-law length dependence for the interchain triplet pairing, but decays exponentially for the singlet pairing and the intrachain triplet pairing. The combined results of negative binding energy and pair correlation functions indicating that the bound-state of the system is a spin-triplet superconducting state where two holes sitting on different chains pair via the ring-exchange mechanism.

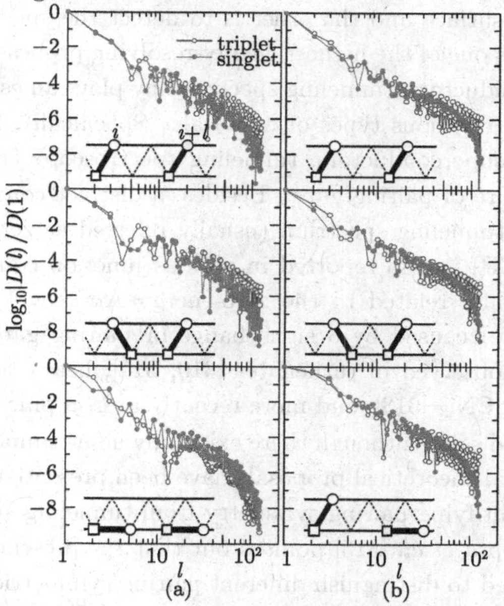

Figure 7.12 Pair correlation functions $D(l)$ calculated at $L = 128$ and $U = 10$ with (a) $|t_1| = 0.25$ and (b) $|t_1| = 0.5$. The correlations are evaluated between two spatially separated pairing operators, which are illustrated as thick solid lines in the insets. From Ref. [48].

8

Tunneling Experiments and Josephson Effect

Towards the ultimate goal of determining the pairing symmetry in the superconducting phase of Bechgaard salts, one can adapt two strategies for experimental attempt: one is to identify the orbital symmetry of the pairing gap along the Fermi surface and the other is to detect the spin part of the gap. Being accepted as one of the highest energy-resolving probes for the electronic states of superconductors, tunneling spectroscopy plays an essential role along both strategies for various types of materials. Specifically, in the context of high-T_c cuprate superconductors, tunneling spectroscopy can help revealing the nodal structure of pairing gap[1]. Besides, a distinct conductance peak at zero bias in the tunneling spectrum (usually referred as zero-biased conductance peak, ZBCP) is also reported in various junction configurations [292–300], and is usually related to the zero-energy Andreev bound state [301] (or mid-gap state) caused by a sign change or pairing gap [302–309]. The ZBCPs are also observed in ruthenates [310, 311], heavy fermion superconductors [312], $MgCNi_3$ [313], and more recently in iron pnictides [314].

For Bechgaard salts, although there exist only a few tunneling experiments up to date, several theoretical proposals have been presented in analyzing the possibility of identifying pairing symmetry from tunneling spectroscopy measurement. Sengupta *et al.* [315] pointed out that the presence and absence of ZBCP can be used to distinguish different pairing symmetries. When tunneling current is applied along the $\boldsymbol{a}$ axis, a ZBCP will be present for the p- and f-wave pairing states, but not for the d-wave state. Tanuma *et al.* [316] further

[1] For detailed information, readers are referred to an excellent review article by S. Kashiwaya and Y. Tanaka [291].

pointed out that the triplet p_x- and f_x-wave states can also be distinguished from the overall shape of the tunneling spectrum, where the fully gapped p_x-wave state has a U-shaped structure and the gapless f_x-wave state acquires a V-shaped spectrum. Subsequent studies by Tanuma *et al.* [317, 318] analyzed the effect induced by the shape of Fermi surface, and suggested that a ZBCP can appear even in the case of d-wave symmetry. In this case, different pairing states can be identified by the way that the ZBCP splits in the presence of a magnetic field.

In addition to the orbital part of the pairing gap, tunneling spectroscopy can also be implemented to investigate the spin symmetry of the superconducting phase. Along this line, Bolech and Giamarchi [319] studied the tunneling process between normal metal and superconductors in the point-contact limit, which is more experimentally relevant to the STM measurement, and suggested that the I-V characteristics in the presence of a magnetic field can be used to distinguish singlet and triplet pairing states. In particular, when a magnetic field is applied perpendicular to the triplet superconducting order parameter $\boldsymbol{H} \perp \boldsymbol{d}$, the tunneling spectrum will be altered from the $\boldsymbol{H} \parallel \boldsymbol{d}$ case. Thus, the spin symmetry of the pairing gap and the direction of the triplet order parameter $\boldsymbol{d}$ can be identified by measuring the I-V characteristics of the junction upon rotation of the magnetic field, provided that the direction of $\boldsymbol{d}$ does not follow the field rotation.

Another type of tunneling process is the Josephson effect where two electrons tunnel as a pair between two superconducting samples. Early proposals to test the symmetry of order parameter in triplet superconductors using the Josephson effect relied on junctions between a singlet and a triplet superconductors in the context of heavy fermion materials [320, 321]. Pals *et al.* [320] showed that the AC Josephson effect with frequency $\omega_{\mathrm{J}} = 2\,\mathrm{eV}$ did not exist between an s-wave superconductor and a triplet superconductor, provided that the pair tunneling matrix element is spin conserving and does not break time-reversal symmetry. These results are not applicable if either spin-orbit coupling is important or the tunneling barrier is non spin-conserving (magnetically active), as shown by Geshkenbein and Larkin [321]. Recently, it is also proposed that an imbalance of spin population and net spin accumulation would take place at a singlet superconductor-insulator-triplet superconductor (SSC-I-TSC) junction [322, 323]. This magnetization can then be detected via a local probe such as scanning SQUID. Meanwhile, if one considers another junction geometry consisting of triplet superconductor–insulator–triplet superconductor (TSC-I-TSC), the Josephson effect should be different from the SSC-I-SSC or SSC-I-TSC versions because there are now two vector order parameters. In particular, it is expected that the Josephson current must depend on the directions, as well as the phases of order parameters of superconductors to the left and to the right of a junction [324].

In this chapter, we discuss some theoretical proposals for implementing single electron and Josephson tunneling measurements to distinguish various pairing states in Bechgaard salts. In Sec. 8.1, we focus on the tunneling spectroscopy measurement in a normal metal–superconductor (N–SC) junction, while the discussion on Josephson effect for a superconductor–insulator–superconductor (SC–I–SC) junction is presented in Sec. 8.2. We stress that such tunneling experiments are in general hard to conduct in organic materials as compared to inorganic ones, since the samples are rather irregular in shape, and are fragile upon cutting, grinding and polishing. Besides, it is also difficult to implement a tunneling experiment under high pressure. Therefore, the most possible candidate for such a measurement is $(TMTSF)_2ClO_4$, which is superconducting at ambient pressure.

8.1 Tunneling spectroscopy and zero-bias conductance peak

The tunneling spectroscopy measurement can be performed in different types of experimental setups, including scanning tunneling microscopy or scanning tunneling spectroscopy (STM or STS), thin-film insulator–superconductor junctions, point contact junctions, break junctions, and squeezable junctions. All of these techniques share common advantages of high energy resolution, flexibility for environments and sensitivity to surface states [291]. One of the central features one can read from the resulting data is the overall shape of the tunneling spectrum, which can help revealing the nodal structure of the superconducting gap. In fact, for a fully gapped superconductor, the spectrum would show a U-shaped structure with a flat bottom, for no quasiparticle can be excited within the full gap. On the other hand, a gapless superconductor would acquire a V-shaped spectrum showing almost linearly increasing behavior around the origin.

Another unambiguous feature one can obtain from the tunneling results is the presence or absence of ZBCP, which can be attributed to zero-energy states (ZESs) confined at surface boundaries. We start our discussion in this section by giving a brief introduction to the physical origin of ZES. We consider a simplified model where electrons are injected from a normal metal to a superconductor, as schematically depicted in Fig. 8.1. The normal state–superconductor (N–SC) interface is assumed to be perfectly flat and along the y-z plane, and the spatial-dependent superconducting order parameter takes a step-function form

$$\Delta(\hat{\boldsymbol{k}}, \boldsymbol{r}) = \Delta_0(\hat{\boldsymbol{k}})e^{i\phi}\Theta(x), \tag{8.1}$$

where $\Theta(x)$ is the Heaviside function and ϕ represents the global phase of the pairing order parameter. At the N–SC interface, a barrier potential is present as $V_0\delta(x)$ with $V_0 > 0$. For simplicity, we consider only one-dimensional process with electrons moving along the x axis, and the Fermi energies in both the N and SC parts are assumed to be equal.

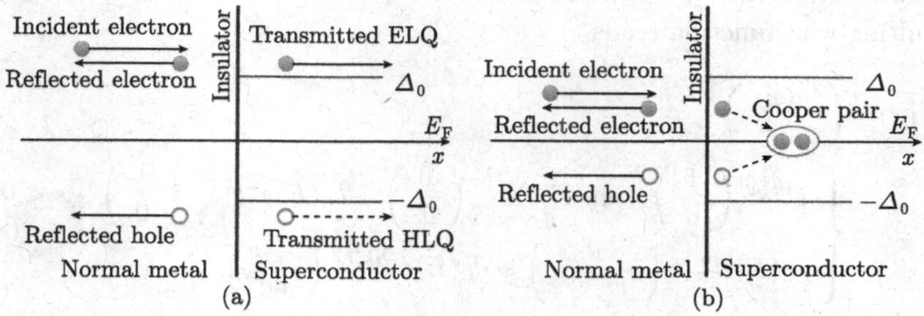

Figure 8.1 Schematic illustration of the electron injection from a normal metal to a superconductor with energy (a) $|E| > \Delta_0$ and (b) $|E| < \Delta_0$. The superconducting gap is assumed to be in the form of a step function $\Delta_0\Theta(x)$. Open and closed circles represent electron (or electron-like quasiparticle, ELQ) and hole (or hole-like quasiparticle, HLQ), respectively. For case (a), four possible trajectories of quasiparticle exist, while for case (b), the ELQ and HLQ cannot exist in the superconductor. Adapted from Ref. [291].

When the incident energy of an electron E is larger than the pairing gap Δ_0 [see Fig. 8.1(a)], there are four possible trajectories for the electron after hitting the interface potential barrier

(i) *normal reflection*: reflected as an electron;

(ii) *electron injection*: transmitted to the superconductor as an electron-like quasiparticle;

(iii) *Andreev reflection*: reflected as a hole;

(iv) *hole injection*: transmitted to the superconductor as a hole-like quasiparticle;

In the case of Andreev reflection, the incident electron is transmitted into the superconductor as a component of a Cooper pair, while a hole is created at the surface boundary and reflected back into the normal metal. Meanwhile, as the incident energy $E < \Delta_0$ [see Fig. 8.1(b)], no quasiparticles can be transmitted into the superconductor and the current can only be carried by Cooper pairs. Notice that the Andreev reflection will not happen unless a finite pairing gap is present, so it is a unique feature resulting from Cooper pairing and the condensation of these pairs.

The probability amplitude of each process can be calculated by solving the Bogoliubov-de Gennes (BdG) equation together with proper boundary conditions. In the case of a conventional superconductor of $\Delta_0(\hat{\boldsymbol{k}}) = \Delta_0$, the resulting wavefunction reads

$$\Psi(x) = \begin{pmatrix} u(x) \\ v(x) \end{pmatrix} = \begin{cases} e^{ik_N^+ x}\begin{pmatrix} 1 \\ 0 \end{pmatrix} + a_0(E)e^{ik_N^- x}\begin{pmatrix} 0 \\ 1 \end{pmatrix} + b_0(E)e^{-ik_N^+ x}\begin{pmatrix} 1 \\ 0 \end{pmatrix}, & x < 0; \\ c_0(E)e^{ik_S^+ x}\begin{pmatrix} u \\ ve^{-i\phi} \end{pmatrix} + d_0(E)e^{-ik_S^- x}\begin{pmatrix} v \\ ue^{-i\phi} \end{pmatrix}, & x \geqslant 0, \end{cases} \tag{8.2}$$

where the terms associated with coefficients $a_0(E)$, $b_0(E)$, $c_0(E)$ and $d_0(E)$ correspond to Andreev reflection, normal reflection, electron injection and hole injection, respectively. The parameters are defined as

$$k_N^\pm = \sqrt{2m(E_F \pm E)}, \quad k_S^\pm = \sqrt{2m(E_F \pm \Omega)},$$
$$u = \sqrt{\frac{E+\Omega}{2E}}, \quad v = \sqrt{\frac{E-\Omega}{2E}}, \quad \Omega = \sqrt{E^2 - |\Delta_0|^2}. \tag{8.3}$$

At the N–SC interface $x = 0$, the wavefunction are connected by the continuity conditions $\Psi(x = 0^+) = \Psi(x = 0^-)$ and $\Psi'(x = 0^+) - \Psi'(x = 0^-) = 2mV_0\Psi(x = 0^-)$. Focusing on the low energy limit with $E \ll E_F$ and in the condition of weak coupling $|\Delta_0| \ll E_\mathrm{F}$, the wave vectors $k^{\pm}_{N,S}$ can be approximated by the Fermi wave vector k_F. Under this circumstance, the coefficients in Eq. (8.2) can be determined as

$$a_0(E) = \frac{uve^{-i\phi}}{(1+Z^2)u^2 - Z^2v^2}, \tag{8.4a}$$

$$b_0(E) = \frac{Z(i+Z)(v^2-u^2)}{(1+Z^2)u^2 - Z^2v^2}, \tag{8.4b}$$

$$c_0(E) = \frac{(1-iZ)u}{(1+Z^2)u^2 - Z^2v^2}, \tag{8.4c}$$

$$d_0(E) = \frac{iZv}{(1+Z^2)u^2 - Z^2v^2}, \tag{8.4d}$$

where $Z \equiv mV_0/k_\mathrm{F}$. From this result, we can easily find that in the case of $E < |\Delta_0|$ and $V_0 = 0$, (i.e., no potential barrier at the N-SC interface, $|a_0(E)|^2 = 1$ and $|b_0(E)|^2 = 0$. This indicates that the incident electron will be totally reflected as a hole.

If the gap function $\Delta_0(\hat{\boldsymbol{k}})$ is anisotropic as in an unconventional superconductor, all processes involving the gap function will be momentum dependent. As a consequence, the wavefunction becomes

$$\Psi(x) = \begin{cases} e^{ik_N^+ x}\begin{pmatrix}1\\0\end{pmatrix} + a_0(E)e^{ik_N^- x}\begin{pmatrix}0\\1\end{pmatrix} + b_0(E)e^{-ik_N^+ x}\begin{pmatrix}1\\0\end{pmatrix}, & x < 0; \\ c_0(E)e^{ik_S^+ x}\begin{pmatrix}u_+\\ v_+e^{-i\phi_+}\end{pmatrix} + d_0(E)e^{-ik_S^- x}\begin{pmatrix}v_-\\ u_-e^{-i\phi_-}\end{pmatrix}, & x \geqslant 0. \end{cases} \tag{8.5}$$

Here, the parameters are defined as

$$u_\pm = \sqrt{\frac{E+\Omega_\pm}{2E}}, \quad v_\pm = \sqrt{\frac{E-\Omega_\pm}{2E}}, \quad \Omega_\pm = \sqrt{E^2 - |\Delta_\pm|^2},$$
$$\Delta_\pm = \Delta_0(\pm\hat{\boldsymbol{k}}_S^\pm), \quad e^{i\phi_\pm} = \frac{\Delta_\pm}{|\Delta_\pm|}. \tag{8.6}$$

Under the same assumptions of low energy and weak interaction, i.e., $\max[|\Delta_0(\hat{\boldsymbol{k}})|, E] \ll E_\mathrm{F}$, all the wave vectors can be approximated by the Fermi wave vector $k_N^\pm = k_S^\pm = k_\mathrm{F}$. This means that the reflected hole bears the same momentum as the incident one, as is often referred as "retro-reflectivity of the Andreev reflection". By connecting the wavefunctions at the N–SC interface, the coefficients can be obtained as

$$a_0(E) = \frac{u_- v_+ e^{-i\phi_+}}{(1+Z^2)u_+u_- - Z^2 v_+ v_- e^{i(\phi_- - \phi_+)}}, \tag{8.7a}$$

$$b_0(E) = \frac{Z(i+Z)[v_+ v_- e^{i(\phi_- - \phi_+)} - u_+ u_-]}{(1+Z^2)u_+u_- - Z^2 v_+ v_- e^{i(\phi_- - \phi_+)}}, \tag{8.7b}$$

$$c_0(E) = \frac{(1-iZ)u_-}{(1+Z^2)u_+u_- - Z^2 v_+ v_- e^{i(\phi_- - \phi_+)}}, \tag{8.7c}$$

$$d_0(E) = \frac{iZ v_+ e^{i(\phi_- - \phi_+)}}{(1+Z^2)u_+u_- - Z^2 v_+ v_- e^{i(\phi_- - \phi_+)}}. \tag{8.7d}$$

Notice that in the case of zero barrier potential $V_0 = 0$, the normal reflection is also suppressed to zero, but the Andreev reflection depends only on Δ_+ and has no Δ_- dependence.

Next, we move to the case where the normal metal has a finite thickness d_N, while the superconductor is still assumed to be semi-infinite for simplicity. In the limit of $d_N \to 0$, this structure resembles the surface of a superconductor. In this case, the electron-hole trajectory can form a closed loop within the N layer via two normal reflections at the N–vacuum (N–V) surface and

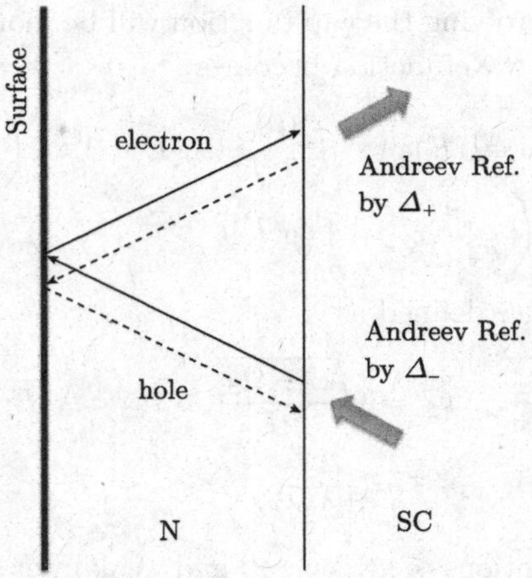

Figure 8.2 Trajectories of bound states in a N–SC bilayer. The quasiparticles follow closed paths throughout two Andreev reflections at the N–SC interface and two specular reflections at the surface and by changing their form between electrons and holes. These states are stable even if the thickness of the N-layer becomes zero. Adapted from Ref. [291].

two Andreev reflections at the N–SC interface, as schematically depicted in Fig. 8.2. The energy of the surface bound state E_b satisfies the equation

$$1 - \exp\left(\frac{4id_N E_b}{v_\mathrm{F}}\right)\left(\frac{v_+ v_- e^{\pm i(\phi_- - \phi_+)}}{u_+ u_v}\right)_{E=E_\mathrm{b}} = 0, \tag{8.8}$$

provided that $|E_\mathrm{b}| < \min(|\Delta_+|, |\Delta_-|)$. For conventional s-wave pairing superconductors, the condition $\Delta_+ = \Delta_-$ ensures $\phi_+ = \phi_-$, such that Eq. (8.8) can be simplified as

$$1 - \frac{E_\mathrm{b} - \Omega}{E_\mathrm{b} + \Omega}\exp\left(\frac{4id_N E_\mathrm{b}}{v_\mathrm{F}}\right) = 0, \tag{8.9}$$

and the resulting E_b is sensitive to the thickness of the N layer d_N. However, in the case of anisotropic unconventional superconductors, the surface bound state energy acquires an additional angle dependence via the phase difference between Δ_+ and Δ_-. In particular, if the N–SC interface is placed in a proper way such that the gap functions Δ_+ and Δ_- have different signs, the phase difference $\phi_- - \phi_+ = \pm\pi$. Thus, one can easily find that $E_b = 0$ is automatically a solution of Eq. (8.8) regardless the value of d_N. This zero-energy surface bound state contributes to resonance tunneling, which leads to the presence of a barrier-height-insensitive conductance peak at zero energy (ZBCP).

The discussion above gives us an idea regarding the physical origin of ZBCP and its relation to the pairing gap function anisotropy. However, the implementation of this idea to Bechggard salts with realistic spectrum and junction structure needs a more careful analysis. In order to analyze the features of tunneling spectroscopy in Bechgaars salts with both singlet and triplet pairing possibilities, Sengupta *et al.* [315] considered quasi-1D systems with separated Fermi sheets labeled by $\alpha = +(\mathrm{R}), -(\mathrm{L})$ and wrote down the following BdG equation [325, 301] for the electron eigenenergies $\mathscr{E}_\lambda$ in the superconducting state

$$\begin{pmatrix} -i\alpha v_\mathrm{F}\partial_x & (\hat{\sigma}\cdot\hat{\boldsymbol{d}})\,\Delta_\alpha(x) \\ (\hat{\sigma}\cdot\hat{\boldsymbol{d}})\,\Delta^*_\alpha(x) & i\alpha v_\mathrm{F}\partial_x \end{pmatrix}\begin{pmatrix} u^\alpha_\lambda \\ v^\alpha_\lambda \end{pmatrix} = \mathscr{E}_\lambda \begin{pmatrix} u^\alpha_\lambda \\ v^\alpha_\lambda \end{pmatrix}, \tag{8.10}$$

where the BdG wavefunctions

$$\Psi_\lambda = e^{i\boldsymbol{k}_\mathrm{F}\cdot\boldsymbol{r}}\left[u_{\lambda,\sigma}(x), i(\sigma_y)_{\sigma'\bar{\sigma}'} v_{\lambda,\bar{\sigma}'}(x)\right] \tag{8.11}$$

with Pauli matrix σ_y. The amplitudes $u_{\lambda,\sigma}$ and $v_{\lambda,\bar{\sigma}}$ correspond to the electron-like and hole-like components, respectively. Here, σ and σ' represent spin indices and $\bar{\sigma}' = -\sigma'$. The coefficient $(\hat{\sigma} \cdot \hat{\boldsymbol{d}})$ within the off-diagonal terms, where $\hat{\boldsymbol{d}} = \boldsymbol{d}/|\boldsymbol{d}|$, is present only for triplet superconductivity and absent for the singlet case. The BdG equation above can be further simplified by choosing the quantization axis to be along $\hat{\boldsymbol{d}}$. In this case, the 4×4 matrix equation (8.10) decouples into two 2×2 matrix equations for the wavefunctions $(u_\sigma, i(\sigma_y)_{\sigma\bar{\sigma}} v_{\bar{\sigma}})$:

$$\begin{pmatrix} -i\alpha v_{\mathrm{F}}\partial_x & \sigma\Delta_\alpha(x) \\ \sigma\Delta_\alpha^*(x) & i\alpha v_{\mathrm{F}}\partial_x \end{pmatrix} \begin{pmatrix} u^\alpha_{\lambda,\sigma} \\ s v^\alpha_{\lambda,\bar{\sigma}} \end{pmatrix} = \mathscr{E}_\lambda \begin{pmatrix} u^\alpha_{\lambda,\sigma} \\ s v^\alpha_{\lambda,\bar{\sigma}} \end{pmatrix}. \tag{8.12}$$

Furthermore, Sengupta *et al.* [315] made an assumption that the $\boldsymbol{d}$-vector is real and pointing along the $\boldsymbol{a}$ axis (parallel to the chains).

The tunneling process is analyzed within the same geometry as in Fig. 8.1, where the superconductor occupies the semi-infinite space $x \geqslant 0$ with a potential barrier at $x = 0$. When electrons reflect from the edge specularly, its k_x momentum changes sign, while other components remain unchanged. The electrons can be looked as scattered from the point $\boldsymbol{k}_{\mathrm{F}}^-$ on the left sheet of the Fermi surface to the point $\boldsymbol{k}_{\mathrm{F}}^+$ on the right sheet, as indicated in the left panel of Fig. 8.3. In this case, the wavefunction Ψ_λ for the BdG equation is described as superposition of the R(+) and L(−) terms:

$$\psi_\lambda = \frac{1}{\sqrt{2}}\left[e^{i\boldsymbol{k}_{\mathrm{F}}^+\cdot \boldsymbol{r}} \begin{pmatrix} u^+_{\lambda,\sigma}(x) \\ v^+_{\lambda,\bar{\sigma}}(x) \end{pmatrix} - e^{i\boldsymbol{k}_{\mathrm{F}}^-\cdot \boldsymbol{r}} \begin{pmatrix} u^-_{\lambda,\sigma}(x) \\ v^-_{\lambda,\bar{\sigma}}(x) \end{pmatrix} \right] \tag{8.13}$$

with boundary conditions

$$u^+(0) = u^-(0), \qquad v^+(0) = v^-(0). \tag{8.14}$$

The BdG equation (8.12) should be solved together with the self-consistency condition

$$\Delta_\alpha(x) = g \sum_\lambda u^\alpha_\lambda(x) v^{\alpha *}_\lambda(x), \tag{8.15}$$

where g is the effective coupling constant, and the sum is taken over all occupied states with $\mathscr{E}_\lambda < 0$ at zero temperature. However, one can first assume a step-function form for $\Delta_\alpha(x)$ to gain some insight into the solutions. Suppose $\Delta_\alpha(x) = \Delta_0 \Theta(x)$ where $\Theta(x)$ is the Heaviside function, then plane waves $\Psi \propto e^{ik_x x}$ are eigenfunctions of Eq. (8.12) with eigenenergies $\mathscr{E} = \pm\sqrt{(v_F k_x)^2 + |\Delta_0|^2}$. Notice that the energy is also real when k_x is purely imaginary $k_x = i\kappa$, and $\mathscr{E} = \pm\sqrt{|\Delta_0|^2 - (v_F\kappa)^2}$. For $\kappa > 0$, this solution describes an electron eigenfunction localized near the edge at $x = 0$: $\Psi \propto e^{-\kappa x}$. Furthermore, because $u^\alpha/v^\alpha = \Delta_\alpha/(\mathscr{E} + i\alpha v_F\kappa)$, the boundary condition expressed in Eq. (8.14) can be satisfied only for the p_x-wave state $\Delta_+ = -\Delta_-$, but not for s-wave with $\Delta_+ = \Delta_-$. Thus, in the p_x-wave case, there exists an edge electron state with energy in the middle of the superconducting gap $\mathscr{E} = 0$, and the localization length $\ell = 1/\kappa$ is equal to the coherence length $\xi_0 = v_F/\Delta_0$.

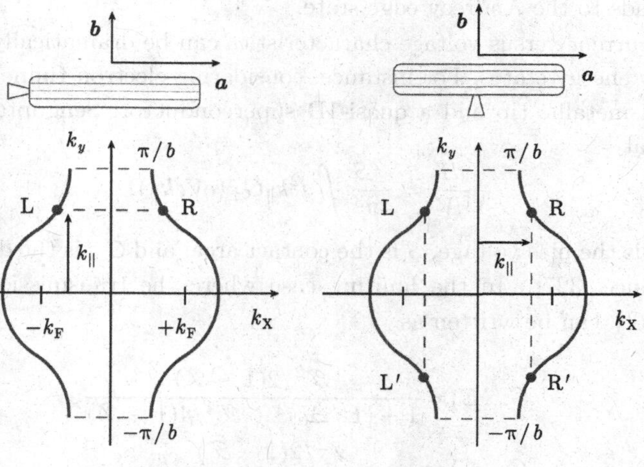

Figure 8.3 Top: $(TMTSF)_2X$ samples with the lines indicating 1D chains. The left and right panels sketch tunneling along the $\boldsymbol{a}$ and $\boldsymbol{b}$ axes, respectively. Bottom: The Fermi surface of $(TMTSF)_2X$, sketched with a greatly exaggerated warping in the k_y direction. Reflection from the edge perpendicular (parallel) to the chains changes electron momentum from L ($\alpha = -$) to R ($\alpha = +$) (R′ to R and L′ to L), as shown in the left (right) panel. From Ref. [315].

Taking into account self-consistency condition (8.15), the BdG equation (8.12) can be solved by extending the wavefunction to the whole space [315]. Sengupta *et al.* noticed that this modified equation coincides with the exactly solvable model describing one-dimensional CDW in polyacetylene [244, 326], and adopted the corresponding solutions. In s-wave case there are no bound-state solutions, while in p_x-wave state the general solutions are given by

$$\begin{pmatrix} u_0(x) \\ v_0(x) \end{pmatrix} = \frac{\sqrt{\kappa}}{2\cosh(\kappa x)} \begin{pmatrix} 1 \\ -1 \end{pmatrix}, \quad \text{for } \mathscr{E}_0 = 0;$$

$$\begin{pmatrix} u_{k_x}(x) \\ v_{k_x}(x) \end{pmatrix} = \frac{e^{ik_x x}}{2\mathscr{E}_{k_x}\sqrt{L_x}} \begin{pmatrix} \mathscr{E}_{k_x} + v_\mathrm{F} k_x + \Delta(x) \\ \mathscr{E}_{k_x} - v_\mathrm{F} k_x - \Delta(x) \end{pmatrix}, \text{ for } \mathscr{E}_{k_x} = \pm E(k_x), \quad (8.16)$$

where $E(k_x) = \sqrt{v_\mathrm{F}^2 k_x^2 + |\Delta_0|^2}$, and L_x is the length of the sample along the chains. In particular, when k_x is purely imaginary, i.e., κ is real, the solution with energy $\mathscr{E}_0 = 0$ represents a localized electron state at $x = 0$, and corresponds to the Andreev edge state.

The current versus voltage characteristics can be dramatically affected by these low energy states. For instance, considering electron tunneling between a normal metallic tip and a quasi-1D superconductor, Sengupta *et al.* [315] found that

$$\frac{dI}{dV} = \frac{e^2 S}{4\pi^3} \int d^2 k_\parallel G_\pm(eV, \boldsymbol{k}_\parallel), \quad (8.17)$$

where V is the bias voltage, S is the contact area, and $G_\pm$ is the dimensionless conductance [327]. In the limiting case where the transmission coefficient $\mathscr{T} \ll 1$, $G_\pm$ can be written as

$$G_+(E) = \frac{\mathscr{T}^2/2(1-\mathscr{T})}{1-(E/\Delta_0)^2 + \mathscr{T}^2/4(1-\mathscr{T})}; \quad (8.18)$$

$$G_-(E) = \frac{\mathscr{T}^2/2(1-\mathscr{T})}{(E/\Delta_0)^2 + \mathscr{T}^2/4(1-\mathscr{T})} \quad (8.19)$$

when $|E| \leqslant \Delta_0$, where Δ_0 is the energy gap in the bulk. The label $\pm$ in $G_\pm$ corresponds to the product of the signs of the pair potentials for the two branches of BdG quasiparticles involved in the tunneling process. For tunneling along the chains, the two branches correspond to the points L ($\alpha = -$) and R ($\alpha = +$) in the left panel of Fig. 8.3, and the sign is $+$ for s-wave and $-$ for p_x-wave.

Numerical results for $G_\pm(E)$ are shown in Fig. 8.4. The zero-bias conductance peak (ZBCP) depicted in the left panel (the peak at $E = 0$) is a manifestation of the midgap Andreev bound states. These states exist at each chain edges for a p_x-wave quasi-1D superconductor. In fact, the discussion above can be generalized to other symmetries, and Andreev bound states exist at those edges where momentum reflection from the edge connect points on the Fermi surface with opposite signs of the superconducting pair potential. As indicated in the left and right diagrams of Fig. 8.3, reflections from the edge perpendicular to the chains connect Fermi sheets L ($\alpha = -$) to R ($\alpha = +$), and reflections from the edge parallel to the chains connect L to L′ and R to R′. By using the index theorem [328] and comparing the signs of the pair potential at these points for different superconducting symmetries, Sengupta *et al.* [315] determined whether ZBCP is present when tunneling into different edges. In Tab. 8.1, the presence or absence of ZBCP is summarized for a few singlet and triplet symmetries. If such tunneling experiments can be performed in Bechgaard salts, this table provides useful information about the symmetry of order parameter.

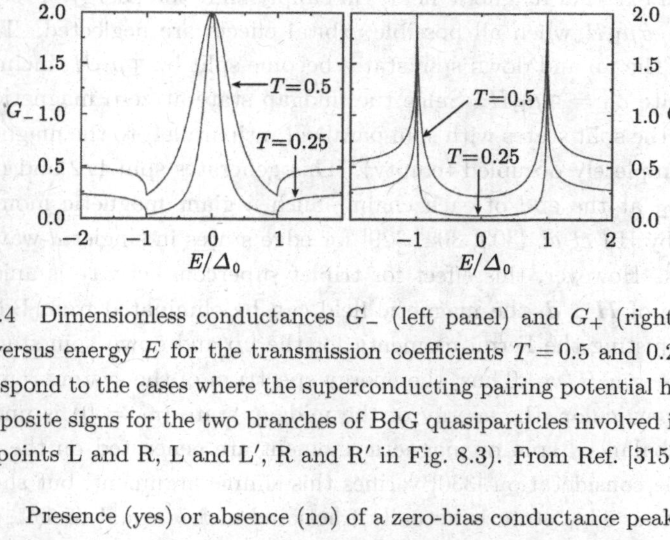

Figure 8.4 Dimensionless conductances G_- (left panel) and G_+ (right panel) are plotted versus energy E for the transmission coefficients $T = 0.5$ and 0.25. G_+ and G_- correspond to the cases where the superconducting pairing potential has the same or the opposite signs for the two branches of BdG quasiparticles involved in tunneling (see the points L and R, L and L′, R and R′ in Fig. 8.3). From Ref. [315].

Table 8.1 Presence (yes) or absence (no) of a zero-bias conductance peak in electron tunneling along the $\boldsymbol{a}$ and $\boldsymbol{b}$ axes for different symmetries of the superconducting pairing potential $\Delta(\boldsymbol{k})$. From Ref. [315].

Symmetry	$\Delta(\boldsymbol{k})$	$\boldsymbol{a}$ axis ZBCP	$\boldsymbol{b}$ axis ZBCP
s	const.	No	No
p_x	$\sin(k_x a)$	Yes	No
p_y	$\sin(k_y b)$	No	Yes
$d_{x^2-y^2}$	$\cos(k_y b)$	No	No
d_{xy}	$\sin(k_x a)\sin(k_y b)$	Yes	Yes

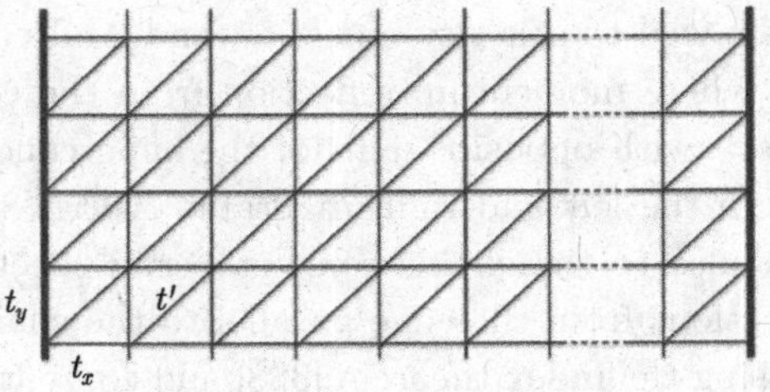

Figure 8.5 Schematic illustration of an anisotropic 2D lattice considered by Tanuma *et al.* [318]. Parameters are taken as $t_y/t_x = 0.1$ to mimic the quasi-1D nature of Bechgaard salts. t' is the next-nearest-neighbor hopping integral.

In addition to their presence in tunneling processes, the midgap edge states also change the system's response to an external magnetic field. In the case of $\boldsymbol{H}\|\boldsymbol{d}$, an external magnetic field will simply shift the energy spectrum Eq. (8.16) by $-\sigma\mu_{\rm B}H$ when all possible orbital effects are neglected. Thus, the energies of the up and down spin states become split by $\mp\mu_{\rm B}H$, including the midgap state $\mathscr{E}_0 = \mp\mu_{\rm B}H$. Since the midgap state at zero magnetic field is half-filled, the split states with spin parallel (antiparallel) to the magnetic field become completely occupied (empty). This generates spin 1/2 and magnetic moment $\mu_{\rm B}$ at the end of each chain. Such a giant magnetic moment was predicted by Hu *et al.* [303, 304, 329] for edge states in singlet d-wave superconductors. However, this effect for triplet superconductivity is anisotropic. In the case of $\boldsymbol{H} \perp \boldsymbol{d}$, the magnetic field can be eliminated from BdG equation by adjusting the Fermi momenta for the up and down spin states to be $k_{{\rm F},\sigma} = k_{\rm F} + \sigma\mu_{\rm B}H/v_{\rm F}$. Thus, the energy spectrum of the system remains the same. In particular, the energy of the midgap state ($\mathscr{E}_0 = 0$) is unchanged, and no unbalanced spin or magnetic moment are generated on the edge. A more subtle consideration [330] verifies this simple argument, but shows that the spin of an edge state is actually fractional and is equal to 1/4, not 1/2. This anisotropic spin response can also be used to distinguish order parameter symmetries in quasi-1D superconductors.

Up to now, we have learned that the presence or absence of an $\boldsymbol{a}$-axis ZBCP, which is related to the zero-energy surface bound state, can be set as an unambiguous feature to identify singlet (either s- or d-wave) states from triplet (p- or f-wave) symmetries. However, it cannot be used to distinguish the orbital symmetries within each category, for the pairing potential bears the same parity upon $k_x \to -k_x$. To solve this issue, an important observation should be made that the s- and p_x-pairing states are fully gapped in quasi-1D Bechgaard salts, while the $d_{x^2-y^2}$- and f_x-wave states have line nodes in the gap function. This difference in nodal structure can manifest itself in the overall structure of tunneling spectrum, and hence can help pinning down the underlying pairing symmetry. Along this line, Tanuma *et al.* [316] analyze the local density-of-states at the surface of superconductors with different pairing symmetries, and conclude that a p_x-wave pairing state should present a U-shaped tunneling spectrum with ZBCP, a $d_{x^2-y^2}$-wave state would acquire a V-shaped structure without ZBCP, and a f_x-wave state should bear a V-shaped spectrum with ZBCP.

Since the surface density-of-states (SDOS) is very sensitive to the shape of Fermi surfaces, Tanuma *et al.* [316] start from an extended Hubbard model on an anisotropic 2D lattice as illustrated in Fig. 8.5, instead of using the tight-binding or effective mass dispersion. The Hamiltonian reads

$$\begin{aligned}\mathscr{H} =& -\sum_{\langle i,j\rangle_x,\sigma}\left(t_x c^\dagger_{i,\sigma}c_{j,\sigma} + \text{H.C.}\right) - \sum_{\langle i,j\rangle_y,\sigma}\left(t_y c^\dagger_{i,\sigma}c_{j,\sigma} + \text{H.C.}\right)\\ &-\frac{V}{2}\sum_{(i,j)_m,\sigma,\sigma'} c^\dagger_{i,\sigma}c^\dagger_{j,\sigma'}c_{j,\sigma'}c_{i,\sigma} - \mu\sum_{i,\sigma}c^\dagger_{i,\sigma}c_{i,\sigma},\end{aligned} \tag{8.20}$$

where $c_{i\sigma}$ ($c^\dagger_{i\sigma}$) is the annihilation (creation) operator of an electron with spin σ at site $\boldsymbol{i} = (i_x, i_y)$, $\langle i,j\rangle_{x,y}$ indicates summation over nearest neighbors along the corresponding direction, $(\boldsymbol{i},\boldsymbol{j})_m$ stands for summation over pairs of sites separated by m lattice spacings along the $\boldsymbol{a}$ axis, and V is the interaction within these pair sites. Here, the quasi-1D nature of Bechgaard salts is employed by assuming $\boldsymbol{a}$ axis to be the easy axis with $t_y/t_x = 0.1$. The chemical potential μ is fixed to make sure the system is quarterly filled.

Within a mean-field approximation, the superconducting order parameter is defined as

$$\Delta^{\sigma\sigma'}_{ij} = \frac{V}{2}\langle c_{i\sigma}c_{j\sigma'}\rangle\delta_{i_y,j_y}, \tag{8.21}$$

where pairing is assumed to take place between electrons within a single chain ($i_y = j_y$). Thus, each chain along the $\boldsymbol{a}$ axis forms a unit cell with a total number of N_L sites. In the discussion of Tanuma *et al.* [316], the following three pairing states are considered and illustrated in Fig. 8.6:

(i) p_x-wave: triplet pairing between sites separated by $m = 2$ lattice spacings. The bulk order parameter takes the form $\Delta_p \sin(2k_x a)$, hence changes sign as $+-$ along the Fermi surface;

(ii) $d_{x^2-y^2}$-wave: singlet pairing between sites separated by $m = 2$ lattice spacings; The bulk order parameter takes the form $\Delta_d \cos(2k_x a)$, hence changes sign as $+-+-$ along the Fermi surface;

(iii) f_x-wave: triplet pairing between sites separated by $m = 4$ lattice spacings; The bulk order parameter takes the form $\Delta_f \sin(4k_x a)$, hence changes sign as $+-+-+-$ along the Fermi surface.

All these three pairing states are consistent with the spin alignment of the $2k_F$ SDW phase with the easy axis along $\boldsymbol{b}'$, hence can be favored by the background spin fluctuations. By Fourier transforming the y coordinate into momentum space k_y, and using i (j) as shorthand notation for i_x (j_x), one obtains the mean-field Hamiltonian as

$$\mathscr{H}_{\mathrm{MF}} = \sum_{k_y,i,j} \hat{\mathbf{C}}_i^\dagger(k_y) \begin{bmatrix} H_{ij}(k_y) & 0 & \Delta_{ij}^{\uparrow\uparrow} & \Delta_{ij}^{\uparrow\downarrow} \\ 0 & H_{ij}(k_y) & \Delta_{ij}^{\downarrow\uparrow} & \Delta_{ij}^{\downarrow\downarrow} \\ \Delta_{ji}^{*\uparrow\uparrow} & \Delta_{ji}^{*\downarrow\uparrow} & -H_{ji}(-k_y) & 0 \\ \Delta_{ji}^{*\uparrow\downarrow} & \Delta_{ji}^{*\downarrow\downarrow} & 0 & -H_{ji}(-k_y) \end{bmatrix} \hat{\mathbf{C}}_j(k_y), \tag{8.22}$$

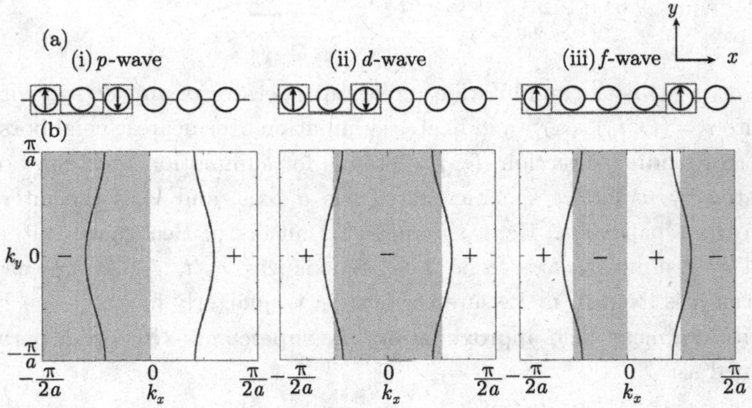

Figure 8.6 (a) An illustration of Cooper pairing in real space and (b) the shape of Fermi surface for $t_y/t_x = 0.1$ at quarter-filling and the pair potential for (i) triplet 'p_x-wave' ($m = 2$), (ii) singlet '$d_{x^2-y^2}$-wave' ($m = 2$), and (iii) triplet 'f_x-wave' ($m = 4$). In (a), the pairs are depicted by dashed lines, and in (b) $+$ $(-)$ denotes the region where the sign of the pair potential is positive (negative). From Ref. [316].

where the wavefunction vector and diagonal elements take the form

$$\hat{\mathbf{C}}_i^\dagger(k_y) = \left[\; c_{i\uparrow}^\dagger(k_y),\;\; c_{i\downarrow}^\dagger(k_y),\;\; c_{i\uparrow}(-k_y),\;\; c_{i\downarrow}(-k_y)\;\right] \tag{8.23}$$

$$H_{ij}(k_y) = -\sum_{\pm}\left[t_x\delta_{i,j\pm1} + 2t_y\cos(k_y b)\delta_{i,j} - \mu\delta_{i,j}\right]. \tag{8.24}$$

Then, Tanuma *et al.* [316] assumed the following forms of the off-diagonal elements for different pairing symmetries

$$p_x\text{-wave: } \Delta_{ij}^{\uparrow\downarrow} = \Delta_{ij}^{\downarrow\uparrow} = \sum_{\pm}\Delta_{ij}^{p}\delta_{i,j\pm2}, \quad \Delta_{ij}^{\uparrow\uparrow} = \Delta_{ij}^{\downarrow\downarrow} = 0, \tag{8.25a}$$

$$d_{x^2-y^2}\text{-wave: } \Delta_{ij}^{\uparrow\downarrow} = -\Delta_{ij}^{\downarrow\uparrow} = \sum_{\pm}\Delta_{ij}^{d}\delta_{i,j\pm2}, \quad \Delta_{ij}^{\uparrow\uparrow} = \Delta_{ij}^{\downarrow\downarrow} = 0, \tag{8.25b}$$

$$f_x\text{-wave: } \Delta_{ij}^{\uparrow\uparrow} = \Delta_{ij}^{\downarrow\downarrow} = \sum_{\pm}\Delta_{ij}^{f}\delta_{i,j\pm4}, \quad \Delta_{ij}^{\uparrow\downarrow} = \Delta_{ij}^{\downarrow\uparrow} = 0. \tag{8.25c}$$

The mean-field Hamiltonian $\mathscr{H}_{\mathrm{MF}}$ is diagonalized via Bogoliubov transformation

$$c_{i\sigma}(k_y) = \sum_{\lambda} u_{i,\lambda}\gamma_\lambda(k_y), \tag{8.26}$$

$$c_{j\sigma'}(-k_y) = \sum_{\lambda} u_{N_L+j,\lambda}^{*}\gamma_\lambda^\dagger(k_y), \tag{8.27}$$

where λ labels different eigenstates. With the aid of the newly defined operators $\gamma_\lambda(k_y)$ which satisfies the fermionic anti-commutation relation, the mean-field Hamiltonian can be diagonalized as

$$\mathscr{H}_{\mathrm{MF}} = \sum_{k_y,\lambda} E_\lambda(k_y)\gamma_\lambda^\dagger(k_y)\gamma_\lambda(k_y). \tag{8.28}$$

The spatial-dependent pairing gap function is then determined self-consistently as

$$\Delta_{j,j\pm m} = \frac{V}{2}\sum_{k_y,\lambda} u_{j\pm m,\lambda}u_{N_L+j,\lambda}^{*}\{1 - n_{\mathrm{F}}[E_\lambda(k_y)]\} \tag{8.29}$$

with $n_{\mathrm{F}}(x)$ the Fermi distribution function.

By numerically diagonalizing the mean-field Hamiltonian Eq. (8.22) with the self-consistency relation Eq. (8.29) until a convergence condition is met, Tanuma *et al.* [316] obtained the spatial-dependent gap function and SDOS for different pairing symmetries. In their calculation, a total number of $N_y = 50$ chains each containing $N_L = 10^3$ sites are considered. The pairing interaction is $V/t_x = 4.0$, and the bulk gap function and the chemical potential for different symmetries are

$$p_x\text{-wave: } \Delta_p/t_x = 0.280, \quad \mu/t_x = 1.39; \tag{8.30a}$$

$$d_{x^2-y^2}\text{-wave: } \Delta_d/t_x = 0.164, \quad \mu/t_x = 1.39; \tag{8.30b}$$

$$f_x\text{-wave: } \Delta_f/t_x = 0.244, \quad \mu/t_x = 1.38. \tag{8.30c}$$

In order to compare with scanning tunneling microscopy (STM) experiments, Tanuma *et al.* [316] assumed that the STM tip is metallic with a flat density of states, and that the tunneling probability is finite only for the nearest site from the tip. The resulting tunneling conductance spectrum converges to the normalized SDOS

$$\bar{\mathscr{N}}(E) = \frac{\displaystyle\int_{-\infty}^{\infty} d\omega \mathscr{N}_S(\omega) \operatorname{sech}^2\left(\frac{\omega - E}{2T}\right)}{\displaystyle\int_{-\infty}^{\infty} d\omega \mathscr{N}_N(\omega) \operatorname{sech}^2\left(\frac{\omega + 2\Delta}{2T}\right)}, \tag{8.31}$$

where

$$\mathscr{N}_S(\omega) = 2\sum_{k_y,\lambda} |u_{1,\lambda}|^2 \delta(\omega - E_\lambda(k_y)) \tag{8.32}$$

denotes the SDOS at the first site from the surface in the superconducting part and $\mathscr{N}_N(\omega)$ represents the DOS in the normal state.

The results for the spatial-dependent pairing potentials for the p_x-, $d_{x^2-y^2}$-, and f_x-wave symmetries are displayed in the left panels of Fig. 8.7. The plotted quantities are defined as

$$p_x\text{-wave: } \Delta^p_{j,\pm a} \equiv \text{Re}[\Delta^p_{j,j\pm 2}]/\Delta_p, \quad \text{Im}[\Delta^p_{j,j\pm 2}] = 0, \tag{8.33a}$$

$$d_{x^2-y^2}\text{-wave: } \Delta^d_{j,\pm a} \equiv \text{Re}[\Delta^d_{j,j\pm 2}]/\Delta_d, \quad \text{Im}[\Delta^d_{j,j\pm 2}] = 0, \tag{8.33b}$$

$$f_x\text{-wave: } \Delta^f_{j,\pm a} \equiv \text{Im}[\Delta^f_{j,j\pm 4}]/\Delta_f, \quad \text{Re}[\Delta^f_{j,j\pm 4}] = 0. \tag{8.33c}$$

For the p_x-wave case, since triplet pairs are formed between two electrons with two lattice spacings apart, the relation $\Delta^p_{j,a} = -\Delta^p_{j+2,-a}$ is satisfied. The pairing gap functions tend to zero near the surface and approach their bulk values of ± 1 in the middle of superconductors. In the case of $d_{x^2-y^2}$-wave symmetry, since the pair is singlet and formed by two electrons separated by two lattice spacings, one has $\Delta^d_{j,a} = \Delta^d_{j+2,-a}$. The pair potentials show an atomic-scale spatial oscillation near the surface and converge to the bulk values towards the center. For the f_x-wave state, the relation $\Delta^f_{j,a} = -\Delta^f_{j+4,-a}$ is hold for the two composite electrons which are four lattice spacings apart within a triplet Cooper pair. The qualitative behavior of the pairing potentials is similar to that in the p_x-wave case, but with some complex details.

The resulting SDOS for various pairing symmetries obtained from the corresponding pair potentials are shown in the right panels of Fig. 8.7. Notice that a ZBCP is present for both the p_x- and f_x-wave states at the surface normal to the $\boldsymbol{a}$ axis, as a direct consequence of the emergence of a zero-energy surface bound state. The ZBCP is absent at the surface normal to the $\boldsymbol{b}$ axis, since in this geometry the injected and reflected quasiparticles feel the same pairing gap, and a zero-energy bound state is not present at the boundary. On the other hand, the $d_{x^2-y^2}$-wave state acquires no ZBCP at any surface. This observation is in qualitative agreement with the outcome obtained by Sengupta *et al.* [315].

In addition to the presence or absence of ZBCP, Tanuma *et al.* [316] also found qualitative difference in the overall structure of the tunneling spectrum among various pairing states. In particular, the p_x-wave state features a U-shaped spectrum with a flat bottom around origin, indicating a full gap along the entire Fermi surface. Meanwhile, both the $d_{x^2-y^2}$- and f_x-wave symmetries present a V-shaped spectrum due to the existence of gap nodes on the Fermi surface. This qualitative feature, together with the presence or absence of ZBCP, hence are suggested to serve as an indicator on the superconducting pairing symmetry.

Although the idea discussed above is rather intuitive, it is by no means an easy task to actually conduct a successful tunneling spectroscopy measurement in organic superconductors. First, organic single crystals are fragile upon standard cutting and surface treating methodologies, which can induce cracks in the samples and hence compromise the tunneling results. Second, the rather high pressure required for $(TMTSF)_2PF_6$ to become superconducting makes it impractical for conventional tunneling setups, while the anion ordering in $(TMTSF)_2ClO_4$ leads to a double-sheet structure of Fermi surface and brings complexity to the resulting tunneling spectrum.

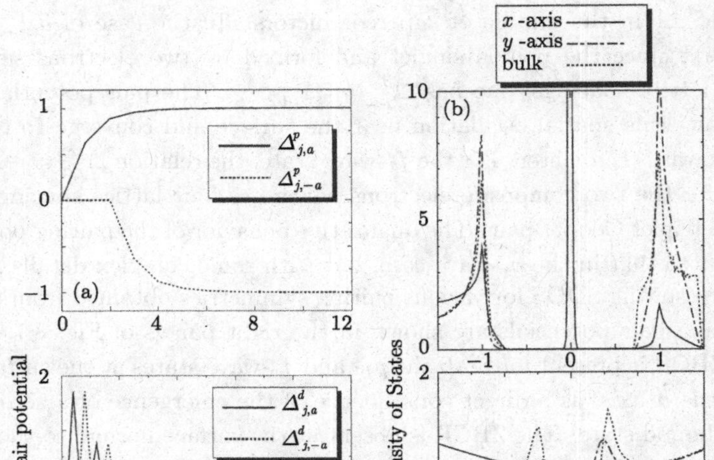

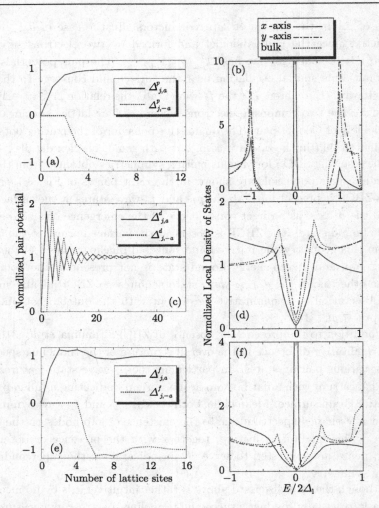

Figure 8.7 The left panels are the spatial dependences of the pair potentials along the ***a*** axis near the surface for different pairing symmetries. The right panels show the SDOS at the surface normal to the ***a*** or ***b*** axis along with the bulk density of states. From Ref. [316].

Despite these difficulties, Arai *et al.* [331] managed to perform an STM spectroscopy study on superconducting $(TMTSF)_2ClO_4$ on the surface perpendicular to the most conducting $\boldsymbol{a}$ axis at temperatures as low as 0.3 K. They observed a V-shaped spectrum and an absence of ZBCP (see Fig. 8.8), which are consistent with the $d_{x^2-y^2}$-wave symmetry as stated above. However, this experiment alone can not set an unambiguous identification of the pairing symmetry. In fact, the presence or absence of the ZBCP is also sensitive to serval other factors in addition to the gap symmetry. First, it is known that in high-T_c cuprate with $d_{x^2-y^2}$ pairing symmetry, the ZBCP may be observed due to atomic-scale roughness in the (100) surfaces where the gap function has the same sign upon reflection [332, 333, 334, 335]. In fact, Iguchi *et al.* [309] observed a ZBCP for Ag/YBCO ramp-edge junctions with incident direction varied from (100) to (110) interfaces. On the other hand, depending on the shape of the Fermi surface and the geometry of the sample boundary, the oscillatory behavior of the wavefunction of the zero-energy bound state can induce an interference effect which locally destroys the ZBCP [333, 334]. Considering the fact that a well-oriented cleavage surface is not easy to obtain for organic samples, this effect may lead to nontrivial consequences.

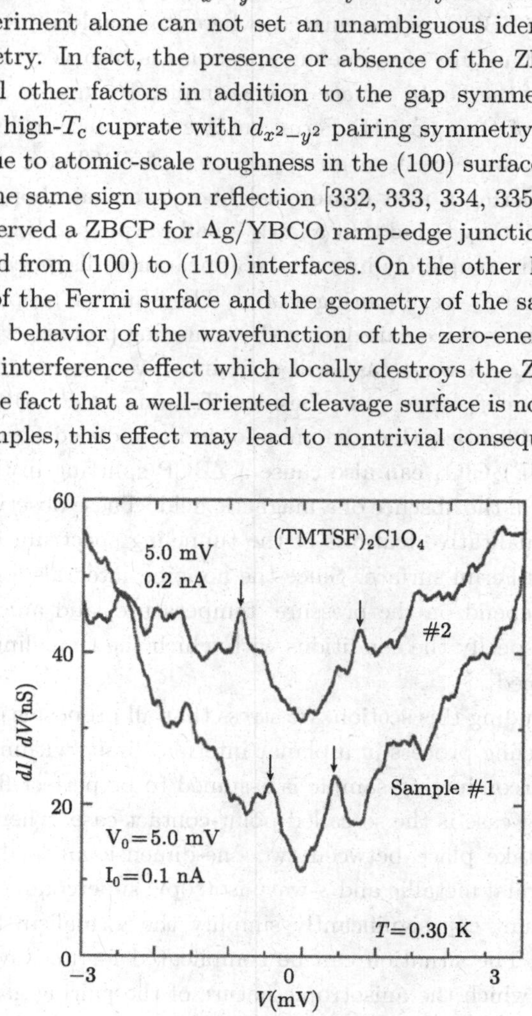

Figure 8.8 Differential conductance dI/dV curves obtained on the surface perpendicular to the $\boldsymbol{a}$ axis of two different $(TMTSF)_2ClO_4$ samples at $T = 0.30$ K. The arrows indicate gap edges. From Ref. [331].

Second, the impurity attached on the sample interface can also be important in the tunneling processes. In particular, it has been shown that impurity scattering near the interface in high-T_c cuprates can induce a splitting or a disappearance of ZBCP on the surfaces where it should be present [336–341].

Third, the structure of tunneling spectrum is also very sensitive to the detailed shape of the Fermi surface. For example, Tanuma *et al.* [317] showed that a warping of the quasi-1D Fermi surface can cause a disappearance of ZBCP for the f_x-wave case, or an appearance of ZBCP for the $d_{x^2-y^2}$-wave state. By considering a next-nearest-neighbor interchain hopping t' as illustrated in Fig. 8.5, they found that a fairly small warping ($t'/t_x = -0.15$) of the Fermi surface can completely inverse the appearance/disappearance of ZBCP in the $d_{x^2-y^2}$- and f_x-wave pairing states. They further proposed to apply a magnetic field along the $\boldsymbol{c}$ axis during the tunneling process, where the ZBCP present in the $d_{x^2-y^2}$-wave case will split into two peaks while no splitting will happen in the f_x-wave case. Besides, Tanuma *et al.* [318] also found in a subsequent publication that the anion-ordering-induced Fermi surface doubling in $(TMTSF)_2ClO_4$ can also cause a ZBCP splitting in the $d_{x^2-y^2}$-wave symmetry, even in the absence of a magnetic field. These observations indicate that even the qualitative features of the tunneling spectrum are sensitive to the shape of the Fermi surface. Since the hopping integrals and the resulting Fermi surface depend on the pressure, temperature, and anions, it is necessary to strictly specify the conditions with which the tunneling experiment is actually performed.

Before concluding this section, we stress that all proposals discussed above model the tunneling process in a planar interface limit, within which the surface of the superconducting sample is assumed to be perfect flat and infinite. Another limiting case is the so-called point-contact case, where the tunneling is assumed to take place between two one-dimensional leads. For tunneling between normal metallic and s-wave isotropic superconducting states, the point-contact limit can significantly simplify the formalism by confining to one dimension. The situation can be complicated for unconventional superconductors, for which the anisotropic nature of the pairing gap function may have nontrivial effect. However, for quasi-1D Bechgaard salts, this complexity may be bypassed by considering only tunneling processes along the $\boldsymbol{a}$ axis and pairing interactions within the same direction. Along this line, Bolech and Giamarchi [319] discussed the point-contact tunneling process between two leads via non equilibrium Keldysh formalism [342], and found different *I-V* characteristics for normal metal–singlet superconductor with *s*-wave symmetry (N-SSC), normal metal–triplet superconductor with *p*-wave symmetry (N-TSC), and singlet superconductor–triplet superconductor (SSC-TSC) junctions.

The numerical results of the *I-V* characteristics for different junction structures are shown in Fig. 8.9. From top to bottom, spectra for (a) N-SSC, (b) N-TSC, and (c) SSC-TSC junctions are displayed in the absence (solid lines) or presence (dashed lines) of a finite magnetic field. The inter-lead tunneling overlap integral is set as $t = 0.2\Delta_0$ and the magnetic field is assumed to be $\mu_B H = 0.2\Delta_0$, where Δ_0 is the amplitude of superconducting order parameter in the bulk. For the case of N-SSC junctions [Fig. 8.9(a)], the *I-V* curve shows a subgap shoulder in the region $|e|V < \Delta_0$ in the absence of magnetic field,

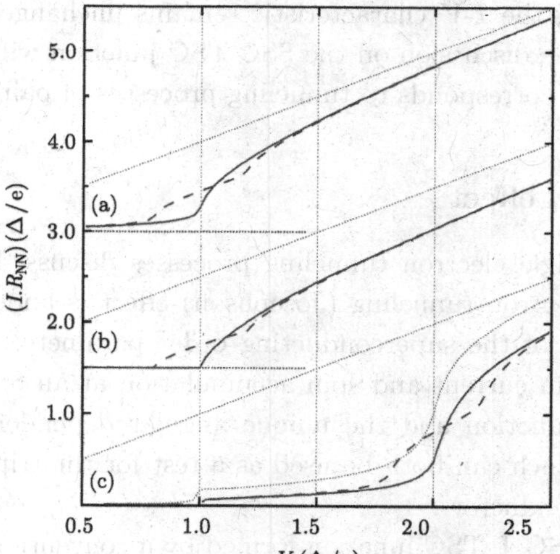

Figure 8.9 The *I-V* characteristics for a one-dimensional lead constructed by (a) N–SSC junction, (b) N–TSC junction, and (c) SSC–TSC junction. The solid (dashed) lines correspond to the results in the absence (presence) of a magnetic field. The diagonal straight dotted lines in all three sets are plotted as reference to the outcome of the N–N case, while the dotted curve in (c) is depicted to show the results for a SSC–SSC junction. In this calculation, only *s*-wave and *p*-wave pairing states are considered for the SSC and TSC, respectively. From Ref. [319].

as a direct consequence of coherent Andreev processes taking place in the junction contact. At the gap edge, the I-V curve bends up rather abruptly, resulting in a differential conductance peak in the dI/dV curve. By applying a magnetic field, the Zeeman effect will induce a splitting of this differential conductance peak, which can be identified on the I-V curve as two kinks. As a comparison, the I-V curves for a N-TSC junction acquire distinct features [Fig. 8.9(b)]. First, there is no subgap shoulder in the absence of magnetic field, since the p_x superconducting state acquires odd parity upon spatial inversion. One should bear in mind that the geometry of this point-contact junction is equivalent to the case of tunneling into a sample surface parallel to the $\boldsymbol{a}$ axis, where no ZBCP is expected. Second, by applying a magnetic field parallel to the triplet $\boldsymbol{d}$-vector ($\boldsymbol{H} \parallel \boldsymbol{d}$), the differential conductance peak also splits into two, as a consequence of the same Zeeman effect as in the N-SSC case. However, if the magnetic field is applied perpendicular to the $\boldsymbol{d}$-vector ($\boldsymbol{H} \perp \boldsymbol{d}$), the I-V characteristic remains unchanged from the zero-magnetic case. The discussion on the SSC-TSC junction will be found in the next section, as it corresponds to tunneling processes of paired electrons.

8.2 Josephson effect

In addition to single electron tunneling processes discussed in the previous section, paired electron tunneling (Josephson) effect is another possible test for the symmetry of the superconducting order parameter. In this section, we discuss the spin current and spin accumulation at an SSC–insulator (I)–TSC Josephson junction and the unique angular dependence in a TSC–I–TSC geometry, which can both be used as a test for the triplet nature of the composite superconductors.

Consider an SSC–I–TSC junction formed by a conventional s-wave superconductor on the left and a p-wave superconductor on the right, separated by a flat plane perpendicular to the so-called x axis. The interface is modeled by the δ-function potential $V(x) = V_0\delta(x)$. The TSC part at $x > 0$ is placed in such a way that the easy conducting axis $\boldsymbol{a}$ is perpendicular to the junction interface, i.e., $\boldsymbol{a} \parallel \boldsymbol{x}$. A schematic illustration of the junction geometry is shown in Fig. 8.10. The pairing potential of the singlet s-wave superconductor on the left (L) side is

$$\Delta_\sigma(x<0) \equiv \Delta_s \mathrm{sgn}(\sigma) = \langle c_{\boldsymbol{k},\sigma} c_{-\boldsymbol{k},\bar{\sigma}} \rangle, \tag{8.34}$$

where σ and $\bar{\sigma} = -\sigma$ indicate spin indices, and Δ_s stands the pairing gap amplitude. Here, the function $\mathrm{sgn}(\sigma) = 1$ and -1 for spin up and down, respectively. The pairing potential of the triplet p-wave superconductor on the right (R) side takes the usual form

$$\Delta_\sigma(x>0,\boldsymbol{k}) \equiv \Delta_p i \left[\sigma_y (\hat{\sigma}\cdot \boldsymbol{d})\right]_{\sigma,\sigma'} \Gamma(\boldsymbol{k}) e^{i\phi_0} = \langle c_{\boldsymbol{k},\sigma} c_{-\boldsymbol{k},\sigma'} \rangle, \tag{8.35}$$

where ϕ_0 is the relative phase between the left and right superconductors of the junction, and $\Gamma(\boldsymbol{k}) = k_x/k_{\mathrm{F}}$ is chosen to account the p_x-wave pairing symmetry which is an important candidate for Bechgaard salts. To illustrate the idea of spin current and spin accumulation with minimum effort, we first neglect the momentum dependence of the $\boldsymbol{d}$-vector and assume $\boldsymbol{d}$ has a uniform orientation [322]. A generalization to the case of $\boldsymbol{k}$-dependent d-vector will be briefly introduced later with details can be found in Ref. [323]. If the spin-orbit coupling in TSC part is weak, the spin quantization axes can be freely rotated such that the $\boldsymbol{z}$ axis is parallel to the $\boldsymbol{d}$-vector. In this case, the triplet pairing potential Eq. (8.35) possesses component only in the $S_z = 0$ channel and takes the form

$$\Delta_\sigma(x>0,\boldsymbol{k}) = \Delta_p \Gamma(\boldsymbol{k}) e^{i\phi_0} = \langle c_{\boldsymbol{k},\sigma} c_{-\boldsymbol{k},\bar{\sigma}} \rangle. \tag{8.36}$$

The electronic wavefunctions are governed by the following BdG equation with a substitution $k_x \to -i\partial_x$

$$\begin{pmatrix} \mathscr{H}_0 + \mu + V(x) & \Delta_\sigma(x,\boldsymbol{k}) \\ \Delta^*_\sigma(x,\boldsymbol{k}) & -\mathscr{H}_0 - \mu - V(x) \end{pmatrix} \Psi_{\lambda\sigma}(x,\boldsymbol{k}) = E_{\lambda\sigma}(x,\boldsymbol{k})\,\Psi_{\lambda\sigma}(x,\boldsymbol{k}). \tag{8.37}$$

where the single particle Hamiltonian

$$\mathscr{H}_0 = \frac{-\partial_x^2}{2m} - t_y \cos(k_y b) - t_z \cos(k_z c) \tag{8.38}$$

is chosen to describe the quasi-1D nature of Bechgaard salts. The wavefunctions on the left and right sides of the junction are connected by the boundary conditions of continuity. Focusing on the subgap bound states with eigenenergies $|E| \leqslant \min[|\Delta|_s, |\Delta_p|]$, the wavefunctions can be rewritten as a superposition of left- and right-moving particles [322, 343],

$$\Psi^\beta_c(x,\boldsymbol{k}) = e^{-\kappa^\beta_c x} \left[A^\beta e^{i\tilde{k}_{\mathrm{F}} x} \begin{pmatrix} u^\beta_{c,+} \\ v^\beta_{c,+} \end{pmatrix} + B^\beta e^{-i\tilde{k}_{\mathrm{F}} x} \begin{pmatrix} u^\beta_{c,-} \\ v^\beta_{c,-} \end{pmatrix} \right], \tag{8.39}$$

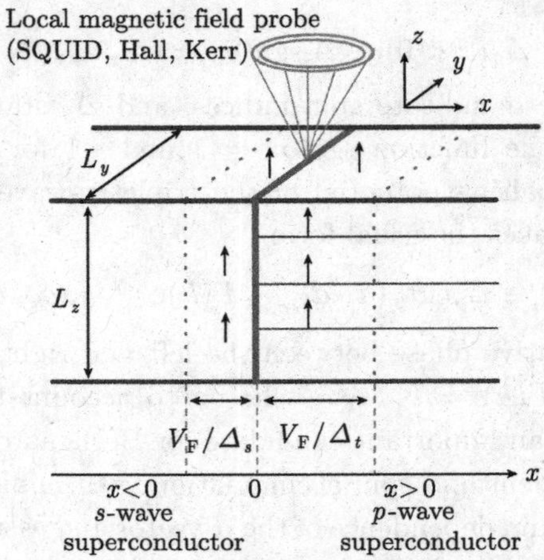

Figure 8.10 A Josephson junction between a singlet and a triplet superconductors. The thin black solid lines represent easily conducting chains for $(TMTSF)_2X$. The blue dotted lines indicate the localization lengths of the Andreev bound states. The cone represents a probe of the local magnetic field produced by spin accumulation at the interface. From Ref. [322].

where $c \equiv (\lambda, \sigma)$ is used to shorten notation, the superscript $\beta = \pm$ labels the wavefunctions on the left ($\beta = -$) and right ($\beta = +$) sides of the junction, and the subscript $\alpha = \pm$ denotes the left ($\alpha = -$) and right ($\alpha = +$) moving quasiparticles. The parameters are defined as

$$\begin{aligned}
\kappa_c^- &= -\sqrt{|\Delta_s|^2 - E_c^2}/v_{\mathrm{F}}, \\
\kappa_c^+ &= \sqrt{|\Delta_p|^2 - E_c^2}/v_{\mathrm{F}}, \\
\tilde{k}_{\mathrm{F}} &= k_{\mathrm{F}} + 2t_y \cos(k_y b)/v_{\mathrm{F}} + 2t_z \cos(k_z c)/v_{\mathrm{F}}.
\end{aligned} \tag{8.40}$$

The Bogoliubov coefficients $u_{c,\alpha}^{\beta}$ and $v_{c,\alpha}^{\beta}$ are solved from Eq. (8.37) away from the junction interface.

Substituting the trial wavefunction Eq. (8.39) into the boundary conditions, we can obtain four linear independent equations for the coefficients A^β and B^β. The condition for the existence of nontrivial solutions translates vanishing of the determinant of the corresponding 4×4 matrix, which takes the form

$$\frac{(\eta_{c,+}^+ - \eta_{c,+}^-)(\eta_{c,-}^+ - \eta_{c,-}^-)}{(\eta_{c,-}^+ - \eta_{c,+}^-)(\eta_{c,+}^+ - \eta_{c,-}^-)} = 1 - \mathscr{T}. \tag{8.41}$$

Here, $\mathscr{T} = v_{\rm F}^2/(v_{\rm F}^2+V_0^2)$ is the transmission coefficient of the insulating barrier, and

$$\eta_{c,\alpha}^{\beta} = \frac{u_{c,\alpha}^{\beta}}{v_{c,\alpha}^{\beta}} = \begin{cases} \dfrac{E_c - i\alpha v_{\rm F}\kappa_c^-}{\Delta_\sigma(x<0)}, & \text{for } \beta = -; \\ \dfrac{E_c - i\alpha v_{\rm F}\kappa_c^+}{\Delta_\sigma(x>0)}, & \text{for } \beta = +. \end{cases} \tag{8.42}$$

Substituting Eq. (8.42) into (8.41), one can obtain the equation for the energies of the Andreev bound states [322],

$$\mathscr{A}\sin(\Phi_{k_y,k_z}) - \mathscr{B}\cos(\Phi_{k_y,k_z}) = \mathrm{sgn}(\sigma)\sin(\phi_0)\mathscr{T}, \tag{8.43}$$

where

$$\Phi_{k_y,k_z} = \frac{1}{2}\left[\phi(\tilde{k}_{\rm F}, k_y, k_z) - \phi(-\tilde{k}_{\rm F}, k_y, k_z)\right] \tag{8.44}$$

with $\phi(k_x, k_y, k_z)$ the phase of the function $\Gamma(\boldsymbol{k})$. The resulting coefficients are

$$\mathscr{A} = (2-\mathscr{T})\mathscr{E}_{c,s}\sqrt{1-\mathscr{E}_{c,t}^2} + \mathscr{T}\mathscr{E}_{c,t}\sqrt{1-\mathscr{E}_{c,s}^2}, \tag{8.45a}$$

$$\mathscr{B} = -\mathscr{T}\mathscr{E}_{c,s}\mathscr{E}_{c,t} + (2-\mathscr{T})\sqrt{(1-\mathscr{E}_{c,s}^2)(1-\mathscr{E}_{c,t}^2)}, \tag{8.45b}$$

where the dimensionless energy of the bound state is determined as

$$\mathscr{E}_{c,i=(s,t)} = \frac{E_c}{|\Delta|_i} \leqslant 1. \tag{8.46}$$

For the specific example of p_x-wave symmetry, the symmetry function satisfies $\Gamma(-k_{\rm F}) = -\Gamma(k_{\rm F})$, which leads to $\Phi_{k_y,k_z} = \pi/2$. In this case, the coefficient is determined to be

$$\mathscr{A} = \mathscr{T}\mathrm{sgn}(\sigma)\sin(\phi_0). \tag{8.47}$$

The eigenenergies E_c are independent of transverse momentum and are plotted as functions of the relative phase factor ϕ_0 in Fig. 8.11(a). Notice that there could be two or four eigenstates depending on ϕ_0, and the eigenenergies corresponding to spin up and spin down have opposite signs $E_\uparrow = -E_\downarrow$. The energy difference between up and down spins can be qualitatively understood in the following way. Since the singlet pairing potential of Eq. (8.34) has opposite signs for spin up and spin down, while the triplet pairing potential Eq. (8.35) has the same sign for both spins, a spin-down quasiparticle would acquire a phase difference $\phi + \pi$ across the junction when a spin-up quasiparticle has a phase difference ϕ. In other words, if one applies a transformation $\sigma \to \bar{\sigma}$, the system must be invariant by requiring $E_\sigma(\phi) = E_{\bar{\sigma}}(\phi + \pi)$, i.e., the bound state energies for $E_\uparrow$ and $E_\downarrow$ are linked by a phase difference of π.

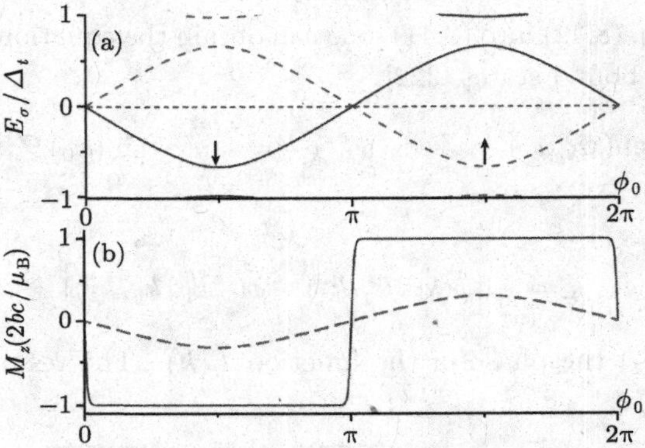

Figure 8.11 (a) The energies $E_\uparrow$ (dashed lines) and $E_\downarrow$ (solid lines) of the Andreev bound states vs. the phase ϕ_0 between the s- and p_x-wave superconductors ($\mathscr{T}$ = 0.8 and $|\Delta_p| = 0.5|\Delta_s|$). (b) Magnetic moment m_z per unit area of the interface, Eq. (8.48), vs. ϕ_0 for $T/|\Delta_p| = 0.01$ (solid line) and 1 (dashed line). From Ref. [322].

The energy difference between spin up and spin down surface bound states can induce a net magnetic moment at the interface. For a given Fermi level at $E = 0$ [dotted line in Fig. 8.11(a)], the magnetic moment accumulated at the junction interface per unit area is give by

$$M_z = \frac{\mu_b}{2bc} \tanh\left(\frac{E_\downarrow(\phi_0)}{2T}\right), \tag{8.48}$$

where μ_B is the Bohr magneton, and the prefactor 1/2 is introduced to compensate double counting [315, 343]. The numerical results for M_z are plotted in Fig. 8.11(b) for two different temperatures. At low temperature of $T/|\Delta_p| = 0.01$ (solid line), the magnetization is close to $\mu_B/2$ per chain, since only the lower bound state is occupied. An elevated temperature of $T/|\Delta_p| = 1$ (dashed line) would populate quasiparticles to excited states above the Fermi level, which provide opposite spin to neutralize the net magnetization. In an open circuit, the phase difference ϕ_0 is self-adjusted to obtain a minimal value of the total energy. For the energy levels shown in Fig. 8.11(a), the minimum is achieved at either $\phi_0 = \pi/2$ or $\phi_0 = 3\pi/2$, which corresponds to negative or positive magnetic moment, respectively. As a consequence, the SSC–I–TSC junction spontaneously breaks time-reversal symmetry and selects one of the two energy minima.

The discussion above for the case of momentum independent $\boldsymbol{d}$-vector can be generalized to the k-dependent cases. Specifically, Lu and Yip [323] considered a $\boldsymbol{d}$-vector in the form of $\boldsymbol{d} \propto k_x\hat{\boldsymbol{y}} - k_y\hat{\boldsymbol{x}}$, and concluded similar findings of spin current and spin accumulation at the SSC–I–TSC junction interface. It also becomes clear that the direction and angle dependences of the spin accumulation can reveal which symmetry is actually broken by the junction. These information hence can help identifying the pairing symmetry of the triplet order parameter.

For possible detection schemes, we stress that the magnetization at the junction interface would manifest itself as a magnetic field which can be determined by local probes such as scanning SQUID or Hall microscope [344, 345]. Assuming a semi-infinite geometry along the z-direction, the magnetic field at the junction interface can be estimated as

$$B \approx \frac{\mu_0\mu_{\mathrm{B}}|\Delta_p|}{4bcv_{\mathrm{F}}}, \tag{8.49}$$

where the condition $|\Delta_s| \gg |\Delta_p|$ is assumed. For $(\mathrm{TMTSF})_2\mathrm{X}$ with $b = 7.711\,\text{Å}$, $c = 13.522\,\text{Å}$, and $v_{\mathrm{F}}/|\Delta_p| = 0.6\,\mu\mathrm{m}$, we have $B \approx 0.3\,\mathrm{G}$. Consider a square scanning SQUID loop of the size $l = 10\,\mu\mathrm{m}$, this magnetic field corresponds to a magnetic flux of $0.06\,\Phi_0$ with Φ_0 the unit flux, which is quite detectable within present technology. Besides, the estimated magnetic field is well above the typical Hall-probe sensitivity of 80 mG at 1 Hz [344].

Next, we discuss another junction geometry consisting of two triplet superconductors with their $\boldsymbol{d}$-vectors both perpendicular to the tunneling axis. If spin-orbit coupling is weak in both superconductors, the Josephson current should depend on the relative orientation of the two $\boldsymbol{d}$-vectors in addition to their relative U(1) phase. This expectation is related to the Malus' law of polarization in optics, although the physics is completely different. In addition, Josephson tunneling can be affected by midgap states, which were discussed in the previous section. These effects were discussed by Vaccarella, Duncan and Sá de Melo (VDS) [324] using a non-local lattice Bogoliubov-de Gennes approach, and are suggested as a probe to distinguish spin and orbital symmetries of the superconducting order parameter in quasi-1D Bechgaard salts.

In anticipation of the existence of surface bound states in particular geometries of quasi-1D systems [315], it is proper to describe the left (L) and right (R) superconductors by the real space mean field Hamiltonian

$$\mathscr{H}_j = \sum_{\boldsymbol{r}_j,\boldsymbol{r}'_j} \Big\{ \Big[t_j(\boldsymbol{r}_j,\boldsymbol{r}'_j) - \mu\delta_{\boldsymbol{r}_j,\boldsymbol{r}'_j} \Big] c^\dagger_{j,\alpha_j}(\boldsymbol{r}_j) c_{j,\alpha_j}(\boldsymbol{r}'_j) + \Big[\Delta_{j,\alpha_j,\beta_j}(\boldsymbol{r}_j,\boldsymbol{r}'_j) c^\dagger_{j,\alpha_j}(\boldsymbol{r}_j) c^\dagger_{j,\beta_j}(\boldsymbol{r}'_j) + \text{H.C.} \Big] \Big\}, \tag{8.50}$$

where α_j, β_j (with j =L, R) are spin indices, $\boldsymbol{r}_j$, $\boldsymbol{r}'_j$ are position labels, and repeated greek indices indicate summation. In the first term, t_j is the transfer integral confined to nearest neighbors. The order parameter matrix in the second term is defined as

$$\Delta_{j,\alpha_j,\beta_j}(\boldsymbol{r}_j,\boldsymbol{r}'_j) = V_{j,\alpha_j,\beta_j,\delta_j,\gamma_j}(\boldsymbol{r}_j,\boldsymbol{r}'_j)\langle c_{j,\delta_j}(\boldsymbol{r}'_j) c_{j,\gamma_j}(\boldsymbol{r}_j)\rangle,$$

where $V_{j,\alpha_j,\beta_j,\delta_j,\gamma_j}(\boldsymbol{r}_j,\boldsymbol{r'}_j)$ is the pairing interaction.

The mean field Hamiltonian on both sides of the junction can be separately diagonalized via lattice Bogoliubov transformations

$$c_{j,\alpha_j}(\boldsymbol{r}_j) = \sum_{N_j} \Big[u_{N_j\alpha_j}(\boldsymbol{r}_j)\gamma_{N_j} + v^*_{N_j\alpha_j}(\boldsymbol{r}_j)\gamma^\dagger_{N_j} \Big],$$

which satisfy the corresponding non-local BdG equations

$$\varepsilon_{N_j} u_{N_j\alpha_j}(\boldsymbol{r}) = \sum_{\boldsymbol{r}'_j} \Big[t_j(\boldsymbol{r}_j,\boldsymbol{r}'_j) - \mu\delta_{\boldsymbol{r}_j\boldsymbol{r}'_j} \Big] u_{N_j\alpha_j}(\boldsymbol{r}'_j) + \sum_{\boldsymbol{r}'_j} \Delta_{j,\alpha_j,\beta_j}(\boldsymbol{r}_j,\boldsymbol{r}'_j) v_{N_j\beta_j}(\boldsymbol{r}'_j), \tag{8.51}$$

$$-\varepsilon_{N_j} v_{N_j\alpha_j}(\boldsymbol{r}_j) = \sum_{\boldsymbol{r}'_j} \Big[t^*_j(\boldsymbol{r}_j,\boldsymbol{r}'_j) - \mu\delta_{\boldsymbol{r}_j\boldsymbol{r}'_j} \Big] v_{N_j\alpha_j}(\boldsymbol{r}'_j) + \sum_{\boldsymbol{r}'_j} \Delta^*_{j,\alpha_j,\beta_j}(\boldsymbol{r}_j,\boldsymbol{r}'_j) u_{N_j\beta_j}(\boldsymbol{r}'_j). \tag{8.52}$$

This set of equations must be solved with the self-consistency condition

$$\Delta_{j,\alpha_j,\beta_j}(\boldsymbol{r}_j,\boldsymbol{r}'_j)=\frac{1}{4}V_{j,\alpha_j,\beta_j,\delta_j,\gamma_j}(\boldsymbol{r}_j,\boldsymbol{r}'_j)\sum_{N_j}(1-2f_{N_j})$$
$$\times\left[v^*_{N_j\gamma_j}(\boldsymbol{r}_j)u_{N_j\delta_j}(\boldsymbol{r}'_j)-v^*_{N_j\delta_j}(\boldsymbol{r}_j)u_{N_j\gamma_j}(\boldsymbol{r}'_j)\right]. \quad (8.53)$$

The tunneling Hamiltonian connecting L and R superconductors is

$$\mathscr{H}_{\mathrm{T}}=\sum_{\boldsymbol{r}_{\mathrm{L}},\boldsymbol{r}_{\mathrm{R}}}\left[\mathscr{T}_{\alpha_{\mathrm{L}}\alpha_{\mathrm{R}}}(\boldsymbol{r}_{\mathrm{L}},\boldsymbol{r}_{\mathrm{R}})c^{\dagger}_{\mathrm{L},\alpha_{\mathrm{L}}}(\boldsymbol{r}_{\mathrm{L}})c_{\mathrm{R},\alpha_{\mathrm{R}}}(\boldsymbol{r}_{\mathrm{R}})+\mathrm{H.C.}\right], \quad (8.54)$$

where the matrix element for tunneling can be separated into spin-conserving ($\mathscr{T}_s$) and spin-dependent ($\mathscr{T}_v$) terms as

$$\mathscr{T}_{\alpha_{\mathrm{L}}\alpha_{\mathrm{R}}}(\boldsymbol{r}_{\mathrm{L}},\boldsymbol{r}_{\mathrm{R}})=\mathscr{T}_s(\boldsymbol{r}_{\mathrm{L}},\boldsymbol{r}_{\mathrm{R}})\delta_{\alpha_{\mathrm{L}},\alpha_{\mathrm{R}}}+\mathscr{T}_v(\boldsymbol{r}_{\mathrm{L}},\boldsymbol{r}_{\mathrm{R}})\cdot\boldsymbol{\sigma}_{\alpha_{\mathrm{L}},\alpha_{\mathrm{R}}}.$$

Following a conventional procedure [346], the pair tunneling current through the junction can be derived to second order in the tunneling matrix element and takes the form

$$J_s(V,T)=2e\mathrm{Im}\Bigg[e^{i\omega_J t}\sum_{\boldsymbol{r}_{\mathrm{L}}\boldsymbol{r}_{\mathrm{R}}\boldsymbol{r}'_{\mathrm{L}}\boldsymbol{r}'_{\mathrm{R}}}\mathscr{T}_{\alpha_{\mathrm{L}}\alpha_{\mathrm{R}}}(\boldsymbol{r}_{\mathrm{L}},\boldsymbol{r}_{\mathrm{R}})\mathscr{T}_{\beta_{\mathrm{L}}\beta_{\mathrm{R}}}(\boldsymbol{r}'_{\mathrm{L}},\boldsymbol{r}'_{\mathrm{R}})$$
$$\times Q^{\alpha_{\mathrm{L}}\beta_{\mathrm{L}}}_{\alpha_{\mathrm{R}}\beta_{\mathrm{R}}}(\boldsymbol{r}_{\mathrm{L}},\boldsymbol{r}_{\mathrm{R}},\boldsymbol{r}'_{\mathrm{L}},\boldsymbol{r}'_{\mathrm{R}},i\omega_n)\Bigg], \quad (8.55)$$

where $\omega_J=2eV$ with V the voltage applied across the junction. The tensor $Q^{\alpha_{\mathrm{L}}\beta_{\mathrm{L}}}_{\alpha_{\mathrm{R}}\beta_{\mathrm{R}}}$ can be written in terms of the anomalous Green's functions as

$$Q^{\alpha_{\mathrm{L}}\beta_{\mathrm{L}}}_{\alpha_{\mathrm{R}}\beta_{\mathrm{R}}}=T\sum_{i\nu_m}F^{\dagger}_{\mathrm{L},\alpha_{\mathrm{L}},\beta_{\mathrm{L}}}(\boldsymbol{r}_{\mathrm{L}},\boldsymbol{r}'_{\mathrm{L}},i\nu_m)F_{\mathrm{R},\alpha_{\mathrm{R}},\beta_{\mathrm{R}}}(\boldsymbol{r}_{\mathrm{R}},\boldsymbol{r}'_{\mathrm{R}},i\omega_n-i\nu_m), \quad (8.56)$$

where ν_m is the fermionic Matsubara frequency, $i\omega_n\to(-eV+i\delta)$, and $F(\boldsymbol{r}_{\mathrm{L}},\boldsymbol{r}'_{\mathrm{L}},i\nu_m)$ is the Fourier transform of

$$F_{j,\alpha_j,\beta_j}(\boldsymbol{r}_j,\boldsymbol{r}'_j,\tau)=\langle T_\tau c_{j,\alpha_j}(\boldsymbol{r}_j,\tau)c_{j,\beta_j}(\boldsymbol{r}'_j,0)\rangle \quad (8.57)$$

with respect to imaginary time τ.

The analysis employed so far is quite general and can be applied to any junction. In the specific problem of quasi-1D Bechgaard salts, we consider a TSC–I–TSC junction where both TSCs are in the weak spin-orbit coupling limit. This limit is quite appropriate for Bechgaard salts because the heaviest element is selenium. In this case, $\Delta_{j,\alpha_j,\beta_j} = i\boldsymbol{d}_j \cdot (\boldsymbol{\sigma}\sigma_y)_{\alpha_j\beta_j} \exp(i\phi_j)$, where $\boldsymbol{d}_j$ is a real vector. VDS [324] assumed that the small spin-orbit coupling weakly pins the $\boldsymbol{d}_j$-vector to the $\boldsymbol{c}_j^*$ axis, as suggested by Knight shift experiments [33, 35]. In the present discussion, we consider only triplet unitary states (time reversal invariant state in the bulk) and spin-conserving tunneling processes ($\mathscr{T}_v = 0$). However, these assumptions are not essential for the polarization effect. As discussed in the previous section, there may exist Andreev bound states at the superconductor–insulator interfaces depending on the order parameter symmetry and the direction of interfaces. These midgap edge states also affect dramatically the pair tunneling current, as they do single electron tunneling [315].

In the case where there are no bound states present, the Josephson current includes only the contribution from scattering states with well defined momentum. Thus the quantum indices N_j in BdG equations (8.51) and (8.52) are represented by momentum $\boldsymbol{k}_j$ and discrete index $s_j = 1, 2$. In this case, the BdG amplitudes become

$$u_{N_j\alpha_j}(\boldsymbol{r}_j) \propto u_{\boldsymbol{k}_j,s_j} \exp(i\boldsymbol{k}_j \cdot \boldsymbol{r}_j); \tag{8.58}$$

$$v_{N_j\alpha_j}(\boldsymbol{r}_j) \propto v_{\boldsymbol{k}_j,s_j} \exp(i\boldsymbol{k}_j \cdot \boldsymbol{r}_j), \tag{8.59}$$

and the tunneling current defined in Eq. (8.55) becomes

$$J_s(V,T) = 2e \left[\mathrm{Im}W \cos(\bar{\phi}) + \mathrm{Re}W \sin(\bar{\phi}) \right]. \tag{8.60}$$

Here, $\bar{\phi} = \phi_{\mathrm{R}} - \phi_{\mathrm{L}} + \omega_J t$, and

$$W = 2 \sum_{\boldsymbol{k}_{\mathrm{L}}\boldsymbol{k}_{\mathrm{R}}} \boldsymbol{d}_{\mathrm{L}}^*(\boldsymbol{k}_{\mathrm{L}}) \cdot \boldsymbol{d}_{\mathrm{R}}(\boldsymbol{k}_{\mathrm{R}}) Q(\boldsymbol{k}_{\mathrm{L}}, \boldsymbol{k}_{\mathrm{R}}) P(\boldsymbol{k}_{\mathrm{L}}, \boldsymbol{k}_{\mathrm{R}}, i\omega_n), \tag{8.61}$$

where $Q(\boldsymbol{k}_{\mathrm{L}}, \boldsymbol{k}_{\mathrm{R}}) = \mathscr{T}_s(\boldsymbol{k}_{\mathrm{L}}, \boldsymbol{k}_{\mathrm{R}})\mathscr{T}_s(-\boldsymbol{k}_{\mathrm{L}}, -\boldsymbol{k}_{\mathrm{R}})$, the function

$$P(\boldsymbol{k}_{\mathrm{L}}, \boldsymbol{k}_{\mathrm{R}}, i\omega_n) = T \sum_{i\nu_m} \left[\nu_m^2 + E_{\boldsymbol{k}_{\mathrm{L}}}^2\right]^{-1} \left[(\omega_n - \nu_m)^2 + E_{\boldsymbol{k}_{\mathrm{R}}}^2\right]^{-1} \tag{8.62}$$

is a sum over Matsubara frequency ν_m, and $E_{\boldsymbol{k}_j} = \sqrt{(\varepsilon_j(\boldsymbol{k}_j) - \mu)^2 + |\boldsymbol{d}_j(\boldsymbol{k}_j)|^2}$ are the excitation energies. Notice that the dot product $\boldsymbol{d}_{\mathrm{L}}^*(\boldsymbol{k}_{\mathrm{L}}) \cdot \boldsymbol{d}_{\mathrm{R}}(\boldsymbol{k}_{\mathrm{R}})$ appears in $J_s(V,T)$, and thus $J_s(V,T)$ is very sensitive to the relative orientation between $\boldsymbol{d}_{\mathrm{L}}^*(\boldsymbol{k}_{\mathrm{L}})$ and $\boldsymbol{d}_{\mathrm{R}}(\boldsymbol{k}_{\mathrm{R}})$ [347]. For instance, if $\boldsymbol{d}_{\mathrm{L}}^*(\boldsymbol{k}_{\mathrm{L}}) \perp \boldsymbol{d}_{\mathrm{R}}(\boldsymbol{k}_{\mathrm{R}})$

then $J_s(V,T)$ vanishes identically for any V and T. Furthermore, $J_s(V,T)$ (for fixed V and T) changes sign depending whether the vectors $\boldsymbol{d}^*_{\mathrm{L}}(\boldsymbol{k}_{\mathrm{L}})$ and $\boldsymbol{d}_{\mathrm{R}}(\boldsymbol{k}_{\mathrm{R}})$ are aligned or anti-aligned.

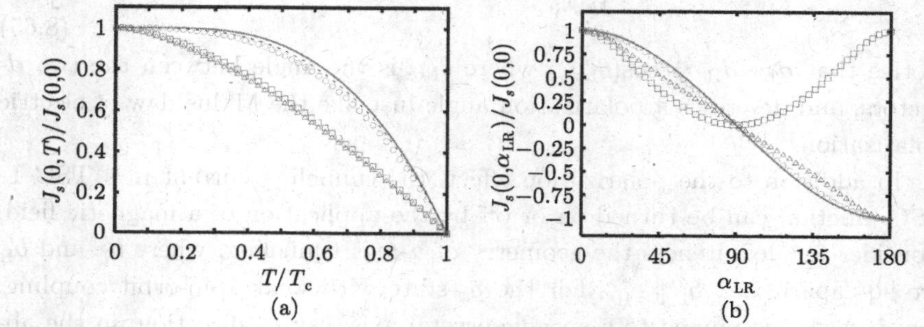

Figure 8.12 Plots of (a) $J_s(0,T)/J_s(0,0)$ versus T/T_c and (b) $J_s(0,\alpha_{\mathrm{LR}})/J_s(0,0)$ versus α_{LR} at $T=0$ [348]. Several $\boldsymbol{a}$-axis tunneling processes are illustrated for p_x and p_y symmetries. *Incoherent processes*: p_x with (i) $Q^{(\mathrm{inc})} \propto X_{\mathrm{L}}X_{\mathrm{R}}$ (solid line) and (ii) $Q^{(\mathrm{inc})} \propto 1_{\mathrm{L}}1_{\mathrm{R}}$ (trivially zero); p_y with (iii) $Q^{(\mathrm{inc})} \propto Y_{\mathrm{L}}Y_{\mathrm{R}}$ (dashed line) and (iv) $Q^{(\mathrm{inc})} \propto 1_{\mathrm{L}}1_{\mathrm{R}}$ (trivially zero). *Coherent processes*: p_x with (i) $Q^{(\mathrm{coh})} \propto X_{\mathrm{L}}X_{\mathrm{R}}$ (circles) and (ii) $Q^{(\mathrm{coh})} \propto 1_{\mathrm{L}}1_{\mathrm{R}}$ (trivially zero); p_y with (iii) $Q^{(\mathrm{coh})} \propto Y_{\mathrm{L}}Y_{\mathrm{R}}$ (triangles) and (iv) $Q^{(\mathrm{coh})} \propto 1_{\mathrm{L}}1_{\mathrm{R}}$ (squares). The L and R superconductors are assumed to be identical quarter-filled systems, with $T_c = 1.5\,\mathrm{K}$, and parameters $t_x = 5800\,\mathrm{K}$, $t_y = 1226\,\mathrm{K}$, $t_z = 48\,\mathrm{K}$. From Ref [324].

The current $J_s(V,T)$ also depends on the tunneling matrix element $\mathscr{T}_s(\boldsymbol{k}_{\mathrm{L}}, \boldsymbol{k}_{\mathrm{R}})$. Assuming an orthorhombic crystal structure, $\mathscr{T}_s$ can be expanded as

$$\mathscr{T}_s(\boldsymbol{k}_{\mathrm{L}}, \boldsymbol{k}_{\mathrm{R}}) = \sum_{\Gamma_{\mathrm{L}}\Gamma_{\mathrm{R}}} \mathscr{T}_{\Gamma_{\mathrm{L}}\Gamma_{\mathrm{R}}}(\boldsymbol{k}_{\mathrm{L}}, \boldsymbol{k}_{\mathrm{R}})\psi^{\Gamma_{\mathrm{L}}}(\boldsymbol{k}_{\mathrm{L}})\psi^{\Gamma_{\mathrm{R}}}(\boldsymbol{k}_{\mathrm{R}}), \tag{8.63}$$

since all $\mathrm{D_{2h}}$ point group representations are one-dimensional and non-degenerate [36, 37]. Here, ψ^{Γ} are the basis functions of $\mathrm{D_{2h}}$ point group defined in Sec. 5.7. For unitary triplet states and weak spin-orbit coupling (see Table 5.4), $\psi^{\Gamma} = X, Y, Z$, or XYZ for p_x, p_y, p_z, and f_{xyz} states, respectively. In this case, the zero bias tunneling current $J_s(0,T)$ simplifies to

$$J_s(0,T) = J_0(T) \times (\hat{\boldsymbol{d}}_{\mathrm{L}} \cdot \hat{\boldsymbol{d}}_{\mathrm{R}}) \times \sin(\bar{\phi}), \tag{8.64}$$

where the U(1) phase difference is $\bar{\phi} = \phi_{\mathrm{R}} - \phi_{\mathrm{L}}$. The prefactor $J_0(T)$ is proportional to the product of the order parameter amplitudes $\Delta_{\mathrm{L}}(T)$ and

$\Delta_{\mathrm{R}}(T)$, and is expressed as

$$J_0(T) = 2e\Delta_{\mathrm{L}}(T)\Delta_{\mathrm{R}}(T)\left[\sum_{\boldsymbol{k}_{\mathrm{L}}\boldsymbol{k}_{\mathrm{R}}} Q(\boldsymbol{k}_{\mathrm{L}}, \boldsymbol{k}_{\mathrm{R}})\psi^{\Gamma_{\mathrm{L}}}(\boldsymbol{k}_{\mathrm{L}})\psi^{\Gamma_{\mathrm{R}}}(\boldsymbol{k}_{\mathrm{R}})P(\boldsymbol{k}_{\mathrm{L}}, \boldsymbol{k}_{\mathrm{R}}, 0)\right] . \tag{8.65}$$

Notice that $\hat{\boldsymbol{d}}_{\mathrm{L}} \cdot \hat{\boldsymbol{d}}_{\mathrm{R}} = \cos(\theta_{\mathrm{LR}})$, where θ_{LR} is the angle between the two $\boldsymbol{d}$-vectors, and describes a polarization angle just like the Malus' law of electric polarization.

In addition to the polarization effect, the tunneling current in a TSC–I–TSC junction can be turned on or off by the application of a magnetic field. Consider for definiteness the geometry of $\boldsymbol{a}$-axis tunneling, where $\boldsymbol{b}'_{\mathrm{L}}$ and $\boldsymbol{b}'_{\mathrm{R}}$ are 90° apart, i.e., $\boldsymbol{b}'_{\mathrm{L}} \parallel \boldsymbol{c}^*_{\mathrm{R}}$. For the p_x-state with weak spin-orbit coupling, the $\boldsymbol{d}_j$ vector is pinned to a specific crystal axis, say $\boldsymbol{c}^*_j$ direction, in the absence of magnetic field. Thus, the above calculation indicates that there is no Josephson current in this geometry. However, this result can be modified if one turns on a strong enough magnetic field along the $\boldsymbol{b}'_{\mathrm{L}}$ direction, provided that H overcomes the spin-orbit pinning effect. With this field, the $\boldsymbol{d}_{\mathrm{R}}$-vector may rotate to be perpendicular to $\boldsymbol{H}$, in order to maximize the susceptibility of the entire system, and minimize the magnetic free energy contribution $F_{\mathrm{mag}} = \sum_j [-H_m \chi^j_{mn} H_n]/2$ with $j =$L, R. The spin-orbit pinning field H_{so} was estimated to be $\sim 0.22\,\mathrm{T}$ in quasi-1D Bechgaard salts, as discussed in Sec. 5.7. Therefore, for magnetic fields applied along the $\boldsymbol{b}'_{\mathrm{L}}$ direction with $H_{\mathrm{so}} < H < H^{c^*}_{\mathrm{c2}}$ ($H^{c^*}_{\mathrm{c2}}$ can reach as high as 0.9 T for $(\mathrm{TMTSF})_2\mathrm{PF}_6$ at given pressure and zero temperature [108]), the $\boldsymbol{d}_{\mathrm{L}}$ and $\boldsymbol{d}_{\mathrm{R}}$ vectors are parallel or anti-parallel, hence can produce finite pair tunneling current.

In the case of zero magnetic field, the zero bias Josephson current depends crucially on the tunneling matrix elements $Q(\boldsymbol{k}_{\mathrm{L}}, \boldsymbol{k}_{\mathrm{R}})$. For purely incoherent tunneling processes, $J^{\mathrm{inc}}_s(0,T)$ is proportional to the $\boldsymbol{k}$-space summation of

$$Q^{(\mathrm{inc})} = 2\mathscr{T}_{1_{\mathrm{L}}1_{\mathrm{R}}}\mathscr{T}_{\Gamma_{\mathrm{L}}\Gamma_{\mathrm{R}}} - 2\mathscr{T}_{1_{\mathrm{L}}\Gamma_{\mathrm{R}}}\mathscr{T}_{\Gamma_{\mathrm{L}}1_{\mathrm{R}}}, \tag{8.66}$$

where 1_j is the identity representation. For the p_x symmetry, the process $Q^{(\mathrm{inc})} \propto X_{\mathrm{L}}X_{\mathrm{R}}$ produces $J_s(0,T) \neq 0$, while $Q^{(\mathrm{inc})} \propto 1_{\mathrm{L}}1_{\mathrm{R}}$ produces a trivially vanishing $J_s(0,T)$. Similarly for the p_y symmetry, $Q^{(\mathrm{inc})} \propto Y_{\mathrm{L}}Y_{\mathrm{R}}$ produces a non-vanishing current but $Q^{(\mathrm{inc})} \propto 1_{\mathrm{L}}1_{\mathrm{R}}$ produces vanishing $J_s(0,T)$. The coherent tunneling processes which conserve parallel momentum can be analyzed analogously [37].

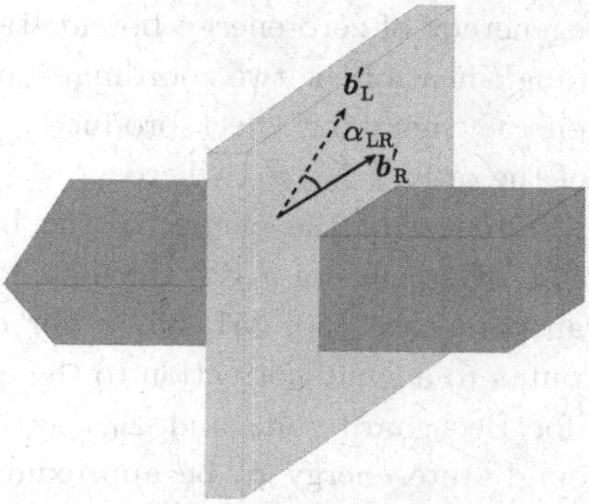

Figure 8.13 A schematic TSC–I–TSC junction with $\boldsymbol{a}$ axes coincide but an angle $\alpha_{\rm LR}$ between $\boldsymbol{b}'$ axes.

Numerical results of $J_s(0,T)$ for $\boldsymbol{a}$-axis tunneling are shown in the left panel of Fig. 8.12, normalized with respect to $J_s(0,0)$. All crystal axes of left and right superconductors are assumed to be parallel in this case. If the requirement is relaxed and rotation about the $\boldsymbol{a}$ axis by some angle $\alpha_{\rm LR}$ is allowed (see Fig. 8.13), the Josephson current acquires an angular dependence $J_s = J_s(V,T,\alpha_{\rm LR})$. This angular dependence is shown in the right panel of Fig. 8.12 for $T=0$. Notice that the Josephson current does not change sign for the p_y symmetry with $Q^{(\rm coh)} \propto 1_{\rm L}1_{\rm R}$ (squares). This is a direct manifestation of the polarization effect of the $\boldsymbol{d}$-vector. Both the temperature and angular dependencies of $J_s(0,T,\alpha_{\rm LR})$ can help distinguish different symmetries, and different (coherent or incoherent) tunneling processes, as seen experimentally for $\boldsymbol{c}^*$ axis tunneling of cuprate superconductors [349].

However, the discussion above in connection to $\boldsymbol{a}$-axis tunneling for p_x and f_{xyz} states is not complete, since surface bound states present at the junction interfaces are neglected. In order to take these states into account, one needs to solve the non-local lattice BdG equations (8.51) and (8.52) together with the self-consistency condition (8.53). VDS [324] considered $\boldsymbol{a}$- ($\boldsymbol{c}^*$-) axis tunneling for p_x (p_z) symmetry and compared the results with standard local quasi-classical continuum approximation [350, 351, 352], as shown in Fig. 8.14. This local quasi-classical approximation is strictly valid only when $k_{{\rm F}_\perp}\xi_\perp \to \infty$, where $k_{{\rm F}_\perp}$ and $\xi_\perp$ are Fermi momentum and coherence length perpendicular to the junction interface, respectively. However, within the non-local lattice

BdG equations the degeneracy of zero energy bound states is lifted near the surface due to the entanglement of the two coordinates of the non-local order parameter. The degeneracy breaking effect produces, perturbatively, finite energy bound states of the order $|\Delta_0|/\gamma_\perp$, where $\gamma_\perp = \xi_\perp/d_\perp$, with $d_\perp$ being lattice spacing. The finiteness of the energy of the bound states cuts off the low temperature $1/T$ divergence of $J_s(T)$ calculated in the quasi-classical approach at small transparencies [350, 351, 352]. For $\boldsymbol{a}$ axis tunneling and p_x symmetry this accounts to a small correction to the quasi-classical results, as $\gamma_x = \xi_x/a \approx 106$ for Bechgaard salts and $\Delta_{0,p_x} \approx 3.73\,\mathrm{K}$. These values lead to the lowest bound state energy to be approximately $T^*_{p_x} \approx 35.2\,\mathrm{mK}$, which means that for $T < T^*_{p_x}$ the quasi-classical results are not reliable (see Fig. 8.14). The quasi-classical results fail in a more dramatic way when $\boldsymbol{c}^*$ axis tunneling for the p_z symmetry is considered. In this case $\xi_z = 21.5\,\text{Å}$, $c = 13.5\,\text{Å}$, $\gamma_z = \xi_z/c \approx 1.59$ and $\Delta_{0,p_z} \approx 3.21\,\mathrm{K}$, resulting in $T^*_{p_z} = 2.02\,\mathrm{K}$. This temperature is even larger than the critical temperature $T_c = 1.5\,\mathrm{K}$ used in this calculation. Thus, perturbative corrections to quasi-classical zero energy bound states are enormous, and the quasi-classical approximation cannot be used except $T \sim T_c$ (see Fig. 8.14).

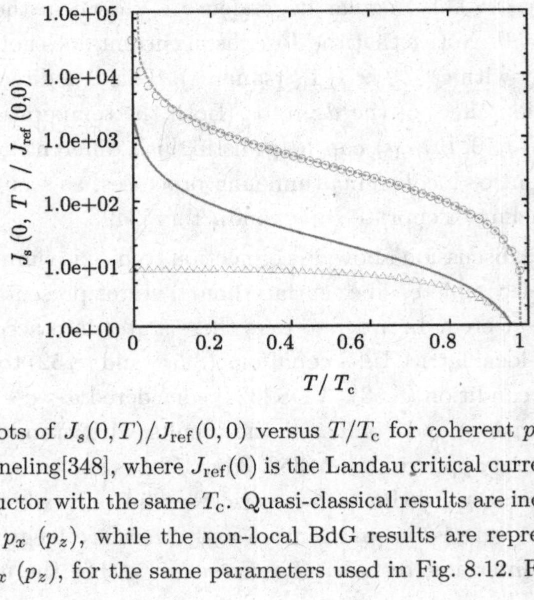

Figure 8.14 Plots of $J_s(0,T)/J_{\mathrm{ref}}(0,0)$ versus T/T_c for coherent p_x-state (p_z-state) $\boldsymbol{a}$- ($\boldsymbol{c}^*$-) axis tunneling[348], where $J_{\mathrm{ref}}(0)$ is the Landau critical current for a singlet s-wave superconductor with the same T_c. Quasi-classical results are indicated by dashed (solid) lines for p_x (p_z), while the non-local BdG results are represented by circles (triangles) for p_x (p_z), for the same parameters used in Fig. 8.12. From Ref. [324].

Lastly, we would like to mention some experimental attempt to investigate Josephson tunneling in quasi-1D Bechgaard salts. Ha *et al.* [353] prepared bi-crystals of $(TMTSF)_2ClO_4$, and observed a large zero-bias conductance peak (ZBCP) during a $\boldsymbol{c}^*$-axis tunneling process, where the $\boldsymbol{a}$ axes of L and R superconductors were about 80° apart. From the discussion in Sec.8.1 (see Table 8.1, the presence of ZBCP in this experimental geometry would suggest a p_z symmetry. However, as shown in Fig. 8.14, there seems to be no zero-energy surface bound states for the p_z symmetry beyond the quasi-classical approximation. Thus, the ZBCP is expected to vanish in such a configuration since the quasi-classical picture cannot be used at low temperatures. In addition, the bi-crystal structure prepared by Ha *et al.* [353] may correspond to TSC–N–TSC or TSC–I–N–I–TSC where N represents a normal metallic system, while our discussion here was confined to spin-conserving processes in a TSC–I–TSC junction. Therefore, a direct comparison between theoretical predictions and experimental results is not presently possible. Further experimental and theoretical works are necessary to clarify these issues.

9 Open Questions

In this chapter, we summarize the main points of this book and discuss briefly several open questions. It is probably fair to say that the experimental evidences discussed in Chap. 3 suggest the existence of a triplet superconducting state in $(TMTSF)_2PF_6$, while the experimental works introduced in Chap. 4 suggest a singlet pairing state in $(TMTSF)_2ClO_4$. Thermodynamical measurements in $(TMTSF)_2ClO_4$ give a rather controversial results regarding the nodal structure of the pairing order parameter. Thus, the true nature of the superconducting phases in Bechgaard salts remains unveiled and needs further investigation.

The possibility of triplet superconductivity in Bechgaard salts is very exciting, but additional phase sensitive experiments are necessary to reveal the orbital and directional structure of the vector order parameter $\boldsymbol{d}$. Unfortunately, these experiments are not easy to perform, and this difficulty has prevented experimentalists from obtaining final answers about the pairing symmetry and the direction of the $\boldsymbol{d}$-vector. Furthermore, Bechgaard salts are triclinic instead of orthorhombic, and the analysis of the order parameter symmetry becomes a bit more complicated.

In the reminder of this chapter we discuss some open questions, which can serve as pointers for directions where more research needs to be done.

- *What is the mechanism for superconductivity in Bechgaard salts?*

It is well known that antiferromagnetic fluctuations can lead to singlet pairing in the d-wave channel [262, 263, 264, 266, 354], and that ferromagnetic fluctuations can lead to triplet superconductivity [355, 356, 357, 358]. However, the phase diagram of $(TMTSF)_2PF_6$ has an SDW and superconducting phase in close vicinity. If the superconducting state of $(TMTSF)_2PF_6$ is indeed triplet as suggested, then it is interesting to ask if AF fluctuations can truly lead to triplet superconductivity in quasi-1D systems.

In strontium ruthenate, neutron scattering experiments [274] have revealed antiferromagnetic peaks at $\boldsymbol{Q} = (Q_x, Q_y, 0)$, for $Q_x a_x = Q_y a_y = \pm 2\pi/3$. This observation leads to proposals that anisotropic AF spin fluctuations may cause spin-triplet superconductivity in strontium ruthenate [43, 275, 277], despite the fact that its parent compound is a ferromagnet, and that strong ferromagnetic fluctuations were observed by NMR measurements [239, 278, 359].

The pairing mechanism of Bechgaard salts is quite possibly very different from that of Sr_2RuO_4, since they have only one band, as opposed to three in the case of ruthenate. In addition, the $\boldsymbol{d}$-vector is possibly perpendicular to the easy axis for spins ($\boldsymbol{b}'$) in quasi-1D systems [23, 33, 35], while $\boldsymbol{d}$ is parallel to the easy axis for spins ($\boldsymbol{c}$) in strontium ruthenate [360]. Furthermore, according to the review by Mackenzie and Maeno [361], triplet superconductivity in Sr_2RuO_4 is still unconfirmed.

The possibility that AF fluctuations can lead to triplet superconductivity in quasi-1D systems was preliminarily investigated [43, 45, 46]. This issue certainly deserves more attention, and should be investigated further, especially after the experiments by Vuletić *et al.* [76], Kornilov *et al.* [77], and Lee *et al.* [79], which suggest the interplay/coexistence of SDW order and triplet superconductivity in $(TMTSF)_2PF_6$. On the other hand, the study of this issue can also help clarify the symmetry of the superconducting state. For instance, it is shown that the conditions for a superconducting instability in the triplet p_x-wave channel is more stringent than in the f-wave case [43, 45, 46]. Finally, it would be quite useful to have neutron scattering experiments performed in Bechgaard salts, if experimental difficulties can be overcomed, since the role of magnetic effects on superconductivity may be ellucidated.

- *What are the effects of the triclinic structure?*

Most of the theoretical calculations discussed in this review assume directly, or indirectly that the crystal structure of quasi-1D Bechgaard salts is orthorhombic. This assumption is based only on convenience. The calculations are a lot easier to perform for various properties of this system, because of the assumed orthogonality of the principal crystallographic axes. The most apparent effect of the triclinic structure is on the orbital symmetry of the superconducting order parameter. The lattice group symmetry changes from D_{2h} (orthorhombic) to C_i (triclinic), as a result the $\boldsymbol{d}$-vector that characterizes the triplet superconducting state also changes to conform with the irreducible representations of the C_i group [210]. This corresponding change in the symmetry of the order parameter $\boldsymbol{d}$-vector affects several tensorial properties, including spin susceptibility $\chi_{mn}(\boldsymbol{q}, \omega, T)$, thermal conductivity $\kappa_{ij}(T)$, and superfluid density $\rho_{ij}(T)$. The upper critical fields may also be affected qualitatively since some mixing of orbital effects is inevitable if crystallographic axes are not orthogonal.

- *How does a Fermi to non-Fermi liquid crossover in the normal state affects the superconducting state?*

The non-Fermi liquid behavior in the normal state of $(TMTSF)_2PF_6$ and $(TMTSF)_2ClO_4$ are discussed in Secs. 3.4 and 4.5. These results indicate the existence of a crossover from Fermi liquid (FL) to non-Fermi liquid (NFL) normal state in both compounds as a function of pressure and magnetic field. This crossover has also been discussed in the normal state of high-T_c superconductors [362], and has a direct counterpart in the superconducting state. It has been shown that singlet d-wave [363, 364, 365] and triplet p-wave superconductors [365, 366] in the continuum limit can exhibit a topological phase transition from state with gapless quasiparticle excitations at lower electronic densities to a state with fully gapped quasiparticle excitations at higher electronic densities. A similar effect should also be present in quasi-1D systems [367] as a function of pressure or density if the density of carriers could be controlled using a field effect transistor (FET) or a ferroelectric field effect transistor (FFET).

The experimental suggestion that the normal state of Bechgaard salts at high pressures is NFL indicates that the system is likely to be very strongly interacting and therefore highly non-degenerate. If this is indeed the case, a picture that was developed to describe the FL to NFL crossover in high-T_c superconductors [213, 214] may be also applicable to quasi-1D systems. In this picture, the normal state of the cuprates is a highly degenerate Fermi liquid in the overdoped regime, while pre-formed pairs (long lived Cooper pairs) exist in the underdoped regime producing a highly interacting non-degenerate system (NFL). The existence of these pre-formed pairs leads to an unusual superconducting state. In particular, the low temperature properties within the superconducting state can change dramatically as a function of carrier density for a d-wave or p-wave superconductor, provided that the interaction strength is strong enough. In $(TMTSF)_2PF_6$, the interaction strength and the

FL to NFL crossover may be tuned by pressure. Thus, it may be possible that the normal state of $(TMTSF)_2PF_6$ evolves from a highly degenerate state (FL) at lower pressures to a highly non-degenerate state (NFL) at higher pressures, and that such evolution in the normal state is paralleled by an evolution in the superconductor from a BCS-like state at lower pressures (where pairing occurs in momentum space) to a BEC-like state at higher pressures (where pairing occurs in real space). This suggestion immediately implies that there is a strong renormalization of the electron-electron interaction as a function of pressure as long as the system is superconducting.

This possible BCS (FL) to BEC (NFL) evolution should have clear signatures on thermodynamic quantities. For instance, consider the triplet superconducting state $^3\mathrm{B}_{1u}(a)$ (p_z) (see Tab. 5.4 in Sec. 5.7) of an orthorhombic crystal in the limit of weak spin-orbit coupling, there is no gap in the excitation spectrum in the BCS regime ($\tilde{\mu} > 0$), but there is a full gap in the excitation spectrum in the BEC regime ($\tilde{\mu} < 0$). Therefore, in analogy with Ref. [364], it is expected that the evolution from BCS to BEC is not smooth and that a topological phase transition takes place without changing the symmetry of the order parameter [367]. The excited states of the systems should also be largely affected and a Lifshitz transition [368] should occur in the superconducting state. However, if the system is in the $^3\mathrm{B}_{3u}(a)$ (p_x) state, then the evolution should be smooth, but thermodynamic properties between the BCS and BEC limits would have major quantitative differences.

This FL to NFL evolution may also occur as a function of carrier density if FET or FFET can be made to work for quasi-1D superconductors. At higher electronic densities the normal state of these systems should be FL-like and the corresponding superconducting state is BCS-like, while at lower densities the normal state should exhibit strong NFL behavior with a corresponding non-BCS superconducting state. Thus, this situation in quasi-1D systems may be similar to that found in high-T_c superconductors.

- *What are the effects of fluctuations beyond mean field theories?*

The superconducting state of quasi-1D superconductors in low and high magnetic fields was analyzed in several theories [16, 18, 34, 49] (see, e.g., Secs. 5.1 and 5.3). However in these efforts mean field approaches were used, and the effects of fluctuations were completely ignored. Phase fluctuation effects only began to be investigated at high magnetic fields parallel to the y

axis, as discussed in Sec. 5.4. These phase fluctuations can melt the predicted magnetic field induced Josephson vortex lattice [18, 20] at a melting temperature $T_m(H) < T_{\rm MF}(H)$, and hence can modify the nature of the predicted reentrant phase expected from mean field calculations. However, phase fluctuations alone cannot describe quantitatively (and perhaps even qualitatively) the observed upper critical fields. Even though the effects of phase fluctuations alone led to a melting transition of the same universality class of the Berezinskii–Kosterlitz–Thouless transition [197, 198], the inclusion of amplitude and quantum (temporal) fluctuations may change the universality class because the system is quasi-one-dimensional. Therefore, the theoretical H–T phase diagram of quasi-1D superconductors at high magnetic fields may be dramatically altered when these effects are included.

Another important aspect of fluctuations concerns the difference in behavior between the superconducting states of $(TMTSF)_2ClO_4$ and $(TMTSF)_2PF_6$. It is important to investigate what is the role of the nearby SDW phase in the case of $(TMTSF)_2PF_6$. An initial experimental study indicated that the upper critical field of $(TMTSF)_2PF_6$ was very sensitive to how close the superconducting and the SDW phases were in the phase diagram[108]. It may be possible that there is some kind of SDW fluctuation feedback mechanism, which would be quite different from the spin fluctuation feedback of Anderson and Brinkman [50, 356] for the case of ^{3}He. Spin fluctuations from the nearby SDW phase can renormalize the attractive interaction that leads to superconductivity, and may lead to an effective interaction that can be very sensitive to applied magnetic fields.

- *What are the effects of alloying in the phase diagram of* $(TMTSF)_2PF_6$?

One way of exploring the proximity effect of SDW into superconducting states in $(TMTSF)_2PF_6$ is by adding ClO_4 into $(TMTSF)_2PF_6$ systems, i.e., by varying δ in the compound $(TMTSF)_2(PF_6)_{1-\delta}(ClO_4)_\delta$. When $\delta = 0$ one has a pure $(TMTSF)_2PF_6$, which has an SDW phase at ambient pressure, while if $\delta = 1$ one has a superconducting $(TMTSF)_2ClO_4$. Therefore there should be a critical value δ_c beyond which the SDW would be replaced by the superconducting state at zero pressure. In particular, beyond the critical value of δ, where the superconducting state should appear for slowly cooled samples, the effects of alloying (which introduces non-magnetic scattering centers) can also aid in the determination of the symmetry of the order parameter in these systems, both at low and high magnetic fields.

- *Is there a coexistence region of SDW and superconducting orders*?

Experiments discussed in Sec. 3.3 suggest coexistence regions between SDW and superconducting (SC) or SDW and metallic (M) states. It has been suggested [76, 77, 79] that the ordered state in these regions are highly inhomogeneous, with macroscopically separated SDW and SC or SDW and M domains. It is possible that this coexistence region is only a consequence of the inhomogeneity of the pressure medium used in experiments, since the liquid used as the pressure medium solidifies at low temperatures. However, when more homogeneous pressure cells become available, the interplay and/or coexistence between SDW and SC orders in such quasi-1D systems should be addressed.

The competition between two uniform and scalar order parameters can be described by a homogeneous ϕ_1–ϕ_2 model [369], where the orders can either exist independently or coexist. This model seems to be over-simplified to describe the interplay between SDW and triplet superconducting (TSC) states in quasi-1D Bechgaard salts for the following reasons: i) the SDW and TSC order parameters are vectors, ii) the coexistence region between SDW and TSC is expected to be inhomogeneous, and iii) the spin anisotropy pins the SDW order parameter vector $\boldsymbol{N}$ mostly along the $\boldsymbol{b}'$ axis [52], while it is expected that the TSC order parameter vector $\boldsymbol{d}$ is either along the $\boldsymbol{a}$ or $\boldsymbol{c}^*$ axis due to spin-orbit effects.

The presence of SDW order in the superconducting state will definitely affect the superconducting properties, especially in magnetic fields. This occurs because SDW order is magnetic in nature and responds to a magnetic field in a different way as a superconductor does. In particular, vortex cores and vortex-core states should be dramatically affected, since the SDW should try to take advantage of the suppression of superconductivity in the vortex core. Furthermore, the upper critical magnetic fields and the vortex lattice structure with a coexisting SDW may also be dramatically modified from their solutions in the absence of SDW order. These effects need to be fully addressed in order to understand experimental results close to the SDW/SC phase boundaries.

- *What are the vortex core states of a quasi-1D triplet superconductor*?

The study of vortex core states in BCS singlet s-wave superconductors has a very long history [215–220, 370–375]. STM experiments were performed by

Hess *et al.* [376] on a layered hexagonal compound 2H–$NbSe_2$, and revealed detailed spatially resolved electronic structure around a vortex. More recently vortex core states have also been analyzed for singlet d-wave superconductors in connection with high-T_c superconductors [222, 377, 378]. STM experiments were also performed in some high-T_c superconductors [379, 380] and revealed the local density of electronic states near and inside a vortex. Thus, it is desirable to generalize these results from the literature for the case of quasi-1D triplet superconductors.

For this purpose, the Bogoliubov-de Gennes equations for a triplet superconductor

$$\begin{aligned}\varepsilon_N u_{N\alpha}(\boldsymbol{r}) =& \mathscr{H}_0(\boldsymbol{r}) u_{N\alpha}(\boldsymbol{r}) \\ &+ \int d^3\boldsymbol{r}' \left[U_{\alpha\beta}(\boldsymbol{r},\boldsymbol{r}') u_{N\beta}(\boldsymbol{r}') + \Delta_{\alpha\beta}(\boldsymbol{r},\boldsymbol{r}') v_{N\beta}(\boldsymbol{r}')\right], \end{aligned} \tag{9.1}$$

$$\begin{aligned}-\varepsilon_N v_{N\alpha}(\boldsymbol{r}) =& \mathscr{H}_0^*(\boldsymbol{r}) v_{N\alpha}(\boldsymbol{r}) \\ &+ \int d^3\boldsymbol{r}' \left[U^*_{\alpha\beta}(\boldsymbol{r},\boldsymbol{r}') v_{N\beta}(\boldsymbol{r}') + \Delta^*_{\alpha\beta}(\boldsymbol{r},\boldsymbol{r}') u_{N\beta}(\boldsymbol{r}')\right] \end{aligned} \tag{9.2}$$

must be solved self-consistently with both the order parameter and the number equations, and in the presence of an external magnetic field.

The solutions of these equations allow the determination of the energies and eigenfunctions of vortex core states for a triplet quasi-1D superconductor, as well as local density of quasiparticles. These vortex core states need to be investigated for different symmetries compatible with a triplet order parameter. Theoretical results can then be compared with STM measurements when and if they become available.

- *What is the thermal conductivity tensor* $\kappa_{ij}(T)$?

Thermal conductivity measurements in $(TMTSF)_2ClO_4$ indicate that these systems are most likely gapped at low temperatures [39]. Additional experiments are required to determine the tensor structure $\kappa_{ij}(T)$ in $(TMTSF)_2ClO_4$. Furthermore, thermal conductivity measurements are still lacking in $(TMTSF)_2PF_6$, perhaps because they may be difficult to perform under pressure. The thermal conductivity tensor can be calculated for various symmetries of the order parameter, and may be used to narrow down the number of possibilities through a comparison with experimental results.

By using the linear response method, the thermal conductivity is

$$\kappa_{ij}(T) = -\frac{1}{T}\lim_{\omega\to 0}\left\{\frac{d\mathrm{Im}\left[K_{ij}^{(\mathrm{R})}(\boldsymbol{q}=0,\omega)\right]}{d\omega}\right\}_{\omega=0}, \tag{9.3}$$

where $K_{ij}^{(\mathrm{R})}(\boldsymbol{q}=0,\omega)$ is the retarded function of the Fourier transform of the thermal flux correlation

$$K_{ij}(\boldsymbol{r}_1,\tau_1;\boldsymbol{r}_2,\tau_2) = \langle \boldsymbol{T}_\tau Q_i(\boldsymbol{r}_1,\tau_1)Q_j(\boldsymbol{r}_2,\tau_2)\rangle. \tag{9.4}$$

In the limit where there is particle-hole (quasiparticle-quasihole) symmetry, typical of the BCS regime, a simple expression can be obtained for the thermal conductivity tensor

$$\kappa_{ij}(T) = -\frac{2}{T}\int \frac{d^3\boldsymbol{k}}{(2\pi)^3}\frac{\partial f(E_{\boldsymbol{k}})}{\partial E_{\boldsymbol{k}}}E_{\boldsymbol{k}}^2\left(\frac{\partial E_{\boldsymbol{k}}}{\partial_{\boldsymbol{k}}}\right)_i\left(\frac{\partial E_{\boldsymbol{k}}}{\partial_{\boldsymbol{k}}}\right)_j\tau_{\boldsymbol{k}}. \tag{9.5}$$

From this approximated expression it is clear that κ_{ij} depends strongly on the symmetry of the order parameter through $E_{\boldsymbol{k}} = \sqrt{\xi_{\boldsymbol{k}}^2 + |\boldsymbol{d}(\boldsymbol{k})|^2}$, and thus can serve as an independent way of measuring the nodal structure of the magnitude of the $\boldsymbol{d}$-vector.

- *Is there a transition from singlet to triplet superconductivity?*

Another interesting possibility arises when studying the competition between singlet and triplet superconductivity in quasi-1D systems. This possible transition was pointed out by Sá de Melo several years ago [63], and is still not fully understood. Imagine that singlet and triplet superconductivity are competing ground states, and assume that the critical temperature for singlet superconductivity $T_{c,s}(0)$ is larger than that for triplet superconductivity $T_{c,t}(0)$ at zero magnetic field, i.e., $T_{c,s}(0) > T_{c,t}(0)$. Therefore, the singlet state is more stable than the triplet state at low magnetic fields. The situation is reversed at high magnetic fields when $\omega_c \sim t_z$ due to the magnetic field induced dimensional crossover [381]. As a consequence, a first order transition line emerges at fields $H = H_t$, when the critical temperatures for singlet and triplet superconductivity are identical $T_{c,s}(H_t) = T_{c,t}(H_t)$. This possibility is illustrated in Fig. 9.1, and the critical field H_t is schematically indicated. A direct experimental signature of this effect should be present in the low temperature spin susceptibility $\chi_\eta(\mathrm{T})$, where $\eta = s,t$. If this magnetic field induced transition happens at low enough magnetic fields, the spin susceptibility at zero temperature changes discontinuously from $\chi_s(0) = 0$ in the singlet case to $\chi_t(0) = \chi_n$ at the field $H = H_t$, where χ_n is the normal state susceptibility.

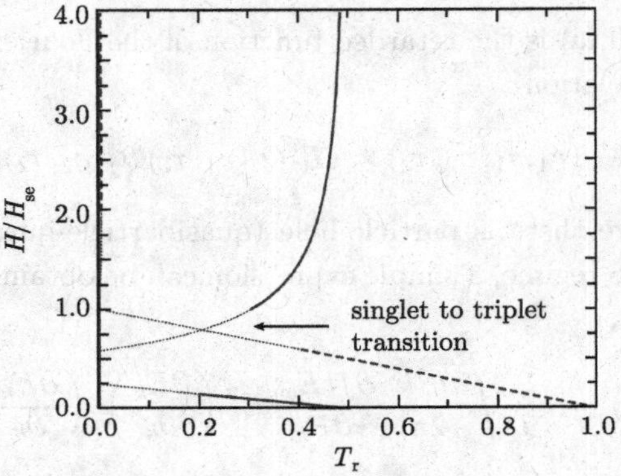

Figure 9.1 The upper critical fields using analytical expressions from Chap. 2 for low and high magnetic fields. The choice of parameters is such that $T_{c,s}(0) = 2T_{c,t}(0)$. The solid lines represent the triplet solution, the dashed lines represent the singlet solution, the thin lines are extrapolations of the solid and dashed curves. The arrow indicates a magnetic field induced phase transition from singlet to triplet superconductivity. The magnetic field H is in units of $H_{se} = 13.271 T_{c,s}^2(0)/(t_z|e|v_F c)$, and $T_r = T/T_{c,s}(0)$. From Ref. [63].

- *Is there magnetic field induced reentrant superconductivity in quasi-two-dimensional superconductors?*

Lastly, it is important to point out that the magnetic field induced dimensional crossover (FIDC), and pairing in the extreme quantum limit may also be achieved in quasi-two-dimensional (quasi-2D) systems like strontium ruthenate or high-T_c Copper oxide superconductors [63, 235].

A simple approximation for the dispersion relation of layered quasi-2D systems is

$$\varepsilon(\boldsymbol{k}) = \frac{k_x^2}{2m} + \frac{k_y^2}{2m} - t_z \cos(k_z c), \tag{9.6}$$

where c is the interplane spacing and t_z is the interplane transfer integral. The x-y planes are assumed to be continuous, but this restriction is not essential. The semiclassical equations of motion for a magnetic field applied along the y

direction are

$$\frac{d^2x(t)}{dt^2} + \omega_c^2 \sin\left[k_z c + Gx(t)\right] = 0, \tag{9.7}$$

and

$$\frac{dz(t)}{dt} = t_z c \cos\left[k_z c + Gx(t)\right], \tag{9.8}$$

when the Landau gauge $\boldsymbol{A} = (0, 0, -Hx)$ is used. Notice that in this gauge, the momenta k_y and k_z are constants of motion. The equation for $x(t)$ is the well known *pendulum* equation, which has open and closed orbit solutions. The characteristic frequency of the motion is ω_c, and the characteristic amplitude of motion along the z direction is proportional to t_z/ω_c, thus a magnetic field dimensional crossover is also possible in quasi-two-dimensional systems. (These open and closed orbits in real space have a counterpart in momentum space as illustrated in Figure 9.2.)

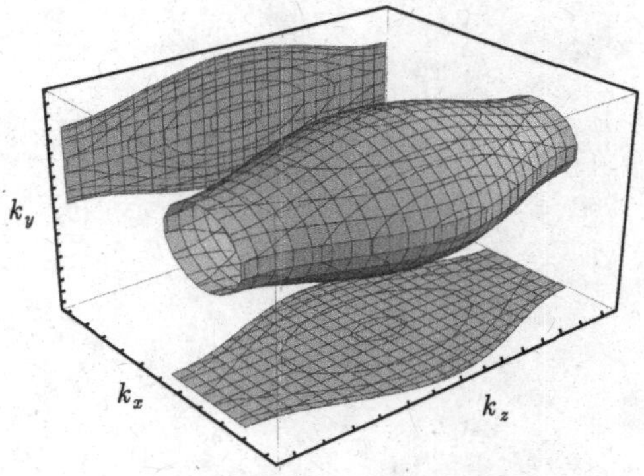

Figure 9.2 The Fermi surface for a quasi-two-dimensional superconductor. The contour plots indicate the semiclassical orbits in a magnetic field. Notice that these orbits can be open or closed. The orbits are closed for small values of k_x and k_y, and open for large values of k_x and k_y.

The quantum level structure also confirms this semiclassical description, since the eigenfunctions of the non-interacting Hamiltonian

$$H_0(\boldsymbol{k} - e\boldsymbol{A}) = \frac{k_x^2}{2m} + \frac{k_y^2}{2m} - t_z \cos(k_z c + Gx) \tag{9.9}$$

also become confined along the z direction, and the extreme quantum limit can be reached when $\omega_c \sim t_z$. The eigenvalue equation for this Hamiltonian is the Mathieu equation, and the eigenfunctions are Mathieu functions.

It may be interesting for experimentalists to look for FIDC and pairing in the extreme quantum limit in Sr_2RuO_4 (a clean, low critical temperature quasi-2D superconductor), specially after theoretical [382, 383] and experimental [384] reports that this system might be a triplet superconductor.

References

[1] D. Jérome, A. Mazaud, M. Ribault, K. Bechgaard, J. Phys. (Paris) Lett. **41**, L95 (1980).

[2] M.Y. Choi, P.M. Chaikin, S.Z. Huang, P. Haen, E.M. Engler, R.L. Greene, Phys. Rev. B **25**, 6208 (1982).

[3] R.L. Greene, P. Haen, S.Z. Huang, E.M. Engler, M.Y. Choi, P.M. Chaikin, Mol. Cryst. Liq. Cryst. **79**, 183 (1982).

[4] S. Bouffard, M. Ribault, R. Brusetti, D. Jérome, K. Bechgaard, J. Phys. C: Solid State Phys. **15**, 2951 (1982).

[5] S. Tomić, D. Jérome, D. Mailly, M. Ribault, K. Bechgaard, J. Phys. (Paris) Colloq. **44**, C3 (1983).

[6] C. Coulon, P. Delhaes, J. Amiell, J.P. Manceau, J.M. Fabre, L. Giral, J. Phys. (Paris) **43**, 1721 (1983).

[7] A.A. Abrikosov, J. Low Temp. Phys. **53**, 359 (1983).

[8] L.P. Gorkov, D. Jérome, J. Phys. (Paris) Lett. **46**, L643 (1985).

[9] R. Brusetti, M. Ribault, D. Jérome, K. Bechgaard, J. Phys. (Paris) **43**, 52 (1982).

[10] T. Ishiguro, K. Kajimura, H. Bando, K. Murata, H. Anzai, Mol. Cryst. Liq. Cryst. **119**, 19 (1985).

[11] A.M. Clogston, Phys. Rev. Lett. **9**, 266 (1962).

[12] B.S. Chandrasekhar, Appl. Phys. Lett. **1**, 7 (1962).

[13] M. Takigawa, H. Yasuoka, G. Saito, J. Phys. Soc. Jpn. **56**, 873 (1987).

[14] L.C. Hebel, C.P. Slichter, Phys. Rev. **113**, 1504 (1959).

[15] Y. Hasegawa, H. Fukuyama, J. Phys. Soc. Jpn. **56**, 877 (1987).

[16] A.G. Lebed, JETP Lett. **44**, 114 (1986).

[17] L.I. Burlachkov, L.P. Gorkov, A.G. Lebed, Europhys. Lett. **4**, 941 (1987).

[18] N. Dupuis, G. Montambaux, C.A.R. Sá de Melo, Phys. Rev. Lett. **70**, 2613 (1993).

[19] N. Dupuis, Phys. Rev. B **50**, 9607 (1994).

[20] C.A.R. Sá de Melo, Physica C **260**, 224 (1996).

[21] I.J. Lee, A.P. Hope, M.J. Leone, M.J. Naughton, Synth. Metals **70**, 747 (1995).

[22] O. Fischer, H.W. Meul, M.G. Karkut, G. Remenyi, U. Welp, J.C. Piccoche, K. Maki, Phys. Rev. Lett. **55**, 2972 (1985).

[23] I.J. Lee, M.J. Naughton, G.M. Danner, P.M. Chaikin, Phys. Rev. Lett. **78**, 3555 (1997).

[24] R.A. Klemm, A. Luther, M.R. Beasley, Phys. Rev. B **12**, 877 (1975).

[25] R.V. Coleman, G.K. Eiserman, S.J. Hillenius, A.T. Mitchell, J.L. Vicent, Phys. Rev. B **27**, 125 (1983).

[26] G. Danner, W. Kang, P.M. Chaikin, Phys. Rev. Lett. **72**, 3714 (1994).

[27] M. Tinkham, *Introduction to Superconductivity*, 2nd edn. (McGraw-Hill, New York, 1996).

[28] I.J. Lee, M.J. Naughton, in *The Superconducting State in Magnetic Fields: Special Topics and New Trends*, ed. by C.A.R. Sá de Melo (World Scientific, Singapore, 1998), chap. 14, p. 272.

[29] N. Dupuis, G. Montambaux, Phys. Rev. B **49**, 8993 (1994).

[30] A.I. Larkin, Y.N. Ovchinnikov, Sov. Phys. JETP **20**, 762 (1965).

[31] P. Fulde, R.A. Ferrell, Phys. Rev. **135**, A550 (1964).

[32] A.G. Lebed, Phys. Rev. B **59**, R721 (1999).

[33] I.J. Lee, D.S. Chow, W.G. Clark, M.J. Strouse, M.J. Naughton, P.M. Chaikin, S.E. Brown, Phys. Rev. B **68**, 092510 (2003).

[34] A.G. Lebed, K. Machida, M. Ozaki, Phys. Rev. B **62**, R795 (2000).

[35] I.J. Lee, S.E. Brown, W.G. Clark, M.J. Strouse, M.J. Naughton, W. Kang, P.M. Chaikin, Phys. Rev. Lett. **88**, 017004 (2002).

[36] R.D. Duncan, C.D. Vaccarella, C.A.R. Sá de Melo, Phys. Rev. B **64**, 172503 (2001).

[37] R.D. Duncan, R.W. Cherng, C.A.R. Sá de Melo, Physica C **391**, 98 (2003).

[38] J. Shinagawa, Y. Kurosaki, F. Zhang, C. Parker, S.E. Brown, D. Jérome, J.B. Christensen, K. Bechgaard, Phys. Rev. Lett. **98**, 147002 (2007).

[39] S. Belin, K. Behnia, Phys. Rev. Lett. **79**, 2125 (1997)

[40] S. Yonezawa, Y. Maeno, K. Bechgaard, D. Jérome, Phys. Rev. B **84**, 140502 (2012).

[41] S. Barisic, S. Brazovskii, in *Recent Developments in Condensed Matter Physics*, ed. by J.T. Devreese (Plenum, New York, 1981), chap. 6, p. 142.

[42] S. Baristic, J. Phys. C **"C3"**, 911 (1983).

[43] K. Kuroki, R. Arita, H. Aoki, Phys. Rev. B **63**, 094509 (2001).

[44] H. Shimahara, Phys. Rev. B **61**, R14936 (2000).

[45] J.C. Nickel, R. Duprat, C. Bourbonnais, N. Dupuis, Phys. Rev. Lett. **95**, 247001 (2005).

[46] J.C. Nickel, R. Duprat, C. Bourbonnais, N. Dupuis, Phys. Rev. B **73**, 165126 (2006).

[47] A.V. Rozhkov, A.J. Millis, Phys. Rev. B **66**, 134509 (2002).

[48] Y. Ohta, S. Nishimoto, T. Shirakawa, Y. Yamaguchi, Phys. Rev. B **72**, 012503 (2005).

[49] C.D. Vaccarella, C.A.R. Sá de Melo, Phys. Rev. B **63**, 180505(R) (2001).

[50] A.J. Leggett, Rev. Mod. Phys. **47**, 331 (1975).

[51] W. Zhang, C.A.R. Sá de Melo, Adv. Phys. **56**, 545 (2007).

[52] T. Ishiguro, K. Yamaji, G. Saito, *Organic Superconductors*, 2nd edn. (Springer, Berlin, 1998).

[53] J.F. Kwak, J.E. Schirber, R.L. Greene, E.M. Engler, Phys. Rev. Lett. **46**, 1296 (1981).

[54] G. Saito, S. Kagoshima, *The Physics and Chemistry of Organic Superductors* (Springer-Verlag, Berlin, 1987).

[55] T. Osada, A. Kawasumi, S. Kagoshima, N. Miura, G. Saito, Phys. Rev. Lett. **66**, 1525 (1991).

[56] T. Osada, S. Kagoshima, N. Miura, Phys. Rev. Lett. **77**, 5261 (1996).

[57] M.J. Naughton, I.J. Lee, P.M. Chaikin, G.M. Danner, Synth. Metals **85**, 1481 (1997).

[58] G. Danner, N.P. Ong, P.M. Chaikin, Phys. Rev. Lett. **78**, 983 (1997).

[59] S.P. Strong, D.G. Clarke, P.W. Anderson, Phys. Rev. Lett. **73**, 1007 (1994).

[60] C.S. Jacobsen, D.B. Tanner, K. Bechgaard, Phys. Rev. B **28**, 7019 (1983).

[61] N. Thorup, G. Rindorf, H. Soling, K. Bechgaard, Acta. Cryst. B **37**, 1236 (1981).

[62] G. Rindorf, H. Soling, N. Thorup, Acta. Cryst. B **38**, 2805 (1982).

[63] C.A.R. Sá de Melo, in *The Superconducting State in Magnetic Fields: Special Topics and New Trends*, ed. by C.A.R. Sá de Melo (World Scientific, Singapore, 1998), chap. 15, p. 296.

[64] Z. Tešanović, M. Rasolt, L. Xing, Phys. Rev. Lett. **63**, 2425 (1989).

[65] M. Rasolt, Z. Tešanović, Rev. Mod. Phys. **64**, 709 (1992).

[66] H. Akera, A.H. MacDonald, S.M. Girvin, M.R. Norman, Phys. Rev. Lett. **67**, 2375 (1991).

[67] M.R. Norman, H. Akera, A.H. MacDonald, in *Physical Phenomena in High Magnetic Fields*, ed. by E. Manousakis, P. Schlottmann, P. Kumar, K. Bedell, F.M. Mueller (Addison-Wesley, Reading, MA, 1992).

[68] M.R. Norman, A.H. MacDonald, H. Akera, Phys. Rev. B **51**, 5927 (1995).

[69] A.K. Rajagopal, R. Vasudevan, Phys. Lett. **23**, 539 (1966).

[70] L.W. Gruenberg, L. Gunther, Phys. Rev. **176**, 606 (1968).

[71] J. Solyom, Adv. Phys. **28**, 201 (1979).

[72] W.E. Lawrence, S. Doniach, in *Proceedings of the 12*th *international conference on low temperature physics (LT12), Kyoto, Japan*, ed. by E. Kanada (Academic Press, New York, 1970).

[73] D.E. Prober, R.E. Schwall, M.R. Beasley, Phys. Rev. B **21**, 2717 (1980).

[74] M.Y. Choi, P.M. Chaikin, R.L. Greene, Phys. Rev. B **34**, 7727 (1986).

[75] I.J. Lee, A.P. Hope, M.J. Leone, M.J. Naughton, Appl. Supercond. **2**, 753 (1994).

[76] T. Vuletić, P. Auban-Senzier, C. Pasquier, S. Tomić, D. Jérome, M. Héritier, K. Bechgaard, Eur. Phys. J. B **25**, 319 (2002).

[77] A.V. Kornilov, V.M. Pudalov, Y. Kitaoka, K. Ishida, G. Q. Zheng, T. Mito, J.S. Qualls, Phys. Rev. B **69**, 224404 (2004).

[78] N. Kang, B. Salameh, P. Auban-Senzier, D. Jérome, C. R. Pasquier, and S. Brazovskii, Phys. Rev. B **81**, 100509 (2010).

[79] I.J. Lee, S.E. Brown, W. Yu, M.J. Naughton, P.M. Chaikin, Phys. Rev. Lett. **94**, 197001 (2005).

[80] D. Jérome, H.J. Schulz, Adv. Phys. **31**, 299 (1982).

[81] K. Bechgaard, D. Jérome, Scient. Am. **247**, 52 (1982).

[82] A.G. Lebed, *The Physics of Organic Superconductors and Conductors* (Springer, Heidelberg, 2008).

[83] K. Murata, H. Anzai, K. Kajimura, T. Ishiguro, G. Saito, Mol. Cryst. Liq. Cryst. **79**, 283 (1982).

[84] K. Murata, M. Tokumoto, H. Anzai, K. Kajimura, T. Ishiguro, Jpn. J. Appl. Phys. **26**, 1367 (1987).

[85] P.M. Chaikin, T. Tiedje, A.N. Bloch, Solid State Commun. **41**, 739 (1982).

[86] I.J. Lee, M.J. Naughton, Phys. Rev. B **58**, R13343 (1998).

[87] D.U. Gubser, W.W. Fuller, T.O. Poehler, J. Stokes, D.O. Cowan, M. Lee, A.N. Bloch, Mol. Cryst. Liq. Cryst. **79**, 225 (1982).

[88] P.M. Chaikin, M.Y. Choi, R.L. Greene, J. Magn. Magn. Mater. **31-34**, 1268 (1983).

[89] Y.N. Ovchinnikov, V.Z. Kresin, Phys. Rev. B **54**, 1251 (1995).

[90] G. Kotliar, C.M. Varma, Phys. Rev. Lett. **77**, 2296 (1996).

[91] D.A. Wollman, D.J. Van Harlingen, W.C. Lee, D.M. Ginsberg, A.J. Leggett, Phys. Rev. Lett. **71**, 2134 (1993).

[92] D.A. Brawer, H.R. Ott, Phys. Rev. B **50**, 6530 (1994).

[93] A. Mathai, Y. Gim, R.C. Black, A. Amar, F.C. Wellstood, Phys. Rev. Lett. **74**, 4523 (1995).

[94] T. Takahashi, H. Kawamura, T. Ohyama, Y. Maniwa, K. Murata, G. Saito, J. Phys. Soc. Jpn. **58**, 703 (1989).

[95] L.J. Azevedo, J.E. Schirber, E.M. Engler, Phys. Rev. B **29**, 464 (1984).

[96] F. Creuzet, C. Bourbonnais, L.G. Caron, D. Jérome, A. Moradpour, Synth. Metals **19**, 277 (1987).

[97] S. Valfells, P. Kuhns, A. Kleinhammes, J.S. Brooks, W. Moulton, S. Takasaki, J. Yamada, H. Anzai, Phys. Rev. B **56**, 2585 (1997).

[98] M. Miljak, J.R. Cooper, K. Bechgaard, J. Phys. (Paris) Colloq. **44**, C3 893 (1983).

[99] J. Orenstein, A.J. Millis, Science **288**, 468 (2000).

[100] N.D. Mathur, F.M. Grosche, S.R. Julin, I.R. Walker, D.M. Freye, R.K.W. Haselwimmer, G.G. Lonzarich, Nature (London) **394**, 39 (1998).

[101] H. Hegger, C. Petrovic, E.G. Moshopoulou, M.F. Hundley, J.L. Sarrao, Z. Fisk, J.D. Thompson, Phys. Rev. Lett. **84**, 4986 (2000).

[102] S.S. Saxena, P. Agarwal, K. Ahilan, F.M. Crosche, R.K.W. Hasellwimmer, M.J. Steiner, E. Pugh, I.R. Walker, S.R. Julian, P. Monthoux, G.G. Lonzarich, A. Huxley, I. Sheikin, D. Braithwaite, J. Flouquet, Nature (London) **406**, 587 (2000).

[103] C. Pfleiderer, M. Uhlarz, S.M. Hayden, R. Vollmer, H.V. Lohneysen, N.R. Bernhoeft, G.G. Lonzarich, Nature (London) **412**, 58 (2001).

[104] J.H. Chu, J.G. Analytis, C. Kucharczyk, I.R. Fisher, Phys. Rev. B **79**, 014506 (2009).

[105] S. Kasahara, T. Shibauchi, K. Hashimoto, K. Ikada, S. Tonegawa, R. Okasaki, H. Shishido, H. Ikeda, H. Takeya, K. Hirata, T. Terashima, Y. Matsuda, Phys. Rev. B **81**, 184519 (2010).

[106] P.M. Chaikin, J. Phys. I **6**, 1875 (1996).

[107] L.J. Azevedo, J.E. Schirber, E.M. Engler, Phys. Rev. B **27**, 5842 (1983).

[108] I.J. Lee, P.M. Chaikin, M.J. Naughton, Phys. Rev. Lett. **88**, 207002 (2002).

[109] N. Biškup, S. Tomić, D. Jérome, Phys. Rev. B **51**, 17972 (1995).

[110] S. Uji, J.S. Brooks, M. Chaparala, S. Takasaki, J. Yamada, H. Anzai, Phys. Rev. B **55**, 14387 (1997).

[111] J.R. Cooper, W. Kang, P. Auban, G. Montambaux, D. Jérome, K. Bechgaard, Phys. Rev. Lett. **63**, 1984 (1989).

[112] S.T. Hannahs, J.S. Brooks, W. Kang, L.Y. Chiang, P.M. Chaikin, Phys. Rev. Lett. **63**, 1988 (1989).

[113] A.V. Kornilov, V.M. Pudalov, Y. Kitaoka, K. Ishida, T. Mito, J.S. Brooks, J.S. Qualls, J.A.A.J. Perenboom, N. Tateiwa, T.C. Kobayashi, Phys. Rev. B **65**, 060404 (2002).

[114] A.V. Kornilov, V.M. Pudalov, Y. Kitaoka, K. Ishida, T. Mito, J.S. Brooks, J.S. Qualls, Synth. Metals **133-134**, 69 (2003).

[115] A.G. Lebed, JETP Lett. **72**, 141 (2000).

[116] A.G. Lebed, Phys. Rev. Lett. **88**, 177001 (2002).

[117] A. Bezryadin, C.N. Lau, M. Tinkham, Nature (London) **404**, 971 (2000).

[118] M.T. Béal-Monod, C. Bourbonnais, Phys. Rev. B **34**, 7716 (1986).

[119] D. Podolsky, E. Altman, T. Rostunov, E. Demler, Phys. Rev. Lett. **93**, 246402 (2004).

[120] D. Podolsky, E. Altman, T. Rostunov, E. Demler, Phys. Rev. B **70**, 224503 (2004).

[121] W. Zhang, C.A.R. Sá de Melo, J. Appl. Phys. **95**, 10B108 (2005).

[122] W. Zhang, C.A.R. Sá de Melo, Phys. Rev. Lett. **97**, 047001 (2006).

[123] W. Wu, P.M. Chaikin, W. Kang, J. Shinagawa, W. Yu, S.E. Brown, Phys. Rev. Lett. **94**, 097004 (2005).

[124] N. Doiron-Leyraud, P. Auban-Senzier, S. René de Cotret, C. Bourbonnais, D. Jérome, K. Bechgaard, L. Taillefer, Phys. Rev. B **80**, 214531 (2009).

[125] S.E. Brown, P.M. Chaikin, M.J. Naughton, in *The Physics of Organic Superconductors and Conductors*, ed. by A.G. Lebed (Spinger, Heidelberg, 2008), chap. 5, p. 49.

[126] P. Lee, N. Nagaosa, X.G. Wen, Rev. Mod. Phys. **78**, 17 (2007).

[127] Q. Si, F. Steglich, Science **329**, 1161 (2010).

[128] L. Fang, H. Luo, P. Cheng, Z. Wang, Y. Jia, G. Mu, B. Shen, I.I. Mazin, L. Shan, C. Ren, H.H. Wen, Phys. Rev. B **80**, 140508 (2009).

[129] C. Bourbonnais, A. Sedeki, Phys. Rev. B **80**, 085105 (2009).

[130] C. Bourbonnais, A. Sedeki, Comp. Rend. Phys. **12**, 532 (2011).

[131] A.J. Millis, H. Monien, D. Pines, Phys. Rev. B **42**, 167 (1990).

[132] T. Moriya, K. Ueda, Rep. Prog. Phys. **66**, 1299 (2003).

[133] N. Doiron-Leyraud, S.R. de Cotret, A. Sedeki, C. Bourbonnais, L. Taillefer, P. Auban-Senzier, D. Jérome, K. Bechgaard, Eur. Phys. J. B **78**, 23 (2010).

[134] C. Proust, E. Boaknin, R.W. Hill, L. Taillefer, A.P. Mackenzie, Phys. Rev. Lett. **89**, 147003 (2002).

[135] R.A. Cooper, Y. Wang, B. Vignolle, O.J. Lipscombe, S.M. Hayden, Y. Tanabe, T. Adachi, Y. Koike, M. Nohara, H. Takagi, C. Proust, N.E. Hussey, Science **323**, 603 (2009).

[136] N. Joo, P. Auban-Senzier, C. Pasquier, P. Monod, D. Jérome, K. Bechgaard, Eur. Phys. J. B **40**, 43 (2004).

[137] N. Joo, P. Auban-Senzier, C.R. Pasquier, D. Jérome, K. Bechgaard, Europhys. Lett. **72**, 645 (2005).

[138] J.I. Oh, M.J. Naughton, Phys. Rev. Lett. **92**, 067001 (2004).

[139] S. Yonezawa, S. Kusaba, Y. Maeno, P. Auban-Senzier, C. Pasquier, K. Bechgaard, D. Jérome, Phys. Rev. Lett. **100**, 117002 (2008).

[140] G.M. Luke, M.T. Rovers, A. Fukaya, I.M. Gat, M.I. Larkin, A. Savici, Y.J. Uemura, K.M. Kojima, P.M. Chaikin, I.J. Lee, M.J. Naughton, Physica B **326**, 378 (2003).

[141] P.W. Anderson, J. Phys. Chem. Solids **11**, 26 (1959).

[142] L. Zuppirolli, in *Low Dimensional Conductors and Superconductors*, ed. by D. Jérome, L.G. Caron (Plenum Press, New York, 1987), p. 307.

[143] M. Sanquer, S. Bouffard, Mol. Cryst. Liq. Cryst. **119**, 147 (1985).

[144] I. Johannsen, K. Bechgaard, C.S. Jacobsen, G. Rindorf, N. Thorup, K. Mortensen, D. Mailly, Mol. Crys. Liq. Cryst. **119**, 277 (1985).

[145] C. Coulon, P. Delhaes, J. Amiell, J.P. Monceau, J.M. Fabre, L. Giral, J. Phys. France **43**, 1721 (1982).

[146] J.P. Pouget, in *Low Dimensional Conductors and Superconductors*, ed. by D. Jérome, L.G. Caron (Plenum Press, New York, 1987), p. 17.

[147] P. Garoche, R. Brusetti, K. Bechgaard, Phys. Rev. Lett. **49**, 1346 (1982).

[148] S. Tomić, D. Jérome, P. Monod, K. Bechgaard, J. Phys. (Paris) Colloq. **44**, C3 (1983).

[149] G. Kriza, O. Traetteberg, J. Phys. IV **C2-3**, 15 (1993).

[150] A.I. Larkin, JETP Lett. **2**, 130 (1930).

[151] K. Maki, H. Won, S. Haas, Phys. Rev. B **69**, 012502 (2004).

[152] H. Schwenk, K. Andres, F. Wudl, Phys. Rev. B **29**, 500 (1984).

[153] D.U. Gubser, W.W. Fuller, T.O. Poehler, D.O. Cowan, M. Lee, R.S. Potember, L.Y. Chiang, A.N. Bloch, Phys. Rev. B **24**, 478 (1981).

[154] A.G. Lebed, S. Wu, Phys. Rev. B **82**, 172504 (2010).

[155] A.G. Lebed, Phys. Rev. Lett. **107**, 087004 (2011).

[156] M.D. Croitoru, M. Houzet, A.I. Buzdin, Phys. Rev. Lett. **108**, 207005 (2012).

[157] M. Chaparala, O.H. Chung, M.J. Naughton, AIP Conf. Proc. **273**, 407 (1993).

[158] I.J. Lee, P.M. Chaikin, M.J. Naughton, Phys. Rev. B **62**, R14699 (2000).

[159] K. Deguchi, T. Ishiguro, Y. Maeno, Rev. Sci. Instrum. **75**, 1188 (2004).

[160] B. Lussier, B. Ellman, L. Taillefer, Phys. Rev. Lett. **73**, 3294 (1994).

[161] A.D. Huxley, H. Suderow, J.P. Brison, J. Flouquet, Phys. Lett. A **209**, 365 (1995).

[162] B. Lussier, B. Ellman, L. Taillefer, Phys. Rev. B **53**, 5145 (1996).

[163] F. Yu, M.B. Salamon, A.J. Leggett, W.C. Lee, D.M. Ginsberg, Phys. Rev. Lett. **74**, 5136 (1995).

[164] H. Aubin, K. Behnia, M. Ribault, R. Gagnon, L. Taillefer, Phys. Rev. Lett. **78**, 2624 (1997).

[165] L. Taillefer, B. Lussier, R. Gagnon, K. Behnia, H. Aubin, Phys. Rev. Lett. **79**, 483 (1997).

[166] C.L. Kane, M.P.A. Fisher, Phys. Rev. Lett. **76**, 3192 (1996).

[167] J. Bardeen, G. Rickaysen, L. Tewordt, Phys. Rev. **113**, 982 (1959).

[168] S. Yonezawa, S. Kusaba, Y. Maeno, P. Auban-Senzier, C. Pasquier, D. Jérome, J. Phys. Soc. Jpn. **77**, 054712 (2008).

[169] G.E. Volovik, JETP Lett. **58**, 469 (1993).

[170] I. Vekhter, P.J. Hirschfeld, J.P. Carbotte, E.J. Nicol, Phys. Rev. B **50**, 9023 (1999).

[171] D.L. Pevelen, J. Gaultier, Y. Barrans, D. Chasseau, F. Castet, L. Ducasse, Eur. Phys. J. B **19**, 363 (2001).

[172] G.M. Luke, A. Keren, L.P. Le, W.D. Wu, Y.J. Uemura, D.A. Bonn, L. Taillefer, J.D. Garrett, Phys. Rev. Lett. **71**, 1466 (1993).

[173] G.M. Luke, Y. Fudamoto, K.M. Kojima, M.I. Larkin, J. Merrin, B. Nachumi, Y.J. Uemura, Y. Maeno, Z.Q. Mao, Y. Mori, H. Nakamura, M. Sigrist, Nature **394**, 558 (1998).

[174] M. Takigawa, G. Saito, J. Phys. Soc. Jpn. **55**, 1233 (1986).

[175] F. Zhang, Y. Kurosaki, J. Shinagawa, B. Alavi, S.E. Brown, Phys. Rev. B **72**, 060501 (2005).

[176] J. Shinagawa, W. Wu, P.M. Chaikin, W. Kang, W. Yu, F. Zhang, Y. Kurosaki, C. Parker, S.E. Brown, J. Low Temp. Phys. **142**, 227 (2006).

[177] A.A. Abrikosov, Sov. Phys. JETP **5**, 1174 (1957).

[178] L.P. Gorkov, Sov. Phys. JETP **9**, 1364 (1959).

[179] L.P. Gorkov, A.G. Lebed, J. Phys. (Paris) Lett. **45**, 433 (1984).

[180] A.I. Buzdin, V.V. Tugushev, Sov. Phys. JETP **58**, 428 (1983).

[181] G. Montambaux, Phys. Scripta T **35**, 188 (1991).

[182] L.N. Bulaeviskii, Int. J. Mod. Phys. B **4**, 1849 (1990).

[183] M. Miyazaki, K. Kishigi, Y. Hasegawa, J. Phys. Soc. Jpn. **68**, 3794 (1999).

[184] L.P. Gorkov, Sov. Phys. Usp. **27**, 809 (1984).

[185] K. Yamaji, J. Phys. Soc. Jpn. **54**, 1034 (1985).

[186] N. Dupuis, V.M. Yakovenko, Phys. Rev. Lett. **80**, 3618 (1998).

[187] M. Miyazaki, Y. Hasegawa, J. Phys. Soc. Jpn. **65**, 3283 (1996).

[188] V.P. Mineev, K.V. Samokhin, *Introduction to Unconventional Superconductivity* (Gordon and Breach, Amsterdam, 1999).

[189] M. Sigrist, K. Ueda, Rev. Mod. Phys. **63**, 239 (1991).

[190] J. Zinn-Justin, *Quantum Field Theory and Critical Phenomena*, 2nd edn. (Oxford University Press, Oxford, U.K., 1993).

[191] N. Goldenfeld, *Lectures on Phase Transitions and the Renormalization Group* (Perseus Books, MA, 1992).

[192] C.D. Vaccarella, C.A.R. Sá de Melo, Phys. Rev. B **64**, 212504 (2001).

[193] B. Horovitz, Phys. Rev. B **47**, 5947 (1992).

[194] B. Horovitz, Phys. Rev. B **47**, 5964 (1992).

[195] S.E. Korshunov, A.I. Larkin, Phys. Rev. B **46**, 6395 (1992).

[196] J.M. Kosterlitz, J. Phys. C **7**, 1046 (1974).

[197] V.L. Berezinskii, Sov. Phys. JETP **34**, 610 (1972).

[198] J.M. Kosterlitz, D.J. Thouless, J. Phys. C **5**, L124 (1972).

[199] Y. Nagai, H. Nakamura, M. Machida, Phys. Rev. B **83**, 104523 (2011).

[200] N. Kopnin, *Theory of Nonequilibrium Superconductivity* (Clarendon, Oxford, 2001).

[201] Y. Nagai, N. Hayashi, Phys. Rev. Lett. **101**, 097001 (2008).

[202] Y. Nagai, Y. Kato, N. Hayashi, K. Yamauchi, H. Harima, Phys. Rev. B **76**, 214514 (2007).

[203] Y. Nagai, Y. Kato, N. Hayashi, J. Phys. Soc. Jpn. **75**, 043706 (2006).

[204] Y. Nagai, Y. Ueno, Y. Kato, N. Hayashi, J. Phys. Soc. Jpn. **75**, 104701 (2006).

[205] A.S. Mel'nikov, D.A. Ryzhov, M.A. Silaev, Phys. Rev. B **78**, 064513 (2008).

[206] M. Takigawa, M. Ichioka, K. Kuroki, Y. Asano, Y. Tanaka, Phys. Rev. Lett. **97**, 187002 (2006).

[207] M. Ichioka, T. Tsuneto, J. Low Temp. Phys. **96**, 213 (1994).

[208] M. Takigawa, M. Ichioka, K. Machida, J. Phys. Soc. Jpn. **69**, 3943 (2000).

[209] M. Takigawa, M. Ichioka, K. Machida, J. Phys. Soc. Jpn. **73**, 450 (2004).

[210] B.H. Powell, J. Phys.: Condens. Matter **20**, 345234 (2008).

[211] P.W. Anderson, Phys. Rev. B **30**, 4000 (1984).

[212] M. Tinkham, *Group Theory and Quantum Mechanics* (McGraw-Hill, New York, 1964).

[213] C.A.R. Sá de Melo, M. Randeria, J.R. Engelbrecht, Phys. Rev. Lett. **71**, 3202 (1993).

[214] J.R. Engelbrecht, M. Randeria, C.A.R. Sá de Melo, Phys. Rev. B **55**, 15153 (1997).

[215] C. Caroli, P.G. de Gennes, J. Matricon, Phys. Lett. **9**, 307 (1964).

[216] J. Bardeen, R. Kummel, A.E. Jacobs, L. Tewordt, Phys. Rev. **187**, 556 (1969).

[217] L. Kramer, W. Pesch, Z. Phys. **269**, 59 (1974).

[218] S. Ullah, A.T. Dorsey, L.J. Buchholtz, Phys. Rev. B **42**, 9950 (1990).

[219] F. Gygi, M. Schluter, Phys. Rev. B **43**, 7609 (1991).

[220] C.A.R. Sá de Melo, Phys. Rev. Lett. **73**, 1978 (1994).

[221] M. Ichioka, N. Hayashi, N. Enomoto, K. Machida, Phys. Rev. B **53**, 15316 (1996).

[222] M. Franz, Z. Tesanovic, Phys. Rev. Lett. **80**, 4763 (1998).

[223] C.A.R. Sá de Melo, Phys. Rev. B **60**, 10423 (1999).

[224] C.H. Choi, J.A. Sauls, Phys. Rev. Lett. **66**, 484 (1991).

[225] B.S. Shivaram, T.F. Rosenbaum, D.G. Hinks, Phys. Rev. Lett. **57**, 1259 (1986).

[226] R.W. Cherng, C.A.R. Sá de Melo, Phys. Rev. B **67**, 212505 (2003).

[227] V.L. Berezinskii, JETP Lett. **79**, 2125 (1974).

[228] T. Rostunov, E. Demler, A. Georges, Phys. Rev. Lett. **96**, 077002 (2006).

[229] B.J. Powell, J.F. Annett, B.L. Györffy, J. Phys. A: Math. Gen. **36**, 9289 (2003).

[230] P. Garoche, R. Brusetti, D. Jérome, K. Bechgaard, J. Phys. (Paris) Lett. **43**, L147 (1982).

[231] L.P. Gor'kov, P.D. Grigoriev, Europhys. Lett. **71**, 425 (2005).

[232] L.P. Gor'kov, P.D. Grigoriev, Phys. Rev. B **75**, 020507 (2007).

[233] A.G. Lebed, Physica B **407**, 1803 (2012).

[234] J.P. Brison, N. Kellera, A. Vernierea, P. Lejaya, L. Schmidtb, A. Buzdin, J. Flouquet, S.R. Juliand, G.G. Lonzarichd, Physica C **250**, 128 (1995).

[235] A.G. Lebed, K. Yamaji, Phys. Rev. Lett. **80**, 2697 (1998).

[236] L.N. Bulaevskii, Sov. Phys. JETP **38**, 634 (1974).

[237] H. Shimahara, Phys. Rev. B **62**, 3524 (2000).

[238] E. Dagotto, Rev. Mod. Phys. **66**, 763 (1994).

[239] Y. Maeno, H. Hashimoto, K. Yoshida, S. Nishizaki, T. Fujita, J.G. Bednorz, F. Lichtenberg, Nature **372**, 532 (1994).

[240] P.M. Chaikin, T.C. Lubensky, *Principles of Condensed Matter Physics* (Cambridge University Press, Cambridge, 1995).

[241] K. Mortensen, Y. Tomkiewicz, T.D. Schultz, E.M. Engler, Phys. Rev. Lett. **46**, 1234 (1981).

[242] W.P. Su, J.R. Schrieffer, A.J. Heeger, Phys. Rev. Lett. **42**, 1698 (1979).

[243] W.P. Su, J.R. Schrieffer, A.J. Heeger, Phys. Rev. B **22**, 2099 (1980).

[244] S.A. Brazovskii, Sov. Phys. JETP **51**, 342 (1980).

[245] S.A. Brazovskii, N.N. Kirova, Sov. Sci. Rev. A Phys. **5**, 99 (1984).

[246] S.A. Brazovskii, L.P. Gor'kov, J.R. Schrieffer, Phys. Scr. **25**, 423 (1982).

[247] S.A. Brazovskii, L.P. Gor'kov, A.G. Lebed, Sov. Phys. JETP **56**, 683 (1982).

[248] L.P. Gor'kov, A.G. Lebed, J. Phys. Colloq. Suppl. **4**, C3 1531 (1983).

[249] S.A. Brazovskii, S.A. Gordyunin, N.N. Kirova, JETP Lett. **31**, 451 (1980).

[250] K. Sano, Y. Ono, J. Phys. Soc. Jpn. **63**, 1250 (1994).

[251] I.F. Foulkes, B.L. Gyorffy, Phys. Rev. B **15**, 1395 (1977).

[252] A.A. Abrikosov, Physica C **222**, 191 (1994).

[253] A.A. Abrikosov, Physica C **244**, 243 (1995).

[254] J. Friedel, M. Kohmoto, Int. J. Mod. Phys. B **15**, 511 (2001).

[255] I. Chang, J. Friedel, M. Kohmoto, Europhys. Lett. **50**, 782 (2000).

[256] G. Varelogiannis, Phys. Rev. Lett. **88**, 117005 (2002).

[257] H. Shimahara, M. Kohmoto, Europhys. Lett. **57**, 247 (2002).

[258] H. Shimahara, M. Kohmoto, Phys. Rev. B **65**, 174502 (2002).

[259] T. Giamarchi, H.J. Schulz, Phys. Rev. B **39**, 4620 (1989).

[260] M. Kohmoto, M. Sato, Europhys. Lett. **56 (5)**, 736 (2001).

[261] Y. Suginishi, H. Shimahara, J. Phys. Soc. Jpn. **73**, 3121 (2004).

[262] V.J. Emery, Synth. Metals **13**, 21 (1986).

[263] H. Shimahara, J. Phys. Soc. Jpn. **58**, 1735 (1989).

[264] H. Kino, H. Kontani, J. Low Temp. Phys. **117**, 317 (1999).

[265] T. Nomura, K. Yamada, J. Phys. Soc. Jpn. **70**, 2694 (2001).

[266] K. Kuroki, H. Aoki, Phys. Rev. B **60**, 3060 (1999).

[267] E.W. Carlson, D. Orgad, S.A. Kivelson, V.J. Emery, Phys. Rev. B **62**, 3422 (2000).

[268] L.G. Caron, C. Bourbonnais, Physica B+C **143**, 453 (1986).

[269] R. Duprat, C. Bourbonnais, Eur. Phys. J. B **21**, 219 (2001).

[270] C. Bourbonnais, R. Duprat, J. Phys. IV (Paris) **114**, 3 (2004).

[271] Y. Fuseya, Y. Suzumura, J. Phys. Soc. Jpn. **74**, 1263 (2005).

[272] D.J. Scalapino, E. Loh, Jr., J.E. Hirsch, Phys. Rev. B **35**, 6694 (1987).

[273] K. Kuroki, J. Phys. Soc. Jpn. **75**, 051013 (2006).

[274] Y. Sidis, M. Braden, P. Bourges, B. Hennion, S. NishiZaki, Y. Maeno, Y. Mori, Phys. Rev. Lett. **83**, 3320 (1999).

[275] T. Takimoto, Phys. Rev. B **62**, R14641 (2000).

[276] M. Sato, M. Kohmoto, J. Phys. Soc. Jpn. **69**, 3505 (2000).

[277] T. Kuwabara, M. Ogata, Phys. Rev. Lett. **85**, 4586 (2000).

[278] H. Mukuda, K. Ishida, Y. Kitaoka, K. Asayama, Z.Q. Mao, Y. Mori, Y. Maeno, J. Phys. Soc. Jpn. **67**, 3945 (1998).

[279] J.P. Pouget, S. Ravy, J. Phys. (Paris) I **6**, 1501 (1996).

[280] M. Kohmoto, M. Sato, cond-mat/0003211 (2000).

[281] Y. Hasegawa, H. Fukuyama, J. Phys. Soc. Jpn. **55**, 3978 (1986).

[282] L. Ducasse, M. Abderrabba, J. Hoarau, M. Pesquer, B. Gallois, J. Gaultier, J. Phys. C **19**, 3805 (1986).

[283] J.M. Delrieu, M. Roger, Z. Toffano, A. Moradpour, K. Bechgaard, J. Phys. (Paris) **47**, 839 (1986).

[284] T. Takahashi, Y. Maniwa, H. Kawamura, G. Saito, J. Phys. Soc. Jpn. **55**, 1364 (1986).

[285] J.B. Torrance, H.J. Pedersen, K. Bechgaard, Phys. Rev. Lett. **49**, 881 (1982).

[286] C. Bourbonnais, in *High Magnetic Fields: Applications in Condensed Matter Physics and Spectroscopy*, ed. by L.P.L. C. Berthier, G. Martinez (Springer, New York, 2002), chap. 9, p. 235.

[287] H. Tasaki, Phys. Rev. Lett. **75**, 4678 (1995).

[288] K. Penc, H. Shiba, F. Mila, T. Tsukagoshi, Phys. Rev. B **54**, 4056 (1996).

[289] S. Daul, R.M. Noack, Phys. Rev. B **58**, 2635 (1998).

[290] S.D. R. M. Noack, S. Kneer, in *Density-Matrix Renormalization: A New Numerical Method in Physics*, ed. by M.K. I. Peschel, X. Wang, K. Hallberg (Springer, Berlin, 1999), chap. 8, p. 197.

[291] S. Kashiwaya, Y. Tanaka, Rep. Prog. Phys. **63**, 1641 (2000).

[292] L. Alff, H. Takashima, S. Kashiwaya, N. Terada, H. Ihara, Y. Tanaka, M. Koyanagi, K. Kajimura, Phys. Rev. B **55**, R14757 (1997).

[293] J.W. Ekin, Y. Xu, S. Mao, T. Venkatesan, D.W. Face, M. Eddy, S.A. Wolf, Phys. Rev. B **56**, 13746 (1997).

[294] M. Covington, M. Aprili, E. Paraoanu, L.H. Greene, F. Xu, J. Zhu, C.A. Mirkin, Phys. Rev. Lett. **79**, 277 (1997).

[295] J.Y.T. Wei, N.C. Yeh, D.F. Garrigus, M. Strasik, Phys. Rev. Lett. **81**, 2542 (1998).

[296] Y. Dagan, G. Deutscher, Phys. Rev. Lett. **87**, 177004 (2001).

[297] H. Aubin, L.H. Greene, S. Jian, D.G. Hinks, Phys. Rev. Lett. **89**, 177001 (2002).

[298] A. Sharoni, O. Millo, A. Kohen, Y. Dagan, R. Beck, G. Deutscher, G. Koren, Phys. Rev. B **65**, 134526 (2002).

[299] A. Biswas, P. Fournier, M.M. Qazilbash, V.N. Smolyaninova, H. Balci, R.L. Greene, Phys. Rev. Lett. **88**, 207004 (2002).

[300] A. Kohen, G. Leibovitch, G. Deutscher, Phys. Rev. Lett. **90**, 207005 (2003).

[301] A.F. Andreev, Sov. Phys. JETP **19**, 1228 (1964).

[302] L.J. Buchholtz, G. Zwicknagl, Phys. Rev. B **23**, 5788 (1981).

[303] C.R. Hu, Phys. Rev. Lett. **72**, 1526 (1994).

[304] J. Yang, C.R. Hu, Phys. Rev. B **50**, 16766 (1994).

[305] L.J. Buchholtz, M. Palumbo, D. Rainer, J.A. Sauls, J. Low Temp. PHys. **101**, 1097 (1995).

[306] Y.N. an K. Nagai, Phys. Rev. B **51**, 16254 (1995).

[307] S. Kashiwaya, Y. Tanaka, M. Koyanagi, K. Kajimura, Phys. Rev. B **53**, 2667 (1996).

[308] Y.S. Barash, A.A. Svidzinsky, H. Burkhardt, Phys. Rev. B **55**, 15282 (1997).

[309] I. Iguchi, W. Wang, M. Yamazaki, Y. Tanaka, S. Kashiwaya, Phys. Rev. B **62**, R6131 (2000).

[310] F. Laube, G. Goll, H. v. Löhneysen, M. Fogelström, F. Lichtenberg, Phys. Rev. Lett. **84**, 1595 (2000).

[311] Z.Q. Mao, K.D. Nelson, R. Jin, Y. Liu, Y. Maeno, Phys. Rev. Lett. **87**, 037003 (2001).

[312] C. Wälti, H.R. Ott, Z. Fisk, J.L. Smith, Phys. Rev. Lett. **84**, 5616 (2000).

[313] Z.Q. Mao, M.M. Rosario, K.D. Nelson, K. Wu, I.G. Deac, P. Schiffer, Y. Liu, T. He, K.A. Regan, R.J. Cava, Phys. Rev. B **67**, 094502 (2003).

[314] X. Lu, W.K. Park, H.Q. Yuan, G.F. Chen, G.L. Luo, N.L. Wang, A.S. Sefat, M.A. McGuire, R. Jin, B.C. Sales, D. Mandrus, J. Gillett, S.E. Sebastian, L.H. Greene, Supercond. Sci. Technol. **23**, 054009 (2010).

[315] K. Sengupta, I. Zutic, H.J. Kwon, V.M. Yakovenko, S. Das Sarma, Phys. Rev. B **63**, 144531 (2001).

[316] Y. Tanuma, K. Kuroki, Y. Tanaka, S. Kashiwayar, Phys. Rev. B **64**, 214510 (2001).

[317] Y. Tanuma, K. Kuroki, Y. Tanaka, R. Arita, S. Kashiwaya, H. Aoki, Phys. Rev. B **66**, 094507 (2002).

[318] Y. Tanuma, K. Kuroki, Y. Tanaka, S. Kashiwaya, Phys. Rev. B **68**, 214513 (2003).

[319] C.J. Bolech, T. Giamarchi, Phys. Rev. Lett. **92**, 127001 (2004).

[320] J.A. Pals, W. van Haeringen, M.H. van Maaren, Phys. Rev. B **15**, 2592 (1977).

[321] V.B. Geshkenbein, A.I. Larkin, JETP Lett. **43**, 395 (1986).

[322] K. Sengupta, V.M. Yakovenko, Phys. Rev. Lett. **101**, 187003 (2008).

[323] C.K. Lu, S. Yip, Phys. Rev. B **80**, 024504 (2009).

[324] C.D. Vaccarella, R.D. Duncan, C.A.R. Sá de Melo, Physica C **391**, 89 (2003).

[325] P.G. de Gennes, *Superconductivity of Metals and Alloys* (W. A. Benjamin, New York, 1966).

[326] H. Takayama, Y.R. Lin-liu, K. Maki, Phys. Rev. B **21**, 2388 (1980).

[327] G.E. Blonder, M. Tinkham, T.M. Klapwijk, Phys. Rev. B **25**, 4515 (1982).

[328] M. Stone, Ann. Phys. **155**, 56 (1984).

[329] C.R. Hu, X.Z. Yan, Phys. Rev. B **60**, R12573 (1999).

[330] D.A. Ivanov, cond-mat/9911147 (1999).

[331] T. Araia, T. Ishiguroa, T. Mangetsuc, J. Yamadąc, H. Anzai, Synth. Met. **133-134**, 209 (2003).

[332] M. Fogelström, D. Rainer, J.A. Sauls, Phys. Rev. Lett. **79**, 281 (1997).

[333] Y. Tanuma, Y. Tanaka, M. Yamashiro, S. Kashiwaya, Phys. Rev. B **57**, 7997 (1998).

[334] Y. Tanuma, Y. Tanaka, M. Ogata, S. Kashiwaya, Phys. Rev. B **60**, 9817 (1999).

[335] K.V. Samokhin, M.B. Walker, Phys. Rev. B **64**, 172506 (2001).

[336] Y. Asano, Y. Tanaka, Phys. Rev. B **65**, 064522 (2002).

[337] Y. Tanaka, Y.V. Nazarov, S. Kashiwaya, Phys. Rev. Lett. **90**, 167003 (2003).

[338] A.A. Golubov, M.Y. Kupriyanov, JETP Lett. **69**, 262 (1999).

[339] A. Poenicke, Y.S. Barash, C. Bruder, V. Istyukov, Phys. Rev. B **59**, 7102 (1999).

[340] K. Yamada, Y. Nagato, S. Higashitani, K. Nagai, J. Phys. Soc. Jpn. **65**, 1540 (1996).

[341] T. Lück, U. Eckern, A. Shelankov, Phys. Rev. B **63**, 064510 (2001).

[342] L.V. Keldysh, Sov. Phys. JETP **20**, 1018 (1965).

[343] H.J. Kwon, K. Sengupta, V. Yakovenko, Eur. Phys. J. B **37**, 349 (2003).

[344] P.G. Björnsson, Y. Maeno, M.E. Huber, K.A. Moler, Phys. Rev. B **72**, 012504 (2005).

[345] J.R. Kirtley, C. Kallin, C.W. Hicks, E.A. Kim, Y. Liu, K.A. Moler, Y. Maeno, K.D. Nelson, Phys. Rev. B **76**, 014526 (2007).

[346] J.B. Ketterson, S.N. Song, *Superconductivity* (Cambridge Univeristy Press, Cambridge, U.K., 1999).

[347] A. Millis, D. Rainer, J.A. Sauls, Phys. Rev. B **38**, 4504 (1988).

[348] Reprinted from *Triplet superconductors: Josephson effect in quasi-one-dimensional systems*, C. D. Vaccarella, R. D. Duncan, and C. A. R. Sá de Melo, Physica C **391**, 89, Copyright 2003, with permission from Elsevier.

[349] Q. Li, Y.N. Tsay, M. Suenaga, R.A. Klemm, G.D. Gu, N. Koshizuka, Phys. Rev. Lett. **83**, 4160 (1999).

[350] Y. Tanaka, S. Kashiwaya, Phys. Rev. B **53**, R11957 (1996).

[351] Y.S. Barash, H. Burkhardt, D. Rainer, Phys. Rev. Lett. **77**, 4070 (1996).

[352] T. Lofwander, V.S. Shumeiko, G. Wendin, Supercond. Sci. Technol. **14**, 53 (2001).

[353] H.I. Ha, J.I. Oh, J. Moser, M.J. Naughton, Synth. Metals **137**, 1215 (2003).

[354] D.J. Scalapino, E. Loh, Jr., J.E. Hirsch, Phys. Rev. B **34**, 8190 (1986).

[355] N.F. Berk, J.R. Schrieffer, Phys. Rev. Lett. **17**, 433 (1966).

[356] P.W. Anderson, W.F. Brinkman, Phys. Rev. Lett. **30**, 1108 (1973).

[357] N. Bulut, D.J. Scalapino, S.R. White, Phys. Rev. B **47**, 6157 (1993).

[358] I.I. Mazin, D.J. Singh, Phys. Rev. Lett. **82**, 4324 (1999).

[359] T. Imai, A.W. Hunt, K.R. Thurber, F.C. Chou, Phys. Rev. Lett. **81**, 3006 (1998).

[360] K. Ishida, H. Mukuda, Y. Kitaoka, K. Asayama, Z.Q. Mao, Y. Mori, Y. Maeno, Nature **396**, 658 (1998).

[361] A.P. Mackenzie, Y. Maeno, Rev. Mod. Phys **75**, 657 (2003).

[362] R.W. Hill, C. Proust, L. Taillefer, P. Fournier, R.L. Greene, Nature **414**, 711 (2001).

[363] L.S. Borkowski, C.A.R.S. de Melo, Acta Phys. Pol. **99**, 691 (2001).

[364] R.D. Duncan, C.A.R. Sá de Melo, Phys. Rev. B **62**, 9675 (2000).

[365] S.S. Botelho, C.A.R. Sá de Melo, Phys. Rev. B **71**, 134507 (2005).

[366] S.S. Botelho, C.A.R. Sá de Melo, J. Low Temp. Phys. **140**, 409 (2005).

[367] R.W. Cherng, C.A.R. Sá de Melo, Phys. Rev. B **74**, 212505 (2006).

[368] I.M. Lifshitz, Sov. Phys. JETP **11**, 1130 (1960).

[369] P.M. Chaikin, T.C. Lubensky, *Principles of Condensed Matter Physics* (Cambridge University Press, Cambridge, UK, 1997).

[370] C. Caroli, J. Matricon, Phys. Kondens. Mater. **3**, 380 (1965).

[371] W. Pesch, L. Kramer, J. Low Temp. Phys. **15**, 367 (1974).

[372] J.D. Shore, M. Huang, A.T. Dorsey, J.P. Sethna, Phys. Rev. Lett. **62**, 3089 (1989).

[373] F. Gygi, M. Schluter, Phys. Rev. Lett. **65**, 1820 (1990).

[374] F. Gygi, M. Schluter, Phys. Rev. B **41**, 822 (1990).

[375] C.A.R. Sá de Melo, J. Supercond. **12**, 459 (1999).

[376] H.F. Hess, R.B. Robinson, R.C. Dynes, J.M. Valles Jr., J.V. Waszczak, Phys. Rev. Lett. **62**, 214 (1989).

[377] K. Maki, N. Schopohl, H. Won, Physica B **204**, 214 (1995).

[378] M. Ichioka, N. Hayashi, M. Enomoto, K. Machida, J. Phys. Soc. Jpn. **64**, 4547 (1995).

[379] I. Maggio-Aprile, C. Renner, A. Erb, E. Walker, O. Fischer, Phys. Rev. Lett. **75**, 2754 (1995).

[380] C. Renner, I. Maggio-Aprile, O. Fischer, in *The Superconducting State in Magnetic Fields: Special Topics and New Trends*, ed. by C.A.R. Sá de Melo (World Scientific, Singapore, 1998), chap. 12, p. 226.

[381] C.D. Vaccarella, C.A.R. Sá de Melo, Physica C **341-348**, 293 (2000).

[382] T.M. Rice, M. Sigrist, J. Phys.: Cond. Matter **7**, L643 (1995).

[383] D.F. Antergerg, T.M. Rice, M. Sigrist, Phys. Rev. Lett. **78**, 3374 (1997).

[384] Y. Maeno, S. Nishizaki, Z.Q. Mao, J. Supercond. **12**, 535 (1999).

Index

Printed in the United States
By Bookmasters